MAPPING
THE DARKNESS

MAPPING THE DARKNESS

THE VISIONARY SCIENTISTS
WHO UNLOCKED
THE MYSTERIES OF SLEEP

KENNETH MILLER

NEW YORK

Hachette Books
Hachette Book Group
1290 Avenue of the Americas
New York, NY 10104
HachetteBooks.com
Twitter.com/HachetteBooks
Instagram.com/HachetteBooks

First Edition: October 2023

Published by Hachette Books, a subsidiary of Hachette Book Group, Inc. The Hachette Books name and logo is a trademark of the Hachette Book Group.

The Hachette Speakers Bureau provides a wide range of authors for speaking events. To find out more, go to hachettespeakersbureau.com or email HachetteSpeakers@hbgusa.com.

Books by Hachette Books may be purchased in bulk for business, educational, or promotional use. For information, please contact your local bookseller or Hachette Book Group Special Markets Department at: special.markets@hbgusa.com.

The publisher is not responsible for websites (or their content) that are not owned by the publisher.

Print book interior design by Six Red Marbles

Library of Congress Control Number: 2023938626

ISBNs: 978-0-306-92495-8 (hardcover); 978-0-306-92497-2 (e-book)

Printed in the United States of America

LSC-C

Printing 1, 2023

For Julie

Contents

Introduction

At first, no one noticed that Joe Borelli was losing his mind—no one but Borelli himself. The wiry, dark-haired radiologist was forty-three years old; he ran two successful practices, taught at the Medical University of South Carolina, and played a ferocious game of tennis. To most people, he seemed more than competent. Yet he'd begun to have trouble remembering names: his children's friends, his fellow Cub Scout leaders, his receptionist. He forgot to run promised errands for his wife. He got lost driving in his own suburban neighborhood. He would doze off over paperwork and awaken with drool dampening his lab coat.[1]

Borelli feared he had a neurodegenerative disease, perhaps early-onset Alzheimer's. But as a physician, he knew that cognitive problems coupled with fatigue could also indicate obstructive sleep apnea (OSA), a disorder in which sagging tissue periodically blocks the upper airway during slumber. Sufferers stop breathing for seconds or minutes until the brain's alarm centers rouse them. Although the cycle may repeat hundreds of times a night, patients typically remember nothing: they assume they're suffering from ordinary tiredness until the collateral damage becomes impossible to ignore.

One night, Borelli checked into a sleep clinic, where a technician pasted electrodes to his scalp, face, legs, and torso, fastened a sensor belt around his chest, and stuck an airflow monitor beneath his nostrils. He managed to drift off despite these encumbrances while computerized equipment monitored his brain waves, eye movements, body movements, heartbeat, and respiration. His numbers all fell within the normal range. Months later, as his symptoms worsened, he went for a workup with a neurologist, who found nothing amiss but suggested he be retested for apnea. Borelli tried another sleep clinic; this time, he was diagnosed with borderline OSA. The doctor sent him home with a continuous positive airway pressure machine, or CPAP—a contraption that resembled a small gas-station tire pump, designed to keep his throat open by gently inflating it. But the face mask was uncomfortable, and he awoke each morning feeling as exhausted as ever. He quit using the device after a few weeks.

Borelli's fingers soon grew so clumsy that he couldn't button his shirt cuffs. His heart raced wildly when he rose from a chair, and even a short walk winded him. Absorbing new information became more difficult as did following a simple train of thought. Haunted by the specter of his own decline, Borelli became increasingly anxious and depressed. He gave up tennis. He resigned his chairmanship of a national professional committee. His marriage dissolved. He found himself daydreaming about suicide. "One day," he later told me, "I just collapsed in the shower, crying." Then he dried himself off and went looking for the best sleep specialist in America.

He settled on Christian Guilleminault, a venerated clinician and researcher at the Stanford University School of Medicine who pioneered the diagnosis and treatment of sleep apnea in the 1970s. Borelli flew across the country for a consultation at Stanford's sprawling outpatient center in Redwood City, California. After shining a flashlight in Borelli's mouth, and running another overnight test to confirm his hunch, Guilleminault delivered a new diagnosis: upper-airway resistance syndrome, or UARS, a condition in which the airway is partially obstructed during sleep, restricting breathing rather than stopping it.[2]

By coincidence, Guilleminault himself had first identified the syndrome in a 1993 journal report. Like classical sleep apnea, UARS may trigger symptoms ranging from depression and cognitive deficits to hypertension—yet it's far less widely known and more difficult to detect.[3]

"You've had this disorder since you were a little boy," Guilleminault declared in his thick French accent. "The damage has been cumulative."

He ordered Borelli to go back on CPAP at more than twice the air pressure of the previous round.[4]

Borelli was reluctant to don the mask again, but he hoped desperately that the old man could save him. Getting a good night's sleep had become a matter of life and death.[5]

———————

I interviewed Borelli in 2013, when *Discover* magazine assigned me to write an article on recent findings in sleep research.[6] Until then, I had paid little attention to sleep, except when I had to pull an all-nighter to make a deadline. But reporting that story opened my eyes to a number of astonishing things. The first, reflected in Borelli's symptoms, was sleep's central role in governing a vast array of mental and physical processes. Sleep, I learned, acts as a master regulator: of memory, mood, and learning; of our hormones, circulatory systems, and immune defenses; of childhood development and the ravages of aging.

As often happens when I'm working on a medical topic, I began seeing ominous signs everywhere. My wife's snoring: Could it be sleep apnea? (A night in a clinic indicated it wasn't.) My son's absentmindedness about turning in homework: Was he being sabotaged by a school day unsuited for teenage sleep rhythms? (Most likely, judging by his improved ability to focus on similar tasks during vacations.) My own increasing frazzled-ness: Was I kidding myself about needing only six hours to get by? (Undoubtedly, and I began trying to eke out more.) Not long after I filed the piece, my eighty-seven-year-old father fell asleep at the wheel of his Prius and plowed into a tree. He survived, despite serious injuries, but the accident marked the end of his ability to live independently. From my reading, I knew that daytime sleepiness in the elderly typically stems from the decline of sleep quality with advancing age. Sooner or later, sleep trouble touches us all.

My reporting also brought another realization—how far sleep science had traveled since the 1920s, when physiologist Nathaniel Kleitman began transforming it into an independent discipline. Just a century ago, only a handful of scientists studied sleep—and not a single one did so full-time.[7] Most saw slumber as a nonevent, a nightly state of suspended animation. Many considered it a vestige of humanity's primitive past, which could safely be minimized or eliminated altogether. Although experts noted a rising tide of sleeplessness, they didn't bother tracking the statistics. Fewer than a dozen sleep disorders had been identified. No one dreamed that treating them could become the focus of a medical career, let alone the basis for a booming commercial industry.

Today, the quest to understand sleep—and to apply that understanding to our daily lives—has become a global obsession. Sleep research centers can be found at every major university. Over 2,500 sleep clinics operate across America, and 4,000 more in other countries.[8] The World Sleep Society, which represents scientists and health care professionals on every continent but Antarctica, boasts 14,000 members.[9] In a 2020 study, analysts valued the so-called sleep economy (encompassing all sleep-related products and services) at $432 billion and projected that it would reach $585 billion by 2024. More than one-fifth of the current total was for items explicitly related to sleep health: $25 billion for CPAP machines; $18 billion for medications and supplements; $15 billion for diagnostic services and devices; another $15 billion for consumer sleep technology; $11 billion for sleep-improvement services; and $9 billion for "ambience optimization" products, such as white-noise machines and blackout curtains.[10] Although a session in a sleep lab remains the gold standard for medical diagnosis, people who prefer to stay home can track their sleep quality using smartwatches, smart pillows, or a $300 titanium ring equipped with heart rate, oxygen, and activity monitors. Those who have difficulty falling asleep can stream soporific podcasts or plug in wireless earbuds that play soothing sounds. For those who need help *staying* asleep, and can spare $2,000, there's an app-synced mattress that heats up and cools down as the user's body temperature fluctuates through the night.[11]

Over the past ninety-odd years, sleep has gone from an afterthought to a central element in our notions of well-being. It has also become one of our biggest sources of collective anxiety. In the United States, according to the Centers for Disease Control and Prevention, 35 percent of the population gets less than seven hours a night—the tipping point, evidence suggests, for increased risk of obesity, diabetes, cancer, cardiovascular disease, depression, and dementia.[12] Diagnosticians now recognize more than eighty sleep disorders, ranging from common ailments such as OSA and chronic insomnia to rare afflictions like exploding head syndrome (characterized by terrifying blasts of hallucinated noise) or rapid eye movement sleep behavior disorder (whose victims act out their nightmares, sometimes injuring or even killing their bedmates). Around seventy million Americans suffer from such miseries.[13] And the sense of a sleep-health crisis has become endemic in many other lands as well. Surveys show that fewer than half of adults worldwide feel they get enough shut-eye; the share is 47 percent in Germany, 46 percent in Brazil, 45 percent in Australia, 40 percent in the United States and United Kingdom, and just 35 percent in France, Japan, and South Korea.[14]

Scientists have developed effective treatments for many sleep disorders, based on a growing understanding of the biological machinery that drives slumber and its dysfunctions. However, they've also learned that a large proportion of sleep troubles arise not from glitches in our individual physiology or neuropsychology but from socioeconomic factors—shift work, school schedules, the stresses and stimuli of a twenty-four-hour culture—that play havoc with our internal rhythms. Beyond the effects on bodily and mental health, these disruptions can lead to calamities of drowsiness: car crashes, medical errors, industrial accidents, and disasters ranging from the *Exxon Valdez* oil spill to the explosion of the Space Shuttle *Challenger*.[15] The invasion of our bedrooms by tiny electronic screens appears to be adding to the problem.[16] A recent study by the RAND Corporation found that insufficient sleep costs the United States $411 billion a year in lost output, or 2.28 percent of gross domestic product.[17]

To attack this crisis, sleep researchers have ventured from the lab into the public square, becoming organizers and advocates for the defense of slumber. They've testified before the US Congress and local school boards.

They've become regulars on late-night talk shows and published hundreds of books aimed at ordinary readers. They've won recognition of sleep medicine as a subspecialty from the American Medical Association and established a center for sleep research at the National Institutes of Health.[18]

And by doing all these things, they've created a world where someone like Joe Borelli has a chance at getting well.

———————

As Borelli began his second round of CPAP treatment, in 2009, he was no more comfortable with the hissing mask than he'd been before. Nonetheless, he kept on wearing it. After a few months, he was waking up most mornings with a long-lost sensation: feeling rested. Even better, the symptoms that had vandalized his life—the memory lapses, anxiety, clumsiness, heart palpitations, and all the rest—began to vanish. So did the high blood pressure for which he'd taken medication since childhood, as well as the spinal arthritis he'd developed from decades of sleeping on his stomach in an unconscious effort to ease his breathing. When I spoke with him, after four years on the machine, he was developing an app to help other users optimize their therapy. He had a new girlfriend: a chess champion, he boasted, with an IQ of 140. And though his mental acuity wasn't quite what it had been before his sleep disorder brought him down, he wasn't complaining. "I lost fifty percent of my brain to this disease," he said, "but I got eighty to ninety percent back. I'm a happy, happy camper."[19]

In some ways, Borelli was a living illustration of what sleep science had accomplished since Kleitman first set up his lab. But as I reported my story, I realized that a great deal about sleep remains unknown.

The central mystery is why we can't just stay awake. "If sleep does not serve an absolutely vital function," said Allan Rechtschaffen, one of the field's pioneers, "then it is the greatest mistake the evolutionary process has ever made."[20] All the other basic animal functions have a clear and simple purpose: eating provides our cells with fuel; respiration converts that fuel to energy; excretion flushes out waste products; sex enables us to

reproduce. But no one can say precisely what sleep is *for*. Although virtually every species with a nervous system does it (with the possible excep-
tion of certain amphibians and cetaceans), and some species have been
shown to die when continuously deprived of it (including rats, dogs, and
humans), debate continues over why sleep first came into being.[21]

That's because sleep is more complex than those other functions, all of
which can be described in a single phrase. According to current defini-
tions, a resting behavior needs to include several characteristics to qualify
indisputably as sleep:

1. It must be part of a circadian cycle, run by biological clockwork
   tied to the twenty-four-hour day.
2. It must involve a posture specific to the species—typically horizontal for humans, for example, and vertical for sperm whales.
3. It must heighten the animal's threshold for sensory arousal, requiring a stronger-than-normal stimulation to elicit a response.
4. It must display a rebound after deprivation, meaning that the
   animal sleeps longer or more deeply to make up for a period of
   enforced wakefulness.[22]

There are gray areas, however. Dolphins sleep with one brain hemisphere at a time, which allows them to ignore rule 2 and keep swimming.
Migrating sparrows flout rule 4, going sleepless for weeks without needing
to compensate later. And a few creatures with very primitive nervous systems, such as hydras and roundworms, have no use for rule 1; they sleep
whenever they please. Of course, some experts consider such irregular
snoozing to be sleep*like* behavior rather than the real thing.[23]

Many observers have pointed out the inconvenience of slumber in
terms of the Darwinian struggle for survival. A sleeping animal can't for-
age for food, defend itself from enemies, or pursue reproductive opportunities. So why did evolution program even insects to take these mandatory
time-outs? Was it to conserve organisms' energy at regular intervals,
reducing the need for calories? To make them lie low during the hours

when predators might leap out of the darkness? To free their brains—or decentralized nerve clusters, in creatures such as jellyfish—to perform offline maintenance? Experts have offered all these hypotheses and more; no consensus yet exists.[24]

Sleep science touches, too, on some of the most enduring conundrums of psychology and philosophy. Where do the boundaries lie between mind and body? Are we better off submitting to the dictates of nature (by matching our sleep rhythms to our internal clocks, for instance) or those of culture (by taking pills to sleep in sync with our alarm clocks)? What is consciousness, and where do dreams fit in? Sleep researchers have debated these issues since the field's infancy, and they show no sign of stopping.

After my encounter with Borelli and the science that healed him, I became obsessed by such questions. I also began to wonder how sleep science had evolved from its youthful awkwardness and obscurity to its current state of sophistication and power.

This book tells a portion of that sprawling story. I've focused most intently on four sleep explorers, each of whom played a crucial role in mapping the invisible continent, unlocking its secrets, and pushing society to translate those discoveries into meaningful action. By training my lens on this quartet, I hope to throw the science they pioneered—and the deeds of their predecessors, contemporaries, and successors—into sharper relief.

They all started as outsiders. At the helm is Nathaniel Kleitman, a refugee from Russian pogroms who fled to the United States at age twenty and soon began laying the groundwork for a new field of biomedical inquiry. With headline-grabbing experiments that included hundred-hour bouts of wakefulness and a monthlong expedition in a Kentucky cave, he won a place for sleep research in the cultural mainstream. He wrote the monumental book that became the field's bible and blazed a trail for all who followed.[25]

The second explorer is one of Kleitman's students at the University of Chicago: Eugene Aserinsky, a brilliant but troubled Brooklynite who stumbled into the field in 1949 and stormed out four years later. In that brief time, Aserinsky discovered rapid eye movement (REM) sleep—revealing that the slumbering brain is as active as its waking counterpart and opening vast new realms for investigation.[26]

Explorer number three is another of Kleitman's mentees: William Dement, now recognized as the father of sleep medicine. A GI Bill psychiatry student from rural Washington State, Dement yearned to understand the dreaming mind, and he grasped the radical implications of REM as neither Aserinsky nor Kleitman did. He led a series of studies that got the revolution underway, then moved on to Stanford, where he spearheaded decades of seminal research into both normal and disordered sleep. Dement invented the modern sleep clinic, crusaded to raise awareness of sleep's fundamental importance to human health, and spurred the creation of the first government agency on the planet devoted to sleep.[27]

The fourth explorer is Mary Carskadon, an unlikely heir to Kleitman's legacy. In an era when women scientists were rare, Carskadon had no thought of becoming one when she signed on as Dement's lab assistant in 1970. A recent college graduate from small-town Pennsylvania, she was more interested in protesting the Vietnam War than analyzing data sets. But her new job revealed her phenomenal gifts and awakened a hunger to learn how the rhythms of modern life affected children and adolescents. While still a PhD student, Carskadon identified the unique sleep needs of teenagers, devised the first tool for quantifying sleepiness, and defined the principle of "sleep debt"—contributions with huge significance for parenting, medical practice, and social policy. After starting her own lab at Brown University, she uncovered the potentially catastrophic impacts of sleep deprivation in young people and inspired an international movement to combat the epidemic.[28]

While these characters often take center stage in the narrative, they seldom stand alone. Other important players include Guilleminault, Michel Jouvet, and several French compatriots; famous figures such as Freud and Pavlov; and lesser-known ones like Giuseppe Moruzzi, Horace Magoun, Ernst Kohlschütter, and Constantin von Economo, from a dozen countries.

Beyond chronicling the rise of sleep science, my aim is to illuminate the ways in which social forces shape scientists—and how scientists help shape society. On one level, *Mapping the Darkness* is the saga of a few brilliant individuals who chanced upon experiences, mentors, allies, professional networks, and historical conditions that helped them make the most of their talents. On another, it's a tale of passionate curiosity, fierce audacity, and near-superhuman perseverance.

These researchers spent decades working to solve puzzles whose importance few other people could grasp, and seeking evidence for theories that often proved mistaken or nonsensical. When they hit a dead end, they analyzed their errors and set off in a new direction. They risked futility and failure, ostracism and ridicule to advance a field that was widely regarded as scientifically irrelevant—or, when it came to matters of public health, as a threat to powerful interests. They shoveled great chunks of their personal lives into the furnace of their all-consuming cause. In the process, they changed our nights—and days—forever.

PART I

COMING INTO THE TERRITORY

1

Exodus

By the time Pesea Kleitman became pregnant with her first child, her husband was too weak to make it through another Russian winter. Tuberculosis was rotting his lungs and wasting his body, tormenting him with chest pain and fevers; no medication could provide a cure. The young cloth merchant's only hope of recovery, doctors said, was several months of complete rest in a warm, dry climate. So the couple decamped to Egypt, a week's journey from their small town in the province of Bessarabia (now the republic of Moldova): by rail to Odessa; by ship across the Black Sea and the Mediterranean; and again by rail from Alexandria to Cairo.[1]

Soon after their arrival, however, the sick man took his last breath. Pesea, a widow at twenty, made the long return trip alone. With her due date looming, she moved in with her parents and four younger siblings in Kishinev, the provincial capital. On April 26, 1895, she gave birth to a boy and named him after his late father.[2]

From an early age, Nathaniel Kleitman showed signs of the extraordinary intelligence that would someday fuel his fame. The family spoke Yiddish at home, but he learned the Russian alphabet in a single day when he was two and a half years old. By the time he started heder (religious school) at four, he could solve complex arithmetic problems in his head. Pesea remarried two years later, to a widowed fabric store owner with children of his own. Her son remained for a time with his grandparents, Leyb and Leya Galanter, who may also have been in the textile business—and who sent him to the best private elementary school they could find. When

Leyb died two years after that, Leya set aside a portion of the estate to fund the precocious child's further education.[3]

Although their house lacked indoor plumbing, like all but the grandest local residences, the Galanters belonged to the upper-middle class of Kishinev's Jewish community—one of the largest in Europe. These Ashkenazim, whose ancestors had migrated from Germany and France during the Middle Ages, comprised nearly half of the municipality's 125,000 residents. They were laborers and factory owners, shopkeepers and tobacco growers, rabbis and blacksmiths, bankers and beggars. And as elsewhere in the czarist empire, they lived under a deepening shadow of persecution.[4]

Anti-Semitism was nothing new in Russia, where Jews had long been restricted to residence in the cluster of territories known as the Pale of Settlement, banned from various trades, and forbidden to wear traditional dress. Young males faced an added threat: conscription by the Imperial Army, which demanded punitive numbers of Jewish recruits and subjected them to discrimination, abuse, and often forcible conversion. (According to family lore, Kleitman's father contracted tuberculosis after starving himself to avoid the draft.) But conditions worsened sharply in the 1880s, after the assassination of Czar Alexander II, which was widely blamed on an imagined Jewish plot. An outbreak of pogroms swept the land. New laws barred Jews from living in towns of fewer than ten thousand people, from voting in municipal elections, and from professions including teaching and the law. Authorities reduced already stringent quotas for Jewish students in secondary and higher education.[5]

In Kishinev, the atmosphere grew dangerously volatile around the turn of the twentieth century, when the city's government-subsidized newspaper began printing lurid accusations of Jewish perfidy on a daily basis. The explosion came in April 1903, after a Christian boy was murdered in a nearby village (by a relative, investigators later found) and a Christian girl died of self-inflicted poisoning in a Jewish hospital. The paper's editor "laid both tragedies at the doors of the Jews," according to a contemporary account, "declaring emphatically that both were murders committed for ritual purposes." He described the purported atrocities in gruesome detail and called for vengeance.[6]

The two-day rampage, beginning on Easter Sunday—shortly before Kleitman's eighth birthday—was the bloodiest pogrom yet. Nearly 50 Jews were killed, 600 injured, and dozens of women raped; 600 businesses were looted, 1,350 dwellings destroyed. No record remains of how the boy's household escaped the onslaught; as in some other cases, the mob may have found the gate to the family compound too difficult to break down. What is known is that Kleitman's uncle Benzion Galanter was among the dead, leaving behind a widow and six children. He'd tried to protect them by brandishing a pistol loaded with blank cartridges at the attackers—who tackled him, gouged out his eyes, and tore out his tongue.[7]

As an adult, Kleitman told his children that the murder had shaken him deeply, though with characteristic reticence, he never spoke of it in public. "He couldn't remember his father's death, but losing his uncle was a real blow," his daughter Hortense said, her own memory of the story undimmed as she neared ninety. "It's interesting," she added, "how many people who went on to do a lot had that kind of early trauma."[8]

The massacre sparked international outrage—especially in the United States, where thousands signed a petition of protest to the czar. It helped turn countless young Russian Jews into militant Zionists or revolutionary leftists. And along with subsequent bloodbaths—including a second one in Kishinev, in 1905—it accelerated an ongoing wave of emigration. Between 1880 and 1914, more than two million Jews fled Russia, the vast majority to America.[9]

Kleitman's trajectory, by contrast, followed his lust for learning. After graduating in 1912 from a private Realschule (a German-style high school emphasizing science and math), he knew the quota system offered little hope of admission to a Russian university. Instead, he set his sights on an elite engineering school in Paris. On an investigative visit, however, he learned that his odds of passing the specialized entrance exam were slim.

Swiftly adjusting his ambitions, he decided to try for the American-run Syrian Protestant College in Lebanon (now the American University of Beirut), with the eventual aim of practicing medicine in Palestine.[10]

Kleitman and his mother, Pesea, in 1912. *Courtesy of the Hanna Holborn Gray Special Collections Research Center, University of Chicago Library.*

In 1914, having taught himself English with a learn-at-home manual, the nineteen-year-old set out on his personal exodus. A snapshot from the time shows a slim youth with a mop of dark hair, a strong nose, a sensitive mouth, and an air of passionate seriousness. As a draft-age Jewish male, Kleitman was legally prohibited from obtaining a passport, so he took the train to Odessa and bought a fake one on the black market. Or, rather, half of one: The document was for two Lithuanian brothers. (Russian passports at the time required neither photo nor signature.) On August 1—with his older "sibling," whom he'd never previously met—Kleitman boarded a steamer bound for the Levant.[11]

A few hours later, Germany declared hostilities against Russia, and the conflict that would become known as World War I erupted. The next day, after stopping in Constantinople, the vessel hurried back to Odessa to avoid attack by enemy destroyers. Because his fictitious brother, who had first dibs on the passport, opted to return home, Kleitman no longer possessed travel documents. Undeterred, he switched to a Turkish steamer, which bore him across the Sea of Marmara, through the Dardanelles, along the shores of Asia Minor, and down the coast of Syria. When the ship reached Beirut, a fellow refugee snuck off and returned with two more false passports—one for himself, the other for Kleitman.[12]

Kleitman spent the next few weeks cramming for the exam to enter Syrian Protestant College's medical school, at a local rooming house where several other Kishinev boys were staying. He passed easily and began taking basic courses in physiology, anatomy, and other subjects. But that October, the Ottoman Empire—which controlled Lebanon—allied itself with Germany, and the school's Russian students became prisoners of war. Kleitman and his compatriots were officially confined to the campus, though enforcement was lax enough that they could make surreptitious forays into town. Toward the end of the school year, the situation grew more ominous: Word arrived that enemy aliens would soon be sent to concentration camps in the Syrian desert, where massacres of Armenians were already underway. With all local ports under blockade, escape by passenger ship was no longer possible.[13]

The last hope of rescue lay with the American government, which was still neutral. The college's administration made a desperate plea to the US ambassador to Turkey, Henry Morgenthau Sr., whose horror at the nascent genocide (and, perhaps, his background as a German-born Jew) helped stir his sympathy for the students' plight. Morgenthau arranged for the cruiser *Des Moines* to evacuate them from Beirut, along with hundreds of other foreign civilians. To save face, Ottoman officials declared that they were expelling the aliens rather than bowing to diplomatic pressure; Kleitman and his classmates had to sign papers promising never to return. The ship was authorized to transport them only to the nearest port of safety.[14]

In July 1915, the *Des Moines* deposited the refugees on the Greek island of Rhodes. From there, Kleitman hopped a steamer to Athens, where he seized the opportunity to tour the ruins of the Acropolis. Then he boarded the *King Constantine*, an ocean liner heading to New York City. The crossing, in steerage, took two weeks.[15]

———————

Kleitman landed at Ellis Island—then the nation's principal port of entry—on August 23. He faced an uncertain welcome. At the time, US immigration policy was in the midst of a long slide from the openness embodied in the Statue of Liberty's inscription ("Give me your tired, your poor, your huddled masses") to the exclusiveness enshrined in the Immigration Act of 1924, whose quotas based on national origins were designed to shut out people of his ilk along with a range of other ethnic groups.[16]

The restrictionists often regarded eastern European Jews as especially objectionable. Madison Grant, the genteel founder of the New York Zoological Society and a champion of the growing eugenics movement (which aimed to prevent reproduction by those deemed genetically unfit), described these immigrants as "half-Asiatic mongrels" and "a curse... draining into the country [from] the great swamp" of their benighted homelands. A turn-of-the-century newspaper cartoon, captioned THE STRANGER AT OUR GATE, depicts a hook-nosed, bearded man in filthy rags, carrying baggage labeled *poverty, disease, superstition, sabbath desecration,* and *anarchy*. Uncle Sam stands before him, holding his own nose to ward off the stench, beneath a sign reading, "United States of America: Admittance Free." The stranger asks, "Can I come in?" The gatekeeper replies, "I s'pose you can, there's no law to keep you out."[17]

Although ethnically targeted bans had not yet progressed beyond those directed at the Chinese in the 1880s, rules barring other types of undesirables had multiplied by the time of Kleitman's arrival. Entry was denied to anarchists and polygamists, as well as any alien likely to become a "public charge" (most often applied to paupers and the physically handicapped). Also inadmissible was anyone exhibiting "moral turpitude," or suffering from epilepsy, "lunacy," "feeblemindedness," or a "loathsome or contagious disease."[18]

Still, the admissions process was strikingly less onerous than it is today. "No passport, visa, or other identification was required," Kleitman recalled decades later, perhaps conscious that his own forged papers might not have passed close scrutiny. Officials performed a brief examination of his mental and physical fitness. They counted the cash in his wallet, which had been thinned by loans to several shipmates who lacked the recommended $25 minimum (worth about $650 today) to prove solvency. They asked whether he advocated the overthrow of the established order; he answered in the negative. And with that, they ushered him through the golden door.[19]

After a ferry ride to Manhattan, Kleitman checked in to a dormitory operated by the Hebrew Immigrant Aid Society on the Lower East Side.[20] He was twenty years old, just one more greenhorn dragging a battered trunk. But within a decade, despite his vow to the immigration agents, he would begin setting the stage for a revolution that neither they nor he could have imagined.

2

Arrival

When Kleitman landed on Manhattan's Lower East Side in the summer of 1915, he had little interest in sleep other than finding a comfortable place to do it. After a short stint at the Hebrew Immigrant Aid Society shelter, followed by a few nights in a rented room on Rivington Street, he located family friends from Kishinev who lived close by—a Mrs. Reisher, her son, and two daughters—and began boarding with them.[1]

With over half a million residents packed into one and a half square miles, the Lower East Side was one of the most crowded communities in the world. A multiethnic enclave, 60 percent Jewish, its tenements teemed with escapees from every hardscrabble corner of Europe; its streets were jammed with peddlers, pushcarts, horse-drawn wagons, and soot-belching Model Ts. Although the neighborhood was one of the poorest in the city, it was also among the most vibrant—home to swaggering gangsters and crusading political activists, grand synagogues and garlicky delicatessens, thriving Yiddish theaters and some of the first movie houses in New York. It held dozens of garment-industry sweatshops, too, as well as a raucous red-light district. On average, immigrants spent fifteen years there before finding their way to more genteel precincts. Kleitman, however, aimed to stay no longer than was absolutely necessary.[2]

At first, he hoped to revive the plan he'd envisioned before leaving Russia: to become a doctor, preferably in Palestine. He figured he would wait out the war, then return to medical school in Europe or the Near East. In the meantime, he worked odd jobs, earning about a dollar a day, of which ten cents was budgeted for the subway and ten cents for lunch. But by 1916, something had shifted. Perhaps he'd realized that the shooting wasn't stopping anytime soon—or perhaps he'd sensed that America offered him another way forward. That fall, he found a job as a laboratory helper at the Rockefeller Institute (now Rockefeller University), rented a room nearby on East Sixty-Eighth Street, and enrolled in night classes at the College of the City of New York in West Harlem.[3]

Founded in 1847, City College was the oldest free public institution of higher learning in the United States. Known as the "poor man's Harvard" for its distinguished faculty and demanding admission standards, it would graduate more Nobel Prize winners than any other public college— thirteen at last count. By the time Kleitman signed on, alumni included such luminaries as Henry Morgenthau Sr. (the ambassador who'd rescued him from Lebanon), George Washington Goethals (chief engineer on the Panama Canal), and Bernard Baruch (economic adviser to Presidents Woodrow Wilson and Franklin D. Roosevelt).[4] Thanks to courses he'd taken in Kishinev and Beirut, Kleitman entered with advanced standing. After two semesters, he switched to day classes, focusing on physics, biology, and chemistry. Evenings, he worked as a cashier at a pastry shop.[5]

As a resident alien, Kleitman wasn't subject to the draft when the United States entered the war in 1917. But the following year, along with more than a thousand other City College students, he volunteered for the Student Army Training Corps, a program initiated by the War Department on campuses across the country. Cadets were housed in barrack dormitories, performed daily military drills, and received a stipend of $30 a month. In the mess hall, Kleitman tasted bacon for the first time and loved it—an event that he, like many immigrant Jews who'd kept kosher in the old country, later recalled as a turning point in his Americanization. Soon afterward, he decided to exercise his right as a soldier to expedited US citizenship. His application was granted on November 2, 1918, nine days before the Armistice.[6]

Kleitman graduated in 1919 with a BS in chemistry and a Phi Beta Kappa key for outstanding grades. That fall, he moved on to Columbia

University to study for a master's in physiology. (To pay the rent, he worked as a teaching fellow at City College.) By then, he'd determined that his interests lay in research and education rather than in practicing medicine. But he had not yet discovered the topic that would obsess him for seven decades.[7] His master's thesis was entitled "Sugar in the Blood of the Frog."[8]

After receiving his MA in 1920, Kleitman faced a quandary.[9] To pursue his calling, he would have to earn a PhD; meanwhile, he needed to earn a living. His options in both areas, however, were constrained by his ethnicity. Nativism and anti-Semitism were gaining momentum across the country, driven in part by widespread fears of contagion from Europe's burgeoning radical movements. (Many conservatives attributed Bolshevism, newly ascendant in Russia, to a Jewish-led conspiracy.) A resurgent Ku Klux Klan targeted Jews as well as Blacks for cross burning, though far less often for lynching; eugenics advocates urged that both groups be encouraged—through sterilization or other means—to die out. Quieter forms of prejudice persisted as well: In polite Gentile society, even well-assimilated Jews were commonly disdained as pushy, moneygrubbing, clannish, and venal. They were excluded from the best clubs, the best neighborhoods, and many of the best jobs.[10]

In academia, a tidal wave of Jewish undergraduates at northeastern schools was spurring a backlash among "native" American students, faculty, and administrators. Most of the newcomers were the children of eastern European immigrants, who saw education as the surest way to succeed in their adopted country. By 1920, the student body at City College was 90 percent Hebraic (as Jews were often labeled); at Columbia, the figure was 40 percent; at Harvard, 20 percent. A college song reflected the casual bigotry of the era:

Oh, Harvard's run by millionaires,
And Yale is run by booze,
Cornell is run by farmers' sons,
Columbia's run by Jews.

So give a cheer for Baxter Street,
Another one for Pell,
And when the little sheenies die,
Their souls will go to hell.[11]

Several leading universities—including Columbia, Harvard, and Yale—imposed informal quotas on Jewish admissions in the early 1920s, sharply reducing their numbers. (Boards of trustees often voted down proposals for explicit strictures, preferring a more discreet approach.)[12] Likewise, Jews who sought academic jobs faced steep obstacles. When Kleitman completed his master's, the number of Jews in the liberal arts or sciences faculties of US colleges and universities was probably under one hundred.[13]

Yet the barriers to entry were not impassable. For an aspirant of suspect pedigree, a key step toward securing a post was backing by a sponsor—often a faculty member of similar heritage who'd made it past the gatekeepers. Such an ally "would testify to the authorities," as the sociologist Lewis S. Feuer put it, "that this man, though a Jew or an immigrant or from the working class, was not insisting on his Jewishness, was devoid of 'pushing' traits, was courteous, and quiet in disposition."[14] (Women professors were growing more numerous, but they remained a small enough minority to justify Feuer's use of the masculine noun.[15] African Americans, by contrast, were almost never admitted to faculty positions outside historically Black colleges and universities.)[16]

Kleitman's first sponsor was City College biology professor Abraham Goldfarb, who volunteered for the role without being asked. It happened in a roundabout way as these things often did. Goldfarb was approached by William Salant, a fellow Columbia alumnus—and Russian Jewish immigrant—who'd recently been appointed as a professor of pharmacology at the Medical College of Georgia. At fifty, Salant was a decade older and considerably more eminent; he'd previously served as chief of the pharmacological laboratory at the US Department of Agriculture and as a senior scientist at the National Bureau of Standards.[17] Now, he told

Goldfarb, he needed to hire an instructor in physiology and pharmacology. Whatever qualifications may have been discussed, both men likely understood—given their own backgrounds and those of most students at City College—that a candidate who faced certain disadvantages would be preferred. At Goldfarb's urging, Salant wrote to Kleitman, offering him the job.[18]

Soon afterward, the twenty-five-year-old stepped off a train in Augusta, where he rented a room in the private home where Salant was staying. The professor and his underling shared meals at a nearby boardinghouse, developing a friendship despite their differences in age and status. Kleitman spent the next two semesters teaching medical students material that he'd only recently learned, and he did well enough that Salant made him promise to return in the fall of 1921.[19]

That June, to earn credits toward a doctorate, Kleitman headed north for summer courses at the University of Chicago. Although its Oxonian buildings and elm-shaded quadrangles gave it an air of antiquity, the school had been founded only three decades before; thanks largely to a generous endowment from oilman John D. Rockefeller, it was already among the nation's premier research institutions. There, Kleitman found his next sponsor—another fellow immigrant, though from a diametrically different background.[20]

This one would help him discover his path.

———

Anton Julius Carlson, chair of the University of Chicago's physiology department, was one of the leading figures of his field, renowned for his research on the peripheral nervous system. Square-jawed and powerfully built, with steel-rimmed glasses and a thatch of graying hair, Carlson radiated physical as well as intellectual energy. Originally from Sweden, he'd made his way to Illinois at age sixteen; working ten-hour days as a carpenter's helper, he saved enough cash to attend a Lutheran college. After graduating, he served a brief stretch as a minister, then lost his faith and headed to Stanford to study science. A groundbreaking paper on cardiac rhythms landed him a teaching position at Chicago in 1904—and within a decade, he'd taken over the department. When he first encountered Kleitman, Carlson was forty-six, and he may have seen the brilliant and driven young man as a version of himself twenty years earlier.[21]

At the end of the summer session, Carlson offered Kleitman an assistantship, which would enable him to earn a salary and a PhD simultaneously. Conscience-stricken, Kleitman declined, citing his vow to his employer in Georgia. When he told Salant about his decision, the professor called him a "damn fool." In the spring of 1922, after teaching two more semesters, he accepted Carlson's standing invitation and relocated to Chicago.[22]

Before Kleitman could begin his studies in earnest, he had to choose an area of concentration. Again, his origins placed some routes off-limits—and, perhaps, spurred him to seek out uncharted territory. Jewish scholars were unwelcome in many long-entrenched disciplines. "Apart from a few appointments to teach Jewish or Semitic studies," Lewis Feuer observed, "the Jews aspiring to academic work tended toward subjects that were new, not already the preserve of some academic establishment, or to such as mathematics where the most objective standards might tend to prevail.... Novel intellectual capital or intelligence, when it is precluded by monopolistic channels of intellectual investment, will then be directed into non-traditional, high risk, and still open fields." This dynamic, historians have suggested, helps account for the high concentration of Jews in the era's cutting-edge sciences, such as psychiatry, biochemistry, and immunology.[23]

It's also possible, of course, that Kleitman was simply drawn to terra incognita by the same questing impulses that had propelled the earliest stages of his journey, from Kishinev to Beirut to Ellis Island. In any case, Carlson gave him a gift that no one else had ever offered: enough time and space to explore the existing literature and discover what called to him. Kleitman found it later that year, in Henri Piéron's book, *Le Problème Physiologique du Sommeil* (in English, *The Physiological Problem of Sleep*)—the most comprehensive monograph on the topic then available.[24]

Published in 1913, Piéron's 520-page masterwork provided an encyclopedic overview of sleep science's evolution since its birth just a few decades earlier.[25]

Although Western thinkers had pondered the mechanics of sleep since ancient Greece—Aristotle thought it was triggered by vapors rising from the stomach to the heart during digestion—no one had studied the topic seriously until the advent of the Industrial Revolution. Two main forces fueled the surge in interest. The first was the drive for labor efficiency, which spurred researchers to investigate the capacities and limits of the human body in all its functions. Because sleep affected not only a worker's fitness but the timing of the workday, understanding its dynamics was essential. The second impetus was the rise of insomnia—an epidemic that emerged with the arrival of cheap and plentiful artificial lighting, which enabled millions to hitch their routines to the alarm clock and shift-change whistle rather than the rhythms of dawn and dusk. As the disorder grew more common, questions about how sleep worked (or didn't) became increasingly urgent.[26]

Piéron's book cataloged the major lines of inquiry, beginning with the riddle of what triggered sleep. In the mid-nineteenth century, researchers had observed that cerebral blood flow diminished in sleeping animals; they deduced that sleep resulted from such a decline. But this theory left many issues unsettled. For example, if the brain was starved of blood each night, how did it manage to generate dreams? And what started the shut-down process in the first place? Some scientists pointed to new research on muscular fatigue, which showed that a buildup of waste products such as lactic acid could inhibit movement in an overused arm or leg. Perhaps a similar process might immobilize a weary brain.[27]

Thus arose various "chemical" theories of sleep—all based on the premise that toxins accumulated in brain cells over the course of the day, eventually curtailing cerebral activity. After a few hours of complete rest, the thinking went, the poisons would be cleared away, leaving the sleeper alert and refreshed. Yet repeated attempts to identify a somnogenic agent proved futile. Experimenters forced animals to inhale carbon dioxide or injected them with alkaloids known to play a role in muscle fatigue. These substances sometimes induced stupor—or death, at high doses—but, bafflingly, never slumber.[28]

If the mechanisms that turned sleep on and off remained mysterious, so did the behaviors of the brain and body *during* sleep. Among the most basic enigmas was why people slept heavily at some points during their nightly voyage and lightly at others. Did these dips and rises follow predetermined cycles, or did they occur in response to external conditions? How much time did sleepers typically spend at one extreme or the other, and how much in the middle?

In 1862, University of Leipzig medical student Ernst Kohlschütter published the first study to chart the depth of sleep from pillow-time onward. Kohlschütter experimented on six fellow students over eight consecutive nights, using a device known as a *Schallpendel*, or "sound pendulum"—a pendular hammer that smacked a slate slab, generating bangs whose loudness could be adjusted by changing the hammer's elevation. The greater the racket required to awaken the subject, the deeper his sleep was judged to be.[29]

Kohlschütter's study suffered from deficiencies in control: At one point, a subject was inadvertently roused by the researcher lighting a cigar. But a more serious flaw was his decision to discard all data points that failed to support his hypothesis—that sleep depth, after reaching its maximum about an hour after onset, diminished in a smooth curve until morning.[30]

In 1888, an Estonian medical student named Eduard Michelson set out to repeat Kohlschütter's study—but with up-to-date methods and without the analytic bias. Michelson's subjects were four young physicians, including himself. Instead of a hammer, he used an electrical apparatus that dropped brass balls of varying weights onto a wooden board next to each sleeper's head. The operator sat in the next room and ran the machine by remote control; rather than watch for ambiguous signs of awakening, he waited for the subject to press a button.[31]

When Michelson crunched the data, it confirmed Kohlschütter's finding that sleep reached its maximum depth about an hour after onset. But the study's other results were startling. Rather than grow steadily shallower as the night progressed, sleep swung rhythmically between peaks and valleys, bottoming out five times in a typical seven-hour session. Each cycle lasted about eighty-four minutes, with average sleep depth diminishing as the waves rolled on. Michelson regarded these "remarkably regular fluctuations" as a "very strange phenomenon"—and by Piéron's time, scientists still had no grasp of their origin or significance.[32]

Le Problème Physiologique du Sommeil described other fascinating experiments. In France, psychologist Nicolas Vaschide had tried to gain a more granular understanding of what happened during sleep by monitoring subjects' pulse, respiration, and motor activity; when these measurements spiked in unison, he woke the sleepers, who told him they'd been dreaming.[33] Other researchers had observed similar activity in sleeping dogs.[34] These findings put a scientific spin on an ancient puzzle: What are dreams for? Do they perform a necessary biological function, or are they a mere by-product of other processes—perhaps, as Charles Dickens suggested in *A Christmas Carol* (1843), the effect of "an undigested bit of beef, a blot of mustard, a crumb of cheese, a fragment of an underdone potato"?[35] No answers were yet available.

The book also covered an emerging debate: whether sleep was controlled by a particular area of the brain. This possibility was first raised in 1890, by Viennese ophthalmologist Ludwig Mauthner. During an outbreak of sleeping sickness in Italy and Austria—characterized by extreme drowsiness, often followed by coma and death—Mauthner noted that common symptoms included drooping eyelids and paralysis of the eye muscles. He was also struck by reports that deceased patients showed inflammation in an area of the midbrain known to control those muscles. Mauthner suggested that sleeping sickness arose from damage to this region, which likely also held an on-off switch for wakefulness. He went on to propose that normal sleep resulted from a temporary break in the region's circuitry, interrupting transmissions between the cerebral cortex and the rest of the nervous system.[36]

Yet Mauthner's theory, like those of his successors, was little more than an educated guess. Scientists were just beginning to establish that different brain regions oversaw distinct aspects of mental and physical function. By the late 1800s, it was clear that the cortex—the convoluted layer of tissue covering the brain's surface—was responsible for "higher" functions such as sensory perception, voluntary movement, and thought. Some lower regions appeared to govern primitive drives and emotions; structures in the brain stem, just above the spinal cord, controlled involuntary functions such as respiration and heartbeat. Researchers were also learning that small areas of the cortex were crucial to specialized behaviors—movement or sensation in an arm or finger; the production or understanding of spoken language. Evidence for a so-called sleep center, however, remained tenuous.[37]

The relationship between sleep and the brain's cellular mechanisms was equally uncertain. Although the existence of brain cells was discovered in the 1830s, the details of their structure remained hidden for over fifty years until new imaging techniques brought them fully to light. It was now known that in both the brain and the peripheral nervous system, messages were passed back and forth by specialized cells known as *neurons*. Most neurons had branching limbs called *dendrites*, which received nerve impulses, and a threadlike fiber called an *axon*, which transmitted them. Inspired by these findings, researchers began to propose "histological" (or tissue-related) theories of sleep, involving the supposed actions of these anatomical features. But these notions, too, were purely speculative. The tools to observe brain cells in action—whether in sleep or waking—did not yet exist.[38]

One thing that scientists could measure directly was how organisms were affected by a *lack* of sleep. The first to do so was a pioneering woman physician and biochemist—Maria Mikhailovna Manaseina, better known as Marie de Manacéïne. In 1894, the Russian researcher reported that prolonged sleep deprivation could be deadlier than starvation. De Manacéïne had kept ten puppies awake for more than a week. Although dogs could survive twenty to twenty-five days without food, the subjects in her study were "irreparably lost" after four to five days without sleep.[39]

Other investigators replicated and expanded upon De Manacéïne's work. Among them were the American psychologists George Thomas White Patrick and J. Allen Gilbert, who in 1896 conducted the first study of sleep deprivation in humans. The pair kept three of their University of Iowa colleagues awake for ninety hours and found that the effects included visual hallucinations, slowed reaction times, diminished grip strength, and weight gain.[40]

Such results lent credence to the notion that sleep might result from an accumulation of fatigue toxins. Scientists redoubled their efforts to find such substances.[41] The most extensive study was conducted by Piéron himself, one of the founders of French experimental psychology. In search of what he called "hypnotoxins," he deprived dozens of dogs of sleep for extended periods, then injected their blood, puréed brains, or cerebrospinal fluid into the brains, veins, or bellies of normal dogs. He found that injecting the fourth brain ventricle with cerebrospinal fluid from a sleep-deprived animal induced "a more or less irresistible need to sleep" in the recipient.[42]

Yet Piéron was unable to identify the chemical or chemicals involved. Nor was he—or any of his colleagues—committed to solving such enigmas. Sleep was still a sideline for scientists in a range of fields; no one seemed to have thought of making it their life's work.[43]

Kleitman found these explorations electrifying, and he was intrigued by the vast areas that remained uncharted. "I was always interested in the activity of the nervous system in reaction to our environment," he later wrote. "And sleep fascinated me as a temporary, purely physiological

---

* A note to readers: Today, this experiment (and some others that will be described in this book) may strike us as horrifyingly inhumane. However, before the advent of tissue cultures and computer simulations, few alternatives to animal models were available for investigating physiological processes that could be harmful to humans. The prevailing view among philosophers as well as scientists was that the benefits to our species outweighed any agony experienced by members of a different one. See Nuno Henrique Franco, "Animal Experiments in Biomedical Research: A Historical Perspective," *Animals* 3 (2013): 238–273.

suspension of that activity." *Le Problème Physiologique du Sommeil* left him hungry to know more.[44]

Two more elements may have fed that hunger. The first was an epidemic of brain disease initially detected in 1916 among patients in a Viennese neurological clinic, whose horrific constellation of symptoms— which could include vision disorders, convulsive movements, muscular rigidity, psychosis, and catatonia—fit no existing diagnosis. Constantin von Economo, a young neurologist at the clinic, identified abnormal sleep as the common element in all these cases. Patients often dropped off while sitting or standing, or slept constantly for weeks on end. Other victims were unable to sleep at all, though they might eventually lapse into somnolence or coma. Examining the brains of deceased patients, von Economo found distinctive patterns of tissue damage. In 1917, he published his findings in an Austrian medical journal, dubbing the ailment *encephalitis lethargica*. By the early 1920s, tens of thousands of cases had been reported worldwide.[45]

With no known cure, encephalitis lethargica killed 40 percent of patients and left 46 percent partially or fully disabled (often with symptoms resembling severe Parkinson's disease, as described in Oliver Sacks's 1973 classic *Awakenings*). It also revived the notion—until then derided by most experts—that a "center" in the brain might be responsible for sleep. Intriguingly, von Economo had found lesions clustered in the same part of the midbrain that Mauthner had pointed to during the sleeping sickness epidemic of 1890, though damage turned up in several other regions as well.[46]

The other factor that may have spurred Kleitman to focus on sleep was that one of the world's most celebrated scientists—the Russian physiologist Ivan Pavlov—had lately turned his own attention to the topic. Pavlov had won a Nobel Prize in 1904 for his work on digestion, which he studied by observing the gastric secretions of dogs through surgically crafted windows, or fistulas, in the creatures' throats and stomachs. That research led him to formulate one of the bedrock practices of behavioral psychology: classical conditioning, in which an animal learns to associate an "unconditioned" stimulus (say, a bowl of food) with a "conditioned" stimulus (a buzzer that sounds whenever food is presented), responding in the same way to the latter (by salivating) even when the former is absent. Pavlov also found that conditioning could have an inhibiting effect. For example, if a dog heard the mealtime buzzer numerous times without receiving food, it would learn to *stop* salivating at the sound. And when dogs received repeated or excessive stimuli, they often fell asleep.[47]

Pavlov believed this reaction was a form of generalized inhibition, a reflex meant to protect brain cells from being damaged by stress. He reasoned that normal sleep, too, must be an inhibitory response, set off by an accumulation of stimuli over the course of the day. This boldly reductionist theory, which was published around the time Kleitman was choosing his dissertation topic, would have served as a thrown gauntlet for a young scientist looking to make his mark.[48]

Whatever forces set the course of Kleitman's career, he seems to have achieved a key insight at the very start. To fully understand a pathology like encephalitis lethargica or insomnia, or to determine the validity of a theory like Pavlov's, it would be necessary to develop a detailed physiological picture of normal sleep. So far, researchers had barely scratched the surface. To make real progress, they would have to shoulder the task full-time, approach it in a systematic way, and keep at it indefinitely. Kleitman wanted to be the first to go all in.

His mentor, as it happened, was an ideal role model for such a mission: a scientist who placed methodical observation ahead of conceptual speculation, valued careful logic over intuitive leaps, and never lost sight of the clinical impacts of the phenomena he studied. Carlson often repeated a maxim to his students: "Keep your mouth closed and your pen dry until you know the facts." When they did open their mouths, his habitual response was a curt and Swedish-accented, "*Vat iss the effidence?*"[49]

Kleitman would emulate that hardheaded skepticism, along with many of Carlson's other predilections. Perhaps the most crucial was a fascination with the body's rhythmic patterns and how they were affected by brain activity. Early in his career, Carlson had shown that in embryonic horseshoe crabs, the heart begins to beat before it's connected to the nervous system—evidence that cardiac muscle cells produce their own rhythms independently of the brain, which acts as a pacemaker in developed animals.[50] Later, he'd turned to the cycles of the digestive system, recording the motions of his own gut while fasting for up to ten days at a time. Carlson also studied the digestive rhythms of animals, from dogs to frogs and turtles. But for several years, his principal subject was a young man named Fred Vlcek.[51]

As a boy, Vlcek had drunk lye, a caustic chemical that left his esophagus blocked with scar tissue. At twelve, he'd undergone surgery that allowed him to be fed through a fistula in his stomach. The fistula enabled Carlson to study Vlcek's stomach contractions by attaching a balloon to a tube and inserting it into the organ; he measured the movements using a pressure gauge and a kymograph (a device that uses a stylus on a rotating drum to record physiological processes). Because Vlcek could chew food but not swallow it, Carlson could investigate how sensory stimuli affected the stomach without the influence of saliva or the food itself. And because Vlcek could talk, unlike lab animals, the researcher could track the relations between his gastric processes and his mental ones.[52]

Carlson's research had convinced him that the urge to eat resulted from a complex interplay between the automatic rhythms of the stomach and a person's habits and mental states. An empty gut produced a wave of contractions every hour or two, he'd found, generating the signals commonly known as hunger pangs. The degree to which people *felt* hungry, however, depended not only on how long it had been since their last meal but also on the times of day when they were accustomed to eating and whether they were distracted by something else. This finding put Carlson at odds with Pavlov, who preached that gastric secretions and stomach contractions were triggered by the appearance of food or conditioned stimuli, and were thus controlled by the cerebral cortex—the part of the brain in charge of processing sensory information, among many other tasks. Carlson argued that the stomach (like the heart) was governed mainly by rhythms intrinsic to the body, though an individual's response to those rhythms could be affected by environmental, psychological, or other factors.[53]

From Carlson, Kleitman learned the value of monitoring the body's rhythms over an extended period, using a combination of direct observation and automated devices to compile a continuous record. Like his teacher, he relied mainly on human experimental subjects, often including himself. And the quest he undertook mirrored Carlson's: to chart an essential physiological cycle, investigate its origins, and gauge its adaptability to an organism's changing needs and behaviors.[54]

The cycle Kleitman hoped to explore, however, had never received a researcher's undivided attention. Where did sleep's rhythms come from? What were their essential characteristics? To what extent could they be modified? How would such alterations affect human health? Probing these questions, he sensed, could keep a scientist busy for a very long time.[55]

Sometime in the second half of 1922, Kleitman proposed this course of study to the department chair. Carlson encouraged him but warned that he'd be entirely on his own—no one at the university knew enough to supervise him. For a brilliant and fiercely self-motivated scholar, the caveat felt like a blessing. Kleitman would forever be grateful for Carlson's "broad-mindedness," as he later wrote, and for the deep trust that it implied.[56]

As warm-up exercises, Kleitman carried out a series of experiments on the effects of caffeine and narcotics on dogs, and another on normal sleep and forced insomnia in puppies.[57] Then he embarked on a project that heralded his vaulting ambitions: the most systematic, thoroughgoing, and technologically advanced study yet undertaken on the physiology of sleep in humans.[58]

Kleitman's avowed goal was to outdo Patrick and Gilbert's 1896 paper on sleep deprivation, and two similar studies that had followed it. All three, he felt, involved too few subjects (one to three apiece) and too few trials (one each), and measured too few physical and mental functions. Only one study included a fully rested control subject with which to compare results.[59]

The fledgling researcher set out to correct those shortcomings. In his study, he and five male classmates underwent multiple stretches of forced insomnia ranging from 40 to 115 hours. Kleitman himself stayed up the longest; he endured more than a dozen sleepless periods, as well as periods of control testing, in which he slept as long as he liked. He had initially intended to keep conditions uniform by having participants lie awake all night on a cot in the physiology lab. When this proved impractical, due to the somniferous effects of reclining, they were made to walk around the neo-Gothic campus or sit in an all-night café to avoid nodding off. "This method is not then perfect," he admitted, "but it is the best we could devise at this time."[60]

Kleitman's approach to data collection made his predecessors' efforts look lackadaisical. Before, during, and after the insomnia sessions, subjects were given tests of memory, concentration, and reaction time. Their pulse, temperature, respiratory rate, oxygen consumption, blood pressure, blood cell counts, and blood sugar were measured, as were the levels of phosphorous, chlorides, nitrogen, creatinine, and acidity in their urine. The pupillary reflex and position of the eyes were examined by spreading the eyelids with the fingers and shining a beam from a pocket flashlight, and the reflexes of the arms, hands, and feet were observed by tracing a pencil along the skin. The face was stimulated by touching it with sharp pieces of paper. Notes were taken on participants' subjective experience and their general behavior. At the end of each session, the length and depth of slumber was recorded for two nights. The point was to detect new clues about the function of sleep by analyzing the effects of its absence as exhaustively as possible.[61]

The most striking results, Kleitman found, involved the urge to sleep. As Carlson had observed regarding hunger, the sleep drive increased over time, but its progress was not linear. Instead, it fluctuated according to the time of day and the activities in which an individual was engaged. Subjects typically felt wide-awake for most of the first night. They had no trouble studying, reading, or doing lab work, though they grew drowsy between 3:00 and 5:00 a.m. Alertness rebounded later in the morning and was easy to maintain as long as subjects kept physically or mentally active. During the second night, they were afflicted with dry eyes and a "buzzing in the head," and found it difficult to study without falling asleep—but as before, physical exercise usually banished the urge. During the third night, staying awake was much harder. Subjects perked up the next morning, but tasks requiring close attention became impossible. When asked to take their own pulse, most lost count within twenty beats.[62]

Although participants invariably found it easier to stay awake when they were busy, keeping them occupied was more challenging at night. "It was amusing to note that some subjects used ruses to escape the watchful eye of the observer," Kleitman wrote. "After being roused many times they pretended to get up for a stroll in the corridors of the building, but would actually walk over to some corner, sit down and fall asleep almost immediately." The sole individual who was kept awake for more than four nights began to experience brief waking dreams. On his fifth morning, subject N.K. (as Kleitman labeled himself) was doing a calculus problem for a mental-acuity test; instead of calling out the answer, he declared, "it is because they are against the system." When questioned, he explained that he'd thought he was having "a heated argument...on the subject of labor unions."[63]

There were other intriguing findings. For example, the night after an insomnia session, subjects slept for only an hour or two longer. The longer they'd been awake, though, the *deeper* they slept. The physical effects of prolonged sleeplessness also included lowered heart rate and respiration, decreased phosphate excretion, and a lessening of daily variations in body temperature, which would normally cycle a degree or two between afternoon highs and predawn lows.[64]

Kleitman titled his study "The Effects of Prolonged Sleeplessness in Man." Published in September 1923 in the *American Journal of Physiology*, it served as his public debut. (It also formed the centerpiece of his doctoral dissertation.) Though little recognized, the paper would come to be known as one of the seminal studies in sleep science, brimming with observations that generations of researchers would devote themselves to investigating further. Besides its content, however, what was remarkable was its tone: incisive, playfully erudite, and magisterially self-assured. The article was clearly meant not only to describe a set of experiments but to announce its twenty-eight-year-old author's arrival—and his intention to transform his chosen field.[65]

He began with the big picture. "Most investigators of the physiology of sleep, in reporting their findings, remind their readers, by way of apology, of the tremendous importance of the subject for the advancement of our knowledge of physiology as a science, as well as for the rational treatment of insomnia," he wrote. Fewer than a dozen scientists were currently investigating this "great physiological mystery."[66]

Kleitman surveyed an impressive portion of the literature on sleep before describing his study's eye-opening results. But the pièce de résistance was a nine-page section titled "Discussion and Theory." Kleitman began by endorsing Piéron's definition of sleep: "a suspension of the sensori-motor activities that bring the living being into relation with its environment." As to why this suspension occurred, he continued, there were several popular notions. One was that the circulatory system becomes fatigued at the end of the day, sending less blood to the brain. Kleitman's own experiments, he noted, showed no evidence of such a phenomenon; moreover, the fact that physical or mental activity made it easier to stay awake made the theory's premise doubtful. Then there was the idea that sleep was triggered by fatigue toxins or by a reflex designed to prevent such toxins from accumulating. Kleitman had detected no biochemical or physiological changes that would support this theory.[67]

Lastly, he turned to Pavlov's finding that animals invariably fell asleep when experiencing "prolonged action of a uniform excitant." Kleitman, it emerged, had been carrying on a correspondence with the famed physiologist—astonishing for a mere grad student. Although Pavlov's complete data had not yet been published, he had stated "in a personal communication to the writer" that "sleep and the so-called internal inhibition of a conditioned reflex are identical phenomena." Kleitman didn't rule out this possibility, but he ventured to question the great man's logic. A person can fall asleep "when not fatigued at all, and idlers have no difficulty in falling asleep…at any hour," he pointed out. And in humans, excessive stimulation seemed more often to prevent sleep than to promote it.[68]

He then offered an elaborate theory of his own, drawing on findings from an array of other scientists. Sleep, Kleitman suggested, was triggered by muscular relaxation, which commonly occurred after people—or other animals—reduced sensory input by lying still and closing their eyes. The process that followed probably involved a blocking of neural communication between the brain's motor centers and the rest of the body, as Mauthner had proposed three decades earlier. (Kleitman cited his own finding that stroking a subject's foot during deep sleep provoked the same toe-spreading reflex as in a person suffering damage to the motor neurons.) There also seemed to be a partial blockade between areas of the brain responsible for sensory processing and those involved in reason and judgment.[69]

This blockade led to a possible explanation for dreams: They might result from lower brain centers misinterpreting nerve impulses emanating from the body. "In the absence of external stimuli, internal sensations [such as] hunger, thirst, or sexual stimuli may start a dream," Kleitman wrote. "Distended seminal vesicles will give rise to an erotic dream, but the powerful stimulus of ejaculation is necessary to overcome all synaptic resistances and waken the sleeper." He did not comment on whether repressed desires might also be involved as Sigmund Freud's contemporaneous and increasingly popular theory asserted. For Kleitman, that speculation was a step too far.[70]

As for sleep's cyclical patterns, he thought they might be primarily the result of habit. "Why is it easier to sleep at night than in the daytime, and why will a person accustomed to get up at, say, 6 a.m., wake up at the usual hour, whether he goes to bed early or late?" This tendency, he suggested, could be analogous to the rhythms of hunger, which in humans corresponded more to accustomed mealtimes than to the biological need to eat.[71]

After cautioning that his suggestions were "provisional," Kleitman concluded with a quote from Piéron: *Une théorie n'est pas la solution d'un problème, c'est au contraire l'énoncé d'un problème à résoudre.* A theory is not the solution to a problem but the statement of a problem to be resolved.[72]

———————

Kleitman had stated the terms of the problem that he would spend most of his life trying to unravel. But he wouldn't start right away. The newly minted PhD applied for and won a National Research Council fellowship to study physiology in Europe. That fall, he boarded a steamer for the Netherlands, where he spent six months at the University of Utrecht working with Rudolf Magnus, a pioneering researcher on animal reflexes. In February 1924, he moved on to Paris, where he began an apprenticeship with three of his idols: Piéron, with whom he investigated retinal nerve function at the Collège de France, and the trailblazing husband-and-wife neuroscientists Louis and Marcelle Lapicque, with whom he studied electrical impulses in nerve cells at the Sorbonne. These pilgrimages were only tenuously connected with sleep; his aim seems to have been to deepen his knowledge of the nervous system in general.[73]

During the summer holidays, Kleitman was visited by one of his professors—Arno B. Luckhardt, known for his discovery of ethylene anesthesia, who was traveling to the German town from which his own parents had emigrated. He also journeyed to see a stepsister who was studying pharmacology in Italy. Kleitman then reunited with his mother, Pesea, who spent some time in Paris before accompanying him back to Kishinev, where he stayed for several weeks.[74]

His hometown had passed from Russian to Romanian rule after World War I, but life there had grown no easier in the ten years since Kleitman fled. An economic crisis had plunged many merchants and professionals into bankruptcy and deepened the misery of the poor. A fascist party, the

National Christian Defense League, was on the rise; like its counterpart in Germany, it adopted the swastika for its banner and called for the expulsion of the Jews. That fall—with no indications of regret—he left for the last time and headed back to Paris where he remained until the end of December.[75]

Kleitman spent most of 1925 finishing his fellowship at the University of Chicago, then took a train trip to California with two cousins, stopping at several national parks along the way. On October 1, he was hired as an assistant professor of physiology at a salary of $3,000 (worth about $43,000 today).[76]

Although it would take a while for anyone else to notice, the modern era of sleep science had begun.

3

Coming into the Territory

The Physiology Building at the University of Chicago was an ornate Victorian chateau, four stories tall, topped with a triple crown of gables and a glowering gargoyle.[1] Within its ivied walls, Kleitman established the world's first dedicated sleep laboratory in 1925—a milestone that was recognized only retrospectively. At the start, he had no lab area of his own; he simply turned whatever space he occupied into a staging ground for his mission.[2] Nor did he confine his research entirely to sleep. The publication lists that he scribbled or typed at the end of each academic year reveal that he continued to pursue other interests (ranging from the nerves of the eyeball to body-righting behavior in the duck) through the early 1930s.[3] Nonetheless, he set himself one central task: to transform sleep science from an obscure backwater into a thriving and coherent discipline.[4]

For a few hours each week, Kleitman lectured on physiology. The remainder of his time, day or night, was devoted largely to thinking about sleep, reading about sleep, writing about sleep, or conducting sleep experiments. The thirty-year-old lived frugally, socializing occasionally with relatives from Kishinev who'd settled in Chicago, and sending a portion of his modest pay to his stepsister and two cousins who were studying abroad.[5] In photos from this period, his expression has lost its youthful ardor. Above the stiff collar and somber necktie, his dark hair is neatly parted, his face round and placid as a winter moon. His eyebrows have vanished (whether due to stress, illness, or some other factor is unknown), adding to the air of inscrutability.[6] Only his eyes hint at the zeal that burned behind the cool exterior, or the stubbornness that banked the flames.

Kleitman wasn't the only scientist studying sleep in 1925, but he was the only one dedicated primarily to that pursuit—as he would remain for nearly three decades.[7] Aided by a rotating squad of graduate students and junior professors, he attacked the mechanics of slumber from every angle then available: behavioral, biochemical, cardiopulmonary, neurological.[8] Over time, the evidence he gathered would help answer some of the fundamental questions about sleep. His methods would provide models for other researchers, his writings would furnish a framework for the field's most important debates, and his mentorship of younger scientists would enable revolutionary discoveries. Yet what set him apart from the competition, in many cases, was not just what he did but what he *didn't* do: play to the crowd, oversimplify his findings, or present his evolving theories as proven facts.

When Kleitman began his work, the best-known American sleep researcher was probably Donald A. Laird, chair of the psychology department at Colgate University.[9] A native of Angola, Indiana,[10] Laird was two years younger than Kleitman[11] and just as driven. Sleep, however, was not his principal concern; indeed, Laird's main focus seemed to be on displaying the boundlessness of his expertise. Newspapers across the nation quoted his pronouncements on topics including Prohibition ("Recent experiments...have shown that even a weak liquor lowers one's mental and motor control"), criminality (in a study of Ohio prisoners, "three out of four were found to be mentally deficient"), gender differences ("Men live their emotions, while women think theirs"), and labor policy ("A four-hour day in industry would be possible if workers were placed in work for which they are best fitted").[12]

Laird's sleep studies at Colgate began in 1924, funded by a grant from the National Research Council.[13] He set up experiments at a frat house, where teams of assistants used the latest technology to gather a wide range of data. Student volunteers wore gas masks that collected their nocturnal exhalations, whose chemistry was analyzed to determine energy expenditure. Motion detectors kept count of the subjects' movements, and a "somnokinetograph" (the only machine of its kind in the world, Laird boasted) measured their degree of relaxation. Sleepers were awakened at intervals and subjected to physical tests, such as lifting a weight with a finger until muscle fatigue set in. Their quality of sleep was measured after eating different types of foods, after fasting, after drinking coffee or caffeinated tea, and while lying on different kinds of mattresses. They were asked to fill out questionnaires on their dreams and on their alertness or sleepiness during waking hours.[14]

But Laird seemed more interested in collecting facts about slumber and burnishing his reputation than in deepening science's understanding of the phenomenon. Over the next decade, he wrote prolifically for newspapers and magazines on the results of his sleep studies. His findings were divided almost evenly between the obvious and the dubious. On the obvious side, he revealed that heavy people sleep better on a firm mattress, and lighter ones on a softer model; most people, moreover, slept poorly in a room that was too warm or too cold, or when too much light or noise came through the windows. On the dubious side, he determined that women slept better than men because they ate more candy—carbohydrates being conducive to sound slumber. He also declared, based on his questionnaires, that 75 percent of people slept without dreaming.[15]

Laird had a penchant for issuing firm advice based on flimsy premises and for casually contradicting his own proclamations. Because sleep grew shallower toward dawn, he warned, morning slumber was "beneficial neither to your body nor your mind," and was "likely to do you grave harm by helping to build up a pattern of laziness." Instead, it was best to sleep for just six hours a night and to compensate with a forty-five-minute nap around noon, which he claimed was equal to three hours of sleep between 3:00 and 7:00 a.m.[16] Laird later urged against this practice, without mentioning that he had once championed it.[17] He also flip-flopped on the issue of sleeping position. Early on, he insisted that it was essential to lie fully extended and to alternate between sleeping on one's side and one's stomach.[18] In 1932, however, he counseled, "Your body will instinctively pick out the best position....Let your sleeping self make the choice."[19] Countless sleeping selves must have uttered a sigh of relief.

By then, Laird had established a full-scale sleep lab, in a ten-room house on the Colgate campus; his book *Sleep: Why We Need It and How to Get It* (1930) had become a bestseller.[20] (Such manuals, which interspersed commonsense tips with nuggets of the latest scientific information, had been popular for a century.)[21] But he left the field in 1939, after becoming director of a consumer research foundation,[22] and his work is largely forgotten today.[23]

At least Laird recognized that slumber was necessary. Other experts dismissed that idea out of hand. A syndicated newspaper story from 1925, headlined NOW SCIENCE PROPOSES TO ABOLISH SLEEP, featured an interview with Columbia University psychologist H. L. Hollingworth, who insisted that although sleep once helped our ancestors avoid being ambushed by predators during the hours of darkness, it was no longer needed in an age when "we can turn night into day by merely pressing a button." We inherit this "disastrous habit," Hollingworth argued, "just as we inherit our annoying appendix, our eyebrows, our useless tonsils, our dangerous wisdom teeth, the degenerate muscles that wiggle our ears, and a number of other vestigial traces of our lowly origins." By reducing one's sleep by five minutes every two months, he suggested, it should be possible to eliminate the nightly atavism in sixteen years. "I think it would be quite a sensible thing to do," he said.[24]

Kleitman, by contrast, refrained from giving any advice at all. During the first decade of his labors, he was quoted in the press on a dozen or so occasions, but almost always concerning the nature of sleep rather than its practice. He remained focused on basic science—the kind aimed at increasing human knowledge rather than solving practical problems, though transformative solutions can arise from it. He wanted, above all, to understand sleep's rhythms: their characteristics, their origins, and their relationship to other bodily and mental functions. Intent on testing his own and others' theories, he showed little interest in influencing popular opinion. The attention he craved was that of his scientific peers.[25]

Kleitman began cautiously, publishing a short paper in 1925 on excretion and respiration during sleep in humans.[26] He also continued his work on human sleep deprivation, subjecting himself and several graduate students to bouts of experimental insomnia lasting up to 155 hours. Those trials attracted some media coverage, in brief articles with headlines like STUDIES SLEEP BY GOING WITHOUT IT or CHICAGO SCIENTISTS SUFFER AGONIES IN STUDY OF EFFECTS OF WAKEFULNESS. "After the first 24 sleepless hours," the Associated Press noted, Kleitman and his assistants found that "continuous speech or action was necessary to keep [subjects] from dozing off... They went cabareting, played horseshoes, indulged in wordy arguments and stationed guards to keep prodding them at the first sign of slumber."[27] The stories usually included a simplified version of Kleitman's working hypothesis on the conditions necessary for sleep to begin. As the *New York Herald-Tribune* put it: "So long as muscles are in contraction, either by exercise or by merely being held tense, sleep seems impossible. Conversely, if all muscles are successfully relaxed and kept that way, sleep soon ensues."[28]

Yet Kleitman was out for bigger game. Sometime during his first months on the faculty, he witnessed an experiment in Pavlovian conditioning at the University of Chicago's pharmacology lab. Instead of using food as the unconditioned stimulus to make dogs salivate, the researchers employed an injection of morphine. After a few sessions, the animals began salivating the moment an experimenter entered the room. "[It] occurred to us," Kleitman later wrote, "that this reflex might be used to test Pavlov's theory of sleep."[29] As in his debut paper two years earlier, he would take on the Russian Nobel laureate—but this time, more directly.

Pavlov's theory seemed simple at first glance. It was unveiled in a lecture (published in a Swedish physiology journal in 1923) titled "Internal Inhibition and Sleep Are Essentially the Same Physicochemical Process." When neurons in the cerebral cortex received repetitive or excessive stimulation, he posited, they shut down to avoid being damaged. This "internal inhibition" began with isolated groups of cells but could spread across the entire cortex and into the lower regions of the brain, leading to a state that Pavlov called "generalized inhibition." The phenomenon, he wrote, "is a daily occurrence. It is our sleep and the sleep of all animals."[30] From that point, though, the conceptual apparatus grew more baroque.

Pavlov had arrived at the notion of generalized inhibition not by observing natural sleep but by analyzing the behavior of dogs undergoing experiments in his St. Petersburg laboratory—where the relationship between stimulus and somnolence was anything but straightforward. Sometimes the animals dozed off when there was a long delay between being placed in the lab stand and receiving the conditioned stimulus, or between conditioned and unconditioned stimuli. Pavlov developed elaborate explanations for these varying responses. One general rule, he claimed, was that when dogs first learned to salivate in response to a conditioned stimulus, they reacted immediately—but with further training, as excitatory impulses waned, the interval lengthened. The reason animals often fell asleep during this delay, Pavlov argued, was that inhibitory impulses in the part of the cortex that controlled salivation had grown strong enough to spread throughout the brain.[31]

This unlikely hypothesis for why a dog might take a nap during a tedious or stressful experiment got surprisingly little pushback from established physiologists, perhaps due to Pavlov's towering stature in the field. But Kleitman and a graduate student, George Crisler, set out to test the assertion's basic premise: that animals responded more slowly to a conditioned stimulus as their training progressed.[32]

The study involved eight dogs, prepared with fistulas in the jaw enabling the collection of saliva in a test tube. The researchers began by placing each subject in a stand; they then injected the animal with morphine, which induced drooling. After several sessions, the dogs learned to associate the stand with the narcotic and started to salivate in expectation of a shot. Gradually, Kleitman and his assistant increased the delay between the conditioned stimulus (placement in the stand) and the unconditioned one (injection with the drug) to as long as two hours. They recorded the moment that salivation began and ended, how many milliliters were secreted, the timing and duration of sleep episodes, and a variety of other behaviors.[33]

The pair's first report on these experiments was published in the *American Journal of Physiology* in February 1927. Its focus was on salivation rather than sleep, but its implications for Pavlov's theory of generalized inhibition were not positive. Over the course of the yearlong study, some of the subjects did fall asleep before receiving morphine, but none displayed the delays in drooling that the model predicted.[34]

"[Our] results differ from those obtained by the workers in Pavlov's laboratory," Kleitman and Crisler wrote, "in that no delayed conditioned reflex was developed at all. Indeed, just the opposite was true." When the dogs first learned to associate the stand with the morphine, it took them fifteen to twenty minutes to start drooling after being placed in the device. Over the following days, however, "the secretion began to pour out earlier, at a gradually increasing rate of flow, until in the fully established reflex the flow reached a maximum value as soon as the animals were placed in the stock and remained constant for the whole pre-morphine period. Nor did our dogs stop secreting saliva if, at the end of the customary test period in the stock, they did not receive morphine. They only slowed down a little."[35]

The paper went unnoticed by the popular press. But it was the opening volley of an attack that would launch Kleitman to a new level of prominence.

———

February 1927 also marked another pivot point in Kleitman's life. That month, a mutual friend introduced him to the partner who would sustain him and his explorations for the next five decades. Paulena Schweizer was raised in the pre-statehood Oklahoma Territory, the daughter of frontier Jews. Her father, Jacob, was an immigrant from Germany who found his way to the settlement of El Reno, became a successful entrepreneur, and served with Theodore Roosevelt's Rough Riders in the Spanish-American

War. Her mother, Minnie, was a socialite from the cattle town of Wichita, Kansas.[36]

Dark-haired and fine-featured, Paulena was drawn to frontiers of a different sort. After studying acting at the American Academy of Dramatic Arts in New York City and elocution at the Kansas City Conservatory, she had spent some time on that city's amateur theater scene.[37] She'd gone on to earn a bachelor's in sociology at the University of Chicago[38] and to volunteer at Hull-House, run by the progressive reformer Jane Addams, which treated the ills of urban poverty through education and cultural exchange.[39] When the thirty-two-year-old Kleitman met her, she was thirty-three and working at the Jewish Social Service Bureau—an idealistic, independent-minded woman with a playful streak beneath the sober surface. Mutually smitten, they married in a judge's chambers just four months after their initial encounter. The couple honeymooned in Kansas City and Los Angeles, where Kleitman met Paulena's sisters and aunts. Then his new bride replaced his roommate in his furnished apartment off campus.[40]

In the fall of 1928, Kleitman returned to Paris, where he spent another few months working with Piéron and the Lapicques—and where Paulena discovered she was pregnant. The following May, back in Chicago, she gave birth to a daughter, Hortense. Later that year, Kleitman was promoted to associate professor and awarded tenure, and the family moved into a spacious apartment on Drexel Avenue. He'd come a long way from his days as a castaway in Mrs. Reisher's tenement flat.[41]

By then, construction had been completed on a new physiology building (later named Abbott Memorial Hall). Part of the university's just-opened medical school, it was a six-story gothic fortress with an arched gateway and a crenellated parapet.[42] Thanks to his promotion, Kleitman finally got a lab in the traditional sense—a two-room suite behind his small office, equipped with metal shelving, marble-topped tables, industrial sinks, and outlets for Bunsen burners. A cot could be installed whenever an experiment called for it.[43]

The young scientist mounted the second stage of his assault on Pavlov the following year, at the XIIIth International Physiological Congress.

Held at Harvard Medical School, the conference was the first such meeting to convene outside Europe. A milestone for American science, it was also the largest physiology conference yet, drawing nearly 1,700 researchers from forty-one countries. Among them were Pavlov and four other Nobel Prize winners, as well as Kleitman's mentor Piéron. The whole crowd sat for a panoramic portrait outside the assembly hall—a battalion of stern-visaged men in three-piece suits, with a few women in cloche hats (mostly, but not all, wives) scattered among them.[44]

Pavlov, at seventy-nine, was "the most notable figure of that great assemblage," one attendee observed, "[and] the most eager and untiring participant in the long drawn-out program." The white-bearded icon delivered a lecture on the broad topic of inhibition. "Vivid, alert, gesticulating, the old man poured out his phrases…never missing fire," neurosurgeon Harvey Cushing recalled. The audience was enraptured, though not necessarily by the substance of Pavlov's discourse. As biographer Daniel P. Todes later wrote: "Physiologists and psychologists certainly respected his scientific achievements, but very few understood his esoteric terminology, fewer still had examined the details of his experiments and conclusions, and only a small handful had adopted his experimental approach. The consistently enthralled accounts of his appearances were notable for their lack of comment about scientific content. They dwelled, rather, upon the great liveliness and energy of this great old man of physiology."[45]

Meanwhile, Kleitman had examined the details that others glossed over—and his intent was to dismantle them. No manuscript of his talk at the congress survives, but it covered the same ground as a paper he published soon afterward in the *American Journal of Physiology*.[46] According to Pavlov, Kleitman noted, salivation should stop *before* sleep set in as inhibition spread across the cerebral cortex. In his own dog experiments, however, "drowsiness and sleep came first, and the cessation of the salivary flow followed, most probably because the action of the conditioned stimulus could not be asserted on the sleeping animal." Kleitman spent several pages describing his methods and enumerating his results, but the message was straightforward: something besides generalized inhibition must send animals to dreamland.[47]

Kleitman's lecture doubtless shocked many who heard it—an upstart's brilliant takedown of a revered elder. (Whether it shocked Pavlov is unknown: Kleitman mentioned in an autobiographical sketch that they met at the conference, but he discreetly refrained from describing the encounter.) Nonetheless, it established his credentials as a scientist to watch, in the most prestigious arena imaginable.[48]

Kleitman with one of his laboratory dogs in the late 1920s. *Courtesy of the Hanna Holborn Gray Special Collections Research Center, University of Chicago Library.*

The lecture also brought Kleitman his widest press coverage yet, though often as an object of mild ridicule. "Professor Ivan Pavlov, a distinguished

Russian investigator, says that his researches lead him to believe that sleep results when there is some inhibition of reflexes to stimuli," reported a typical item, in the *Honolulu Advertiser*. "The opposing theory comes from Dr. Nathaniel Kleitman of the University of Chicago, who holds that sleep comes first and automatically shuts off the reflex activity. In other words, sleep is the cause and not the effect of cessation of reflexes." To science, the article continued, the question was far from settled. "But to the man in the street, or more literally the man in bed, it does not matter much....If the neighbors upstairs would shut off the radio and the baby would stop crying and flaming youth next door would stop practicing on the saxophone and the old pay check was as big as all the bills that have to be paid on the First we could all go to sleep and leave the physiologists to figure out why any way they please."[49]

The *Advertiser*'s pundit had a point. For people struggling to get a good night's sleep, research like Kleitman's offered no immediate help. Yet the advice of experts like Laird was often equally ineffectual. Since doctors first detected an insomnia epidemic in the 1890s, the plague had continued to spread, despite the publication of countless articles and books touting surefire cures. Many sufferers turned to medication when reading, warm baths, a light snack, or relaxation exercises (the experts' favored prescriptions) failed to work.[50] The available pharmaceuticals, however, had serious drawbacks. By 1930, an estimated one billion doses of barbiturates—the latest sleep aids, introduced around the turn of the century—were consumed each year.[51] Like the bromides that preceded them, these drugs had been introduced as a safer alternative to earlier sleep meds but had turned out to be highly addictive and prone to abuse. Dosages had to be increased over time to achieve the desired result, often leading to potentially fatal overdoses.[52]

And insomnia wasn't the only sleep-related ailment that science couldn't unravel. There was narcolepsy, a debilitating disease characterized by daytime sleep attacks and sudden bouts of temporary paralysis.

There were parasomnias such as sleepwalking and night terrors, which made sleep itself dangerous or harrowing.[53] There was shift work—not a medical condition but a health-sapping misery for many of the millions compelled to practice it. "I wash up, go home, eat, and go to bed," wrote a journalist who worked nights at a steel mill as part of an exposé of the industry's labor practices. "Anything that happens in your home or city that week is blotted out, as if it occurred upon a distant continent."[54]

What was it about modern society that made insomnia so prevalent? Why were both sleeplessness and excessive sleepiness so difficult to treat? Why did night workers often feel exhausted, even when they got eight hours of sleep a day? Such questions were not purely physiological; they also touched on matters of psychology, sociology, technology, and economics. (That "old pay check"; that radio upstairs.) But they could not be fully answered until sleep's mechanics were better understood.

Researchers were making some progress. In 1926, Constantin von Economo published a study in a German journal pinpointing two crucial areas of brain damage in encephalitis lethargica, the disease he'd identified nine years before. Both occupied the hypothalamus, an almond-shaped region just above the brain stem. In patients who exhibited excessive sleep, lesions were found in the posterior hypothalamus and its junction with the midbrain; in those who showed excessive wakefulness, there were pathological changes in the anterior hypothalamus and its junction with the forebrain. Von Economo posited that these areas constituted a sleep-regulation center, divided into a "sleep part," or *Schlafteil*, and a "wake part," or *Wachteil*. He expanded on this idea in a 1929 lecture at Columbia University's College of Physicians and Surgeons. Slumber was a complex process, he theorized, in which the sleep and wake centers worked in coordination. The rhythm of their interplay was likely governed by the endocrine system, whose periodic release of hormones to control bodily functions was just becoming known.[55]*

---

* For a diagram of the brain, see page 289.

Von Economo never had the chance to pursue this proposition further; he died of a heart attack in 1931, at fifty-five.[56] That year, however, the Swiss physiologist Walter Rudolf Hess reported that by electrically stimulating the posterior hypothalamus in cats, he had induced sleep while stimulating the anterior hypothalamus had triggered extreme excitement and fight-or-flight behavior. Hess argued that these responses confirmed the existence of a center governing both sleep and arousal in that tiny portion of the brain.[57]

Kleitman wasn't convinced. He remained attached to the theory he'd proposed in his first paper, nearly a decade earlier: that sleep was triggered by a reduction of sensory input, and that its rhythms were governed primarily by forces native to the cerebral cortex, such as habit, conditioning, and volition. Yet he had never put that notion to the test.[58]

Now, he set about doing so, using a technique that physiologists had long employed when studying the brain—removing a structure to determine what functions it controls. If the cortex was in charge of sleep's rhythms, as Kleitman believed, then getting rid of it would presumably cause sleep to occur in an erratic way.

It would be unethical to perform such an experiment on people, of course, since it would destroy their ability to think, communicate, or otherwise engage with the world. So Kleitman turned once again to dogs, which he considered to be naturally monophasic sleepers (slumbering in one long chunk), like adult humans.[59] This was, in fact, a questionable premise for both species—probably based on his unfamiliarity with cultures that practiced biphasic sleep,* as well as with dogs. In reality, those animals are *poly*phasic, sleeping in sessions averaging about eighty minutes with periods of waking in between. It is true, however, that they sleep mostly at night, napping for only about three hours over the course

---

* Before the Industrial Revolution, evidence suggests, people in many parts of the world slept in two sessions, separated by a period of wakefulness that could be used for chores, prayers, or other purposes. See A. Roger Ekirch, *At Day's Close: Night in Times Past* (New York: W. W. Norton, 2006), 300–311.

of a day.[60] Previous experiments had shown that dogs' waking behavior became aimless and disorganized after the cortex was cut away, but no one had examined whether their sleeping behavior followed suit.[61]

With a young Haitian surgeon named Nelaton Camille (who later became mayor of Port-au-Prince),[62] Kleitman performed the gruesome operation on several dogs, four of which survived long enough to be studied. In February 1932, the pair published the results: Decorticated dogs appeared to sleep normally—"curled up or sprawled on the floor, eyes closed, and breathing quiet and regular"—but they no longer did so in any discernible pattern. Instead, they dozed for anywhere between thirty minutes and several hours; after waking, they would wander in circles for an equally random period of time, then sleep again, without regard to the positions of the sun and moon.[63]

To Kleitman, these findings didn't rule out the existence of a subcortical sleep center, whether in the hypothalamus or elsewhere—but if one existed, he argued, it couldn't work properly on its own. "We see that diurnal sleep, developed around the 24-hour cycle of day and night, is...a definitely cortical phenomenon, has its center or is located in the cerebral cortex, disappears when the cortex is removed," he and Camille wrote. A subcortical center might act as a simple on-off switch, but more sophisticated circuitry was needed for "the special diurnal sleep seen only in higher animals and in man."[64]

Decades later, it would emerge that von Economo's view was closer to the truth: The hypothalamus plays a key role in regulating the rhythms of wakefulness as well as sleep, triggering the release or reduction of hormones and neurotransmitters that promote slumber. Other subcortical centers help orchestrate the process as well.[65] The likeliest explanation for Kleitman's results is that his dogs' cycles were disrupted by the trauma of their surgery; their sleep might have returned to normal had the animals not been euthanized soon afterward.[66]

Although Kleitman had guessed correctly that the cortex plays a role in sleep, its contributions turned out to be vastly different from what he imagined.[67]

Kleitman's work in the early 1930s was increasingly influenced by events outside the lab. One inescapable factor was the Great Depression, whose impact deepened with every passing month.[68] Across the country, banks collapsed, businesses folded, and farms went bust. As unemployment soared to 25 percent, the jobless hopped freight trains to distant towns in search of work. Families queued for meals on breadlines, and shantytowns sprang up in vacant lots and public parks.[69]

At the University of Chicago physiology department, Anton Carlson supported some PhD candidates using his own salary, allowing one of them to live in his lab and eat food intended for the experimental animals. The department chair worked heroically to drum up funding for his researchers, wrangling gifts from wealthy friends and product-testing work from manufacturers. Kleitman oversaw several such studies, including one in which graduate students drank a bottle of low-alcohol beer every fifteen minutes, with or without a liverwurst sandwich.[70] A food conglomerate, the Wander Company, gave him two yearly grants of $3,400 (worth $70,000 apiece today) to test several of its products. Among them was the malted-milk mix Ovaltine, which the company touted as a health tonic and insomnia remedy. Kleitman's trials suggested that three teaspoons might improve sleep slightly, though two—the suggested serving—had no significant effect. Wander promptly launched an advertising campaign proclaiming that University of Chicago researchers had confirmed Ovaltine's soporific power. The ads were pulled after Kleitman and Carlson protested, but the two men came in for some ribbing from their colleagues. On the departmental bulletin board, someone pinned a cartoon depicting a nubile bride glaring at her sleeping husband. Its caption: "Damn that Ovaltine!"[71]

Despite these distractions, Kleitman launched a series of studies aimed at scrutinizing human sleep with a granularity that no one had yet attempted. He had four intertwined goals: to analyze the rhythms of sleep; to determine how those rhythms interacted with the body's other cycles;

to test how different variables in waking behavior affected sleep rhythms; and to investigate how changes in sleep rhythms affected people's daytime performance. By charting factors such as motility, body temperature, and mental acuity at different times of day and night, and under varying conditions and routines, he hoped to map not only the contours of normal sleep but also to learn what distinguished it from abnormal varieties.[72]

Of all the variables Kleitman aimed to examine, only the body temperature cycle had been studied extensively. It had long been known that temperatures in humans fluctuated by a degree or two each day, reaching a minimum sometime after midnight and a maximum in the afternoon. Kleitman hypothesized that the daily cycles of both temperature and sleep were dictated by patterns of habit and behavior (such as food intake, muscular exertion, and brain activity) rather than by rhythms intrinsic to the brain or body. He also posited that both cycles could be changed without harm, though a person might need some time to adjust. But the only way to find out for sure was through observation and experiment.[73]

He began his explorations shortly after the birth of his second daughter, Esther, in 1931.[74] Family members would remember Kleitman as an attentive and nurturing father—but he was a scientist first and foremost.[75] Although it was known that infants slept far more than older children or adults, their sleeping patterns had never been examined in detail. He attached an array of instruments to both little girls' cribs (and later to their beds), recording all the data he could gather.[76]*

Kleitman did the same with his adult study subjects, often using devices of his own invention. To measure the duration of sleep movements, he linked motion detectors to automatically triggered clocks. To gauge the movements' strength, he created a sensor that measured the

---

* When Kleitman didn't have a preprinted form on hand, he recorded his data on any available scrap of paper—including, one day in 1942, eleven-year-old Esther's French homework. On the back of the sheet are Kleitman's entries; on the front is a typewritten list of phrases useful at a Paris hotel. In the margins of the list, the sentences "Esther likes to talk to the boys" and "Esther Kleitman is very smart" are scrawled in girlish cursive. Nathaniel Kleitman Papers, Box 3, Folder 8, Hanna Holborn Gray Special Collections Research Center, University of Chicago Library.

depression of mattress springs. Later, he connected an electric rectal thermometer to these gadgets, enabling him to trace the relationship between motility and body temperature in real time.[77]

The motility experiments showed that sleepers moved frequently throughout the night, though they spent less than a minute per hour doing so. It could be "stated with certainty," Kleitman later wrote, that in normal people, motionless sleep was neither possible nor desirable. He found no significant relationship between motility and sleep quality, though movement did increase during the hours before waking. Nor did he detect any definite connection between motility and body temperature in healthy sleepers.[78]

Kleitman then turned to diurnal variations in cognitive and other types of performance. He tracked subjects' speed and accuracy at mental gymnastics (multiplying complex numbers, translating letters into code), and gauged their physical steadiness (holding a stylus in a hole, standing with eyes closed), at different times of day and night. Meanwhile, he measured their body temperature. On average, Kleitman concluded, performance reached "a maximum in the afternoon and minima early in the morning and late at night," with temperatures running in rough parallel. People's minds were sharper and their reactions faster, it seemed, around the peak of the temperature cycle—a finding that would eventually have a profound impact on his work and that of his scientific heirs.[79]

Next, he studied the effects of alcohol and caffeine on motility and temperature during sleep. Booze, Kleitman found, depressed movement during the first half of the night and increased it during the second half, affecting temperature much the same way. Coffee, by contrast, increased both movements and temperature all night. "N.K., an abstainer from coffee, complained of a 'miserable night' after the consumption of three cups of the beverage," he wrote of himself.[80]

In April 1934, Kleitman addressed a meeting of the American College of Physicians in Chicago, which was covered by the United Press. For the first time, he was able to share experimental findings that were relevant to ordinary readers. The researcher "blasted many a superstition of the man who has learned about sleep only by doing it," the article reported. "For example: You do not sleep better just before dawn. You sleep soundest some 10 minutes after hitting the sheets. Alcohol does not disturb sleep. Quite the opposite.* You are not at your best early in the morning. Most people gain in activity until noon; some reach the peak of efficiency as late as six p.m."[81]

Nor could most humans function efficiently with minimal slumber as theorists like Professor Hollingworth claimed. Not everyone required the standard eight hours, Kleitman told the assembly, but everyone was entitled to "at least that much sleep, because sleep is a pleasure. A little more of it than is needed makes the world a brighter place to live in, and certainly a safer one."[82]

———————

Kleitman was beginning to sense that his research might lead toward some world-changing applications. He was also starting to gather material for a comprehensive book on sleep, a successor to the one that had inspired him to enter the field—Piéron's *Le Problème Physiologique du Sommeil*.[83] (With Paulena's help, he developed a system of color-coded file cards for classifying studies from around the world, which he stored by the hundreds in his office.)[84] For his investigations to reach their full potential, however, he would need to find funding that didn't require him to spend time testing drink mixes. Carlson urged his protégé to reach out to the Rockefeller Foundation, which was financing a planned neuropsychiatry department at the University of Chicago medical school.[85]

In March 1934, Kleitman wrote to the director of the foundation's medical sciences division, Alan Gregg, outlining his research and suggesting that it would fit perfectly with the mission of the new program. His experimental approach, he told Gregg, made him uniquely positioned to

———————

* This would prove to be a misapprehension; later studies showed that alcohol disturbs sleep in the second half of the night when (as Kleitman correctly found) it increases body temperature. See Jordan Gaines Lewis, "The Science of Why Alcohol Is Bad Before Bed," Dreams, https://www.dreams.co.uk/sleep-matters-club/the-science-behind-why-alcohol-is-bad-before-bed.

"settle the controversy concerning the localization of a sleep 'center' in the cerebral cortex or the subcortical structures." Opening a lab in the neurophysiology unit, which was to be located at Albert Merritt Billings Hospital, would enable him to study diurnal rhythms in "mental defectives" (as the intellectually disabled were then known), to determine "the extent to which the cortical center of sleep dominates the subcortical one." He also hoped to study sleep disturbances in encephalitis lethargica, narcolepsy, and insomnia, as well as in "psychopathic individuals." With more cash and research subjects, he promised, his work could achieve clinically useful results. Gregg agreed to a meeting at his office soon afterward, when Kleitman would be in New York City for a conference.[86]

It's easy to imagine Kleitman's excitement before this encounter—and also, almost certainly, his dread. Not long ago, he had been eating chicken necks on the Lower East Side. Although he'd traveled a remarkable distance since then, he remained an outsider in crucial ways—not only as a member of a widely despised minority but also as the champion of a discipline that, strictly speaking, did not yet exist. Gregg, by contrast, was from an old New England family, the Harvard-educated son of a prominent Congregationalist minister who'd followed his calling to Colorado Springs. Just five years Kleitman's senior, he had spent nearly a decade running Rockefeller projects from Brazil to Paris before assuming his current position. He was insiderness personified.[87]

This social chasm may help explain why the meeting unfolded as it did. Perhaps Kleitman found Gregg intimidating. Perhaps his own considerable pride made it painful to beg this philanthropic potentate for favors. Or perhaps the immigrant scientist, conscious of lingering stereotypes about "pushy" Jews, tried too hard to keep his eagerness from showing. In any case, Gregg found Kleitman's manner stiff and awkward, his arguments unconvincing.[88] "I told him we could not do anything on the project before autumn and that I would think it over," Gregg wrote in his diary.[89] Months of silence followed.[90]

In May 1934, Carlson wrote to Gregg, urging him to give this promising scientist a second hearing. Kleitman's work, he explained, was "a

significant part of the problems in neuro-psychiatry that are being pursued on this campus," but was threatened by departmental budget cuts.
Gregg did not respond. Nor did Kleitman succeed in arranging another
meeting when he passed through New York in September, after his annual
trip to Europe with Paulena and the girls (this time to Italy, where they
vacationed with his mother and stepfather and visited relatives who'd emigrated from Kishinev). In January 1935, when there was still no answer,
Carlson sent a more insistent letter. "Dr. Kleitman and I," he wrote, "are
naturally anxious for some word from you concerning this matter."[91]

Still skeptical, but swayed by Carlson's persistence, Gregg traveled to
Chicago. This time, he met Kleitman in his lab, where an astonishing
transformation took place. Surrounded by his jury-rigged instruments—
whose ingenuity Gregg could now grasp for himself—the formerly diffi-
dent researcher was self-assured, enthusiastic, and persuasively articulate.
The interview, Gregg noted in his diary, was a "remarkable illustration of
[the] advantages of seeing a man in his own laboratory." Kleitman, he later
observed, had begun his career as a lab assistant at the Rockefeller Institute, and "must have reverted to his old mood in his first presentation in
our office, for only on second presentation . . . did he do himself justice."[92]

With Gregg's support, Kleitman received nearly $16,000 from the
Rockefeller Foundation (around $285,000 today) over the three-year
period beginning in April 1935. He set up his new laboratory at Billings
Hospital and prepared to expand his explorations.[93]

———————

Around the same time, a Belgian neurophysiologist named Frédéric
Bremer published a sensational paper in a French journal, *Comptes rendus
des séances de la société de biologie*. Bremer hadn't set out to study sleep;
he'd wanted to understand how the cerebellum—a region behind the
brain stem—regulates posture. To explore the workings of an animal's
cerebellum, most researchers removed the forebrain (the upper portion
of the brain). Bremer decided to take a less destructive approach. After

opening a cat's skull, he left the forebrain intact but isolated it from the lower brain by slicing through the brain stem just above a structure called the *pons*.[94]

To Bremer's bewilderment, the anesthetized animal plunged into what appeared to be a deep sleep, marked by contracted pupils and downward-rolled eyeballs. The scientist confirmed that impression using an electroencephalogram—a remarkable new tool that could distinguish states of consciousness by reading brain waves. In his paper, titled *"Cerveau 'isolé' et physiologie du sommeil,"** he drew a seemingly logical conclusion: when stimuli from lower regions were prevented from reaching the upper brain, sleep resulted automatically—just as Kleitman and likeminded researchers had been saying all along.[95]

The *cerveau isolé* study helped strengthen Kleitman's conviction that he was on the right track.[96] But a formidable obstacle soon blocked his path: academic politics. The neuropsychiatry department's chairman, who envisioned an innovative interdisciplinary program focusing on basic science, clashed with more conservative medical-school administrators and quit after a turbulent year. The chairman's replacement was a plodding administrator who had little interest in physiological research. Kleitman made scant headway in this uncongenial environment. In March 1936, having spent little of his Rockefeller funding, he wrote to Gregg apologizing for the delay; their relationship subsequently soured.[97]

In the years covered by the grant, Kleitman succeeded in publishing just a handful of papers.[98] Only one was noteworthy: a study of sleep and temperature rhythms in thirty-five infants at two Chicago hospitals. Although other scientists had charted the development of the human temperature cycle, he and his coauthors did so with far greater thoroughness and precision. Their results showed that the diurnal temperature curve was not fully established until the second year of life, when children had also learned to sleep through the night. Kleitman took this finding as evidence

---

* Translation: "'Isolated' brain and physiology of sleep."

for his theory that rhythms of both temperature and sleep arose mainly from habits and learned behavior—that is, from phenomena whose development followed that of the cerebral cortex.[99]

His paucity of publications, however, did not reflect a lowering of aspirations. In 1937, Kleitman began to steer his studies of diurnal rhythms in a new direction. He'd become obsessed with a question no previous researcher had investigated: What would happen to the temperature curve if sleep were no longer attached to a twenty-four-hour day?[100]

This was not an abstract question. Such a dislocation could happen with shift work, for example, or in people whose sleep schedules were thrown off-kilter by illness or insomnia. Given his earlier findings that the cycles of body temperature and mental performance rose and fell together, it was crucial to investigate how the sleep-wake cycle fit into the picture of the industrializing world.[101]

For a month, Kleitman tried to shift his own sleep-wake rhythm to follow a forty-eight-hour cycle—an effort that involved staying awake for thirty-nine-hour stretches and then sleeping for nine hours. His temperature rhythm stayed stubbornly in place. Later, he placed a student on a twelve-hour cycle, with two sleep periods of three and a half hours per day; the subject's temperature curve refused to follow. Doubling or halving the conventional sleep cycle likewise barely changed the undulations on a monthlong graph. These results suggested that there might be limits to how far sleep rhythms could be modified without putting them permanently out of sync with temperature rhythms.[102]

To determine where those boundaries lay, Kleitman tried less-radical alterations, placing himself and several students on cycles lasting twenty-one or twenty-eight hours. Before long, some of the students were able to acclimatize to these patterns. Kleitman's temperature, however, stubbornly stuck to a twenty-four-hour cycle, and he felt more irritable and exhausted with each altered day.[103]

Why couldn't he adjust? Were his cortical circuits less adaptable than those of other subjects, or were they more sensitive to external stimuli? It was impossible to tell without controlling for such stimuli—the cues of sunset and sunrise; of cooler, quieter nights and warmer, noisier days.[104] Hoping to eliminate those variables, Kleitman planned an expedition to Norway's Arctic hinterlands, to be undertaken during the midsummer period of round-the-clock sunshine. In January 1938, he asked the Rockefeller Foundation to finance the project. Gregg had come to regret his earlier investment, however, and tersely declined the request.[105]

The rejection must have been wrenching. Kleitman was now forty-two; he'd been tilling his arcane field for fifteen years, with little to show besides two dozen small studies that—however provocative their findings—raised far more questions than they answered. He'd unveiled some of sleep's hidden patterns and debunked some misconceptions about its nature, but he'd solved no deep mysteries and made no great mark. He had disappointed the benefactor who could have changed his fortunes, squandering the efforts Carlson had made on his behalf. The funding for Kleitman's work at Billings Hospital would expire that fall and was unlikely to be renewed. Not since his days as a student POW in Beirut had his way forward seemed so uncertain.[106]

Then a colleague suggested a trip to Mammoth Cave.[107]

4

Descent and Transformation

Between ten and fifteen million years ago, in what is now south-central Kentucky, trickles of groundwater began probing the cracks in a fossil seabed. Over the eons, the pockets grew and grew until they'd formed the most extensive cave system in North America—over four hundred miles of underground limestone chambers, canyons, tubes, shafts, and keyhole passageways, interwoven with Stygian rivers.[1] Indigenous people first explored the caverns around 3000 B.C.E., lighting their way with cane torches; they used mussel-shell scrapers to mine gypsum and other minerals (likely used in medicines, pigments, and agriculture), and left behind petroglyphs, pictographs, and mummies. White settlers, carrying oil lamps, followed in the 1790s.[2]

The earliest commercial use of Mammoth Cave was as a source of saltpeter, shipped east by the ton to make gunpowder for the War of 1812. In the 1830s, local entrepreneurs set about transforming the site into a tourist attraction, often enlisting enslaved Black men as explorers and guides.[3] One of the latter, Stephen Bishop, discovered and mapped many of the cave's major landmarks while still in his teens. Known as "the Sable Genius," he became world-famous for his wit, charm, and geological erudition. (Bishop, who described the place as "grand, gloomy, and peculiar," was granted his freedom at thirty-five, only to die of unrecorded causes the following year.)[4] As hotels sprang up around the cave entrances, visitors poured in. In 1926, spurred by a growing movement to put the national wonder in public hands—and by an outbreak of business rivalry known as the "cave wars," featuring arson, dynamite, and clashing gangs of roadside touts—the federal government authorized the creation of Mammoth Cave National Park.[5] By 1938, three years before the transition was completed, more than seventy thousand people flocked to the grottos annually.[6]

When Kleitman told fellow faculty members he was seeking a locale for a monthlong sleep experiment—someplace as isolated from the rhythms of day and night as the Arctic in summer, but less expensive to get to— University of Chicago geologist J Harlen Bretz said he knew just the spot. A renowned cave expert, Bretz volunteered to make introductions and help with logistics.[7]

That February, Kleitman wrote to the property's general manager, W. W. Thompson, describing his requirements for an unprecedented expedition. Unlike the era's other trailblazing voyages (Lindbergh leaping the Atlantic, Byrd bagging the North Pole), this one would go inward: not just below terra firma but deep into the mysteries of the human brain and body. Kleitman needed a space for himself and an assistant, he told Thompson, removed from the flow of foot traffic but accessible to delivery of food and other supplies. The pair would spend thirty-two days and nights keeping to a twenty-eight-hour sleep-wake schedule, and it was imperative that they receive no sensory cues to the cycles of the twenty-four-hour world. He had one more request: that the project be carried out discreetly. "We are anxious to avoid unnecessary publicity," Kleitman cautioned, "at least until we are through with our work in the cave."[8]

Thompson agreed. He offered the use of a chamber called Rafinesque Hall, about a quarter mile from the entrance. The vast space—2,500 feet long by 150 feet wide, with a 40-foot ceiling—sat just beyond the itinerary of the tour groups, off a passage known as Audubon Avenue. Room service would be provided by the Mammoth Cave Hotel, along with the furniture for Kleitman's live-in laboratory. The hotel's carpenters built two iron beds to his specifications, with a shelf below each mattress for the motion-detecting apparatus; to keep out rats and other vermin, the frames stood forty inches off the ground, and the legs were planted in water-filled buckets. Other amenities consisted of a folding dining table, two dining chairs, two rocking chairs, a washstand, and bedside tables for recording equipment. Portable generators powered the scientific instruments and an emergency telephone, but gasoline lamps supplied the only light.[9]

On June 4, Kleitman and his companion—a graduate student named Bruce Richardson—made the short hike to their base camp. The two cast disparate shadows on the boulders that loomed over the outpost; Richardson, a rangy twenty-five-year-old from Simpsonville, South Carolina, stood a head taller than his slight, round-faced professor. Kleitman had chosen him, however, for what they shared in common: Both had found it impossible to adjust to a twenty-eight-hour schedule in previous experiments. The question was whether they would do better in a diurnally unvarying environment, 140 feet beneath Earth's surface. With no clues to the time outside their burrow, would their bodies adapt more easily to the rhythms of an artificial day? If not, did that mean (as a few scientists were beginning to suspect) that some internal clock kept organisms marching to its beat? Whatever the results, there could be profound implications for the way humans ordered their sleep-wake cycles aboveground.[10]

Kleitman and Richardson kept to a strict regimen, attempting to sleep from 12:00 a.m. to 9:00 a.m. Sunday; 4:00 a.m. to 1:00 p.m. Monday; 8:00 a.m. to 5:00 p.m. Tuesday, and so on—nine hours in bed and nineteen hours out for each cycle.[11] A hotel employee, Halver Johnson, brought meals, mail, newspapers, and smoking supplies twice a day, and spirited away the contents of the chamber pots. Encompassed by darkness, the men ate broiled Kentucky ham, fried chicken, and T-bone steak.[12] To pass the time, they organized data, wrote letters, read periodicals and books (both devoured the best-selling swashbuckler *Anthony Adverse*), brushed up on bridge rules, or took brief walks around the cavern. With the air a constant fifty-four degrees Fahrenheit, they wore overcoats, hooded sweaters, galoshes, and two layers of woolen pants to ward off hypothermia. Every two hours, they measured their own temperatures, relying on alarm clocks to rouse them during sleep periods. As they lay dreaming, the motion detectors kept vigil, disgorging ticker tape into a box beside each bunk.[13]

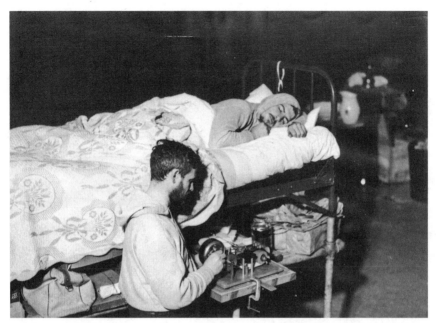

Kleitman (in bed) and Bruce Richardson in Mammoth Cave. *Courtesy of the Hanna Holborn Gray Special Collections Research Center, University of Chicago Library.*

For five twenty-eight-hour cycles, events in the cave went precisely as planned. But despite the property manager's vow of secrecy, the newspapers got wind of the expedition, and a swarm of reporters descended on his office. On June 10, Thompson sent a note with the food delivery, begging Kleitman to allow a visit from a representative of the press, who would be chaperoned by the park's publicity director. "There are... a number of rumors going around concerning your and Mr. Richardson's stay in Mammoth Cave, none of which is accurate," he wrote. "I can assure you that we do not want to impose on you in any way but feel that this would enable us to handle the story in a dignified way, which I know is the very thing you want."[14]

Kleitman was distressed at this development. "Publicity may be welcome in Washington or Hollywood," he later explained, "but it is definitely frowned upon among scientists."[15] He feared that an influx of newshounds could undermine both his academic reputation and the conditions of his experiment.[16] Yet there was danger, too, in refusing the request and letting rumors be reported as fact. He consented to an interview, and correspondent S. V. Stiles—chief of the Louisville bureau of the Associated Press—arrived the next day.[17]

Stiles's seven-hundred-word dispatch turned out to be a humdinger, full of fascinating science and colorful details. "Neither man has shaved in a week," he reported. "There's no sunrise or sunset to disturb them... yet they find themselves still unconsciously guided by the twenty-four-hour cycle that has ruled all their lives." He explained the project's rationale with admirable clarity and described the explorers' living conditions down to the cigarette packs scattered across the breakfast table. The men assured him that their spouses had no reason to worry; the only discomfort was a slight dampness to the bedsheets. "When wives have crazy husbands, they get used to their doing things like this," Kleitman said with a laugh, "and we feel that we may contribute something to the knowledge of men's reactions."[18]

The story was picked up across the country, and it quickly developed legs—whether because of its sheer novelty, or because of the intriguing questions it posed about one of the basic cycles of human existence. A few days later, the AP reported that the duo had used their emergency phone to check on the outcome of the world heavyweight championship fight between Joe Louis and the German boxer Max Schmeling. (Louis, the African American "Brown Bomber," swiftly dispatched Hitler's purported übermensch.) More reporters came calling.[19] Then Kleitman received a request from *News of the Day* to shoot a newsreel on the last day of the expedition. "We will in no way ridicule the experiment," the manager of the film crew wrote. "I have a letter from my New York office guaranteeing this at my request." The scientist again said yes.[20]

The clip, which was shown in thousands of movie theaters, opens with a title frame: STUDYING MYSTERY OF SLEEP, SCIENTISTS LIVE MONTH IN CAVE. "An alarm clock calls time on a unique experiment," the announcer declaims as Richardson awakens and Kleitman extracts a length of tape from the bedside box. "Two Chicago university experts, who in the service of science have been living in the depths of Mother Earth for more than a month, end their test."[21]

Then comes a close-up of Kleitman, his bearded face glowing eerily in the shadows. "We were entirely successful in our undertaking," he says in a reedy voice with a hint of an accent, "but want it known that it is in no way a stunt or an act of endurance or perseverance, but a bona fide scientific experiment."[22]

After Richardson adds a few words ("The most pleasant part of the experiment, to me, has been my ability to sleep"), we see a group of park workers hiking down to escort the explorers back to civilization. "To penetrate the recesses in which the two men dwell," the announcer says, "the visitors cross a subterranean river and work their way through stone jungles of stalactites and stalagmites." The procession arrives in Rafinesque Hall, where each member shakes hands with Kleitman and Richardson. "Well," the voiceover concludes cheerily, "if the experiment helps us to sleep better these hot nights, we vote for science!"[23]

The newsreel is in some ways misleading. There were, in fact, no stalactites or stalagmites along the path to the chamber; that footage had been shot in another part of the cave.[24] The announcer mispronounces Kleitman's name as *Kleet*-man.[25] And contrary to the scientist's declaration, the experiment was not entirely successful. Once again, he'd been unable to acclimatize to a twenty-eight-hour schedule while his student had done so within two weeks. Kleitman's body temperature continued on its twenty-four-hour curve, and he always began feeling sleepy around 10:00 p.m.—even when that time fell during the morning or afternoon of the artificial day. He later suggested that either age or variations among individuals might account for the difference between his and Richardson's adaptability. The sample size, he admitted, was too small to draw conclusions.[26]

Yet such details did nothing to diminish the story's appeal. On July 7, 1938, when the men climbed the seventy-one steps from the cave entrance to the surface, dozens of journalists, park officials, townspeople, and tourists thronged to greet them—a sea of straw boaters and seersucker suits, sparkling with camera flashes. Beneath their beards, one reporter wrote, the men's skin bore "a slightly greenish cast." It was 7:45 on a midsummer evening, and sunset gilded the surrounding woods.[27]

What struck Kleitman most sharply, he told the newspapers, was the fragrance of the forest—a pleasant shock after more than a month with little to smell but dank stone, Southern cooking, tobacco smoke, and bodily excretions. The clamor of the crowd must have been startling, too, as well as the commotion of bats among the oaks and hickories, and the brilliance of the pink-and-orange sky. Sweating in the sudden heat, the two explorers peeled off their coats and climbed into a car, which whisked them to the hotel a few hundred yards away. After taking their first baths in four weeks, they repaired to the dining room for a banquet in their honor. There, Kleitman reunited with Paulena, who'd driven down from Chicago with one of his frequent coauthors, physiologist Francis Joseph Mullin. According to a picture caption, the dinner (flowered dresses, flowered wallpaper, long table of polite strangers) was "a joyful event to all concerned."[28]

The end of the expedition got front-page coverage. In the photo spreads, Richardson smiles raffishly while his mentor—reserved and formal even in his hoodie and unkempt whiskers—often looks as if he's been interrupted mid-lecture by an impertinent remark. Above his sober gaze, the headlines were gleeful: SCIENTIFIC CAVEMEN EMERGE... SCIENTISTS QUIT PLAYING MOLES IN SLEEP STUDY... OLD WORLD LOOKS STRANGE AFTER A MONTH IN CAVERN. Kleitman's return to Chicago was celebrated with a picture of his two young daughters caressing his beard. WELCOME CAVEMAN-SAVANT HOME.[29]

That week, there were other headlines. President Franklin Delano Roosevelt began a campaign to bolster the New Deal. British warships arrived to quell riots in Palestine. Chinese insurgents launched terrorist attacks on Japanese occupiers in Shanghai. Spanish loyalists pushed back fascist rebels in Valencia. In Maple Grove, Minnesota, a farmer, his wife, and a hired man were found murdered in their beds. To many Americans, the cave experiment was simply a welcome distraction from the general grimness.[30]

Its repercussions, however, can be felt to this day. In later decades, Kleitman's eccentric expedition would inspire a spate of experiments that revolutionized our understanding of how sleep-wake rhythms function. More fundamentally, his emergence from Mammoth Cave marked a new era: after nearly a century of germination, the science of sleep had come out from underground.

———

Kleitman was now a national celebrity. Nonetheless, Alan Gregg declined to renew the scientist's Rockefeller funding when it expired that October. "It is not evident," Gregg wrote in his final assessment, "that the grant for Kleitman's work...has provided results proportionate to the total expenditure of $15,583.74." Kleitman's results, he added, were "negligible without being false or useless." Perhaps the problem lay with the topic itself, at least insofar as it could be illuminated by one researcher. Sleep, he asserted, "remains a most common phenomenon, marvelous to the few and attractive as a major interest to almost no one. It is possible that this grant illustrates the error of attacking a large task in a small way."[31]

Although Kleitman never read those slighting words, the foundation's withdrawal of support conveyed their message.[32] Yet the impact of the blow was likely dampened by the sudden change in his circumstances. He no longer needed Gregg to validate the importance of his work. Kleitman had, in fact, convinced millions of people at one stroke that sleep was of "major interest"—and that it was a legitimate scientific field. The self-improvement guru Dale Carnegie dubbed him "probably the greatest authority in the world" on the topic.[33] Newspaper features described how Kleitman's "ten-thousand nights of scientific experiment" had "upset many popular ideas about the technique of sleeping," particularly the notion that people could get by with minimal slumber if they exercised sufficient willpower.[34]

Kleitman cemented his preeminence in 1939, with the publication of *Sleep and Wakefulness as Alternating Phases in the Cycle of Existence.* The first true textbook on sleep science, his 638-page opus (revised and enlarged in 1963) would serve as the field's bible for decades.[35] Kleitman had spent six years writing the tome, which provided a critical digest of virtually every study, including his own, in the twenty-six years since Piéron's *Le Problème Physiologique du Sommeil,* as well as historical background dating back beyond Aristotle.[36] The book was one of just a handful of scholarly entries in a marketplace dominated by titles like Donald Laird's *Sleep: Why We Need It and How to Get It* and Edmund Jacobson's *You Can Sleep Well: The ABC's of Restful Sleep for the Average Person.*[37] "Among a mass of literature on sleep published in recent years," declared the *Journal of the American Medical Association,* the volume represented a "refreshing scientific document in the midst of theory, superstition and suggestion." It offered a more cogent evaluation of "all the facts heretofore accumulated than any other work on the subject thus far available."[38]

Besides its scope and erudition, what made *Sleep and Wakefulness* momentous was its conceptual framework. Kleitman's research had led him to challenge the age-old assumption that sleep was "a periodic temporary cessation, or interruption, of the waking state, which is the prevalent mode of existence for the healthy human adult." Instead, he made it his objective to prove that "sleep is in reality a complement to the waking state...the one related to the other as the trough of a wave is related to the crest." Because of this interdependency, he argued, sleep science couldn't be just about sleep—it had to examine the waking state, too. By the same token, however, sleep was just as important as wakefulness and just as worthy of study.[39] This duality would become a credo for Kleitman's scientific heirs.

Another striking feature of Kleitman's magnum opus was the way it embodied his strengths and limitations as a thinker. The book displayed the evolution of his ideas since the start of his career, as well as the sometimes-contradictory facets of his scientific persona: his devotion to experimental rigor and his openness to evidence that might undercut his own theories, but also his bias toward a hierarchical worldview (very much in keeping with his times) where humans held the apex of the animal kingdom and the brain's highest region dominated our "cycle of existence."

Sleep and Wakefulness was divided into eight sections, each grappling with a different aspect of the cycle.[40] These set the stage for Kleitman's grand finale, titled "An Evolutionary Theory of Sleep and Wakefulness"— the most comprehensive explanation of the topic's *whys* and *hows* that any scientist had ever offered. "It is commonly stated that we know nothing, or practically nothing, as to what causes sleep," the chapter began. "By this type of expression we tacitly admit that wakefulness is our predominating mode of existence and that sleep represents an interruption of that state." Yet the common wisdom, he countered, had it backward. "From the evolutionary standpoint," he wrote, "it is perhaps not sleep that needs to be explained but wakefulness."[41]

Individuals as well as species may exhibit different types of wakefulness at different stages of development, Kleitman argued. In primitive animals, and in infants of more advanced species, sleep is the dominant mode. Such creatures awaken only when external stimuli or internal sensations—hunger, thirst, the need to relieve a distended bladder or move a cramped muscle—force the issue. He called this type of behavior "the wakefulness of necessity." Higher animals, once they've reached maturity, show a different kind of waking behavior: "the wakefulness of choice." Kleitman described this state as "a prolongation and modification of the wakefulness of necessity," in which both phases of the sleep-wake cycle became increasingly consolidated.[42]

Just as there were gradations between sleep and waking, there were gradations between these two types of wakefulness; dogs might show more "choice" than mice, for example. For *Homo sapiens*, however, choosing to stay awake came far more easily. Why? Because our species possessed the most elaborate cerebral cortex in existence. From the moment that humans woke up, the cortex was busy processing not only bodily sensations and external stimuli but also thoughts, memories, and plans. The result, Kleitman wrote, was "a constant stream of impulses" that kept the brain from shutting down. The length of wakefulness increased until early adulthood, when it comprised about 70 percent of the diurnal cycle, or sixteen to seventeen out of twenty-four hours.[43]

Building on this foundation, Kleitman went on to attribute nearly every aspect of the human sleep-wake cycle to choices, learned behaviors, and sensory input rather than "some general properties of protoplasm" (that is, biological rhythms).[44] What makes people feel the need to sleep after an extended period of wakefulness? Fatigue of the cerebral cortex—and of the muscles of the eyes, head, face, and neck, which take up more space in the cortex's motor areas than those of other body parts.[45] Why does sleep depth rise and fall throughout the night? Unlike Eduard Michelson, who'd judged these regular fluctuations a "very strange phenomenon" when he discovered them in 1891, Kleitman offered a prosaic explanation: Sleepers periodically grow uncomfortable lying in one position and come close to awakening while changing to another. What drives diurnal temperature rhythms? Different levels of mental and muscular activity during sleep and waking—though the temperature curve may persist through force of habit even when sleeping patterns change.[46]

Still, Kleitman was willing to admit he could be wrong. He concluded his book with the same quote he'd used in the paper that launched his career—Piéron's *Une théorie n'est pas la solution d'un problème, c'est au contraire l'énoncé d'un problème a résoudre.* A theory is not the solution to a problem but the statement of a problem to be resolved.[47]

———

Sleep and Wakefulness decisively established Kleitman as his field's foremost authority. Not until years later did his grand theory prove to be mistaken—in part due to discoveries concerning the "general properties of protoplasm," as well as to new findings in neuroanatomy and electrophysiology. (He turned out to be wrong, too, about humans having the longest wakefulness period of any animal: research showed that many species outlast us, including sheep, horses, elephants, giraffes, and migrating birds.)[48] For now, he stood alone on the summit. But soon after the book's publication, in July 1939, larger events overshadowed his triumph.

That summer, with the thunderheads of global conflict gathering, Kleitman skipped his usual European trip; instead, he sailed with Paulena and the girls to Bermuda, where he studied diurnal pigment changes in the beach woodlouse *Ligia baudiniana*. After their return to Chicago, his mother, Pesea, arrived on the French liner *Normandie* for a planned six-month visit. Two weeks later, on September 1, Hitler's forces invaded Poland, igniting World War II. Pesea moved in permanently with her son and daughter-in-law.[49] Kleitman's stepfather, Avram Roitberg, and a stepsister, Esther Torban, stayed behind in Kishinev and perished in the Holocaust along with other members of the extended family. (The exact circumstances remain unknown.)[50] Most of the city's seventy thousand Jews suffered the same fate, whether they were starved in the Nazi-imposed ghetto, executed in mass shootings, asphyxiated in mobile gas vans, or deported to brutal concentration camps elsewhere in the region.[51]

Like many other immigrants with loved ones in the killing zones, Kleitman was unable to learn of their circumstances until after VE Day; all he could do was worry. With nonmilitary research largely at a standstill,[52] he channeled his skills and energies—and, perhaps, his anxieties—toward aiding the war effort. He became a special consultant to the US Department of Labor, responsible for advising the government on the optimal timing of shift work for war production.[53]

Based on his earlier studies of the relationship between sleep schedules, body temperature, and mental performance, Kleitman devised a new plan to replace the traditional pattern of eight-hour shifts starting at 7:00 a.m., 3:00 p.m., and 11:00 p.m. Instead, he suggested, shifts should begin at noon, 8:00 p.m., and 4:00 a.m.; that way, no one would have to sleep in the afternoon, when it was typically most difficult, and no one would be forced to change their normal bedtime by more than four or five hours. He also urged that shifts be rotated no more than once every few months, so that workers' temperature rhythms would have time to realign with their sleep-wake schedules.[54]

In June 1942, the Labor Department published Kleitman's proposal as an eight-page pamphlet titled *Arranging Shifts for Maximum Production.* "More machine hours—that is what this country needs for greater production to win the war," the text began. "But while machines must run round the clock, the men and women who man them must work by shifts. With multiple-shift production comes the problem of night work: experience has shown that accidents often go up, and production down, on the graveyard shift." One way to prevent these problems, the pamphlet suggested, might be to adopt Kleitman's extended-rotation policy and to switch to the time blocks that he'd patriotically labeled the Red, White, and Blue shifts.[55]

The idea stirred widespread interest among private- and public-sector employers. Over the following months, Kleitman received dozens of letters requesting further information; the senders ranged from corporate executives to union representatives, from public utility operators to police chiefs.[56] The American Can Company reprinted the proposal in its company newsletter and asked for feedback from workers and managers at its sixty-seven plants.[57] Kleitman was invited to give a presentation to the Industrial Hygiene Foundation and to write an article for the *Mining Congress Journal.*[58] Yet no record exists of how many workplaces actually adopted the Red, White, and Blue plan, or how well it worked where it was implemented. The Labor Department never ran a controlled trial of the schedule, and Kleitman lacked the resources to do so himself. Thanks in part to his efforts, the principle of extended or permanent shift assignments became a lasting feature of enlightened labor policies, but the time slots he championed never took hold on a large scale.[59]

After the war, the navy agreed to let Kleitman try out his shift scheduling ideas on a submarine, the USS *Tusk,* but the experiment was a disaster: the scheme disrupted workflow, led to overconsumption of food supplies, and left the crew no more alert than before.[60] The fiasco reflected the state of sleep science a decade after Mammoth Cave. Kleitman's vision had begun to redefine the discipline. In the pages of his groundbreaking textbook, researchers could find a set of facts, concepts, problems, and techniques to serve as points of departure for new explorations. The discipline was developing a sense of intellectual coherence, as well as a modicum of visibility, respectability, and public influence. Yet its grasp of sleep's underlying mechanisms—and its ability to actually improve people's lives—remained severely limited.

The field's world-altering transformation would begin with an accidental discovery in Kleitman's lab, using a technology that he and other sleep researchers had all but forgotten.

PART II

CHASING DREAMS

5

The Body Electric

The invention that transfigured sleep science was conceived during a brief encounter with the uncanny. Its birth, however, required decades of obsessive labor.

On a spring morning in 1893, just before his twentieth birthday, a German cavalry recruit named Hans Berger was out on maneuvers in the vineyard-covered hills of Würzburg. As he rode along a steep ravine, his horse suddenly reared and flung him from the saddle. Tumbling down the slope alongside the animal, Berger landed beneath the wheels of a massive field gun. He shut his eyes and drew what he expected to be his last breath. But the artillery wagon halted at the crucial moment, and he rose from the dust unscathed.[1]

That evening, he received a telegram from his father, the chief physician of a Bavarian mental hospital, reporting that Berger's sister was frantic with worry; she was certain Hans had met with an accident. Stunned by what he later termed "a case of spontaneous telepathy," the young cadet resolved to change his life. Henceforth, he would devote himself to discovering the physiological basis of "the energy of mind." When his military duties ended, he returned to college, where he'd previously been studying mathematics and astronomy, and switched his focus to neuropsychiatry.[2]

At the time, the study of paranormal phenomena, though not uncontroversial, was closer to the scientific mainstream than it is today; the London-based Society for Psychical Research boasted such luminaries as the physicist Sir Oliver Lodge, the chemist Sir William Crookes, and the psychologist William James.[3] In many fields, the line between the paranormal and the normal was still being worked out. The Danish experimental psychologist Alfred Lehmann, for example, proposed that the brain converted chemical energy into three components: heat, electricity, and "P-energy"—the psychic energy associated with different mental states, of which telepathy might be a rare manifestation. To understand the relationship between brain and mind, Lehmann asserted, it was crucial to measure each of these components precisely.[4]

Berger embraced such quasi-mystical notions, but he was careful to maintain an aura of *Korrektheit und Seriosität*.[5] After earning his doctorate in 1897, he took a junior staff position at the University of Jena's psychiatric clinic. There he remained for forty-one years, eventually becoming the facility's director, as well as dean of the medical school and rector of the university. "Shy, reticent, and inhibited," according to a psychiatrist who'd served under him, "obviously fond of his instruments...and somewhat afraid of his patients," Berger kept his passion to himself. Colleagues found him dull and unimaginative, almost laughably devoted to routine.[6] In secret, however, Berger spent his spare time searching for something as elusive—and potentially transformative—as the philosopher's stone: a way to quantify P-energy in the human brain.[7]

Since existing devices were unable to detect that mysterious force, Berger hoped to trace its fluctuations indirectly, through the electrical currents of the cerebral cortex. For clues to how this might be accomplished, he looked to the work of earlier researchers who'd studied animal brains.[8] In 1870, German physicians Gustav Fritsch and Eduard Hitzig had shown that by electrically stimulating an area of a dog's cortex, they could induce movement in the creature's legs—proof that nerve impulses in a specific part of the brain controlled behavior in a specific part of an animal's body.[9] In 1875, British physician Richard Caton made another landmark discovery: the presence of continuous, *spontaneous* electrical activity in the exposed brains of rabbits and monkeys. When Caton placed electrodes at two points on the surface of the cortex, he reported in the *British Medical Journal*, "feeble currents of varying direction" passed through his measuring instrument.[10]

Caton's findings suggested that such currents behaved in complex ways. Their amplitude decreased in an area of the cortex associated with the eye, for example, when he stimulated the opposite retina with light; the same thing happened in a different area when the animal chewed its food.[11] In follow-up reports, Caton showed that fluctuations also occurred in response to tactile stimulation, and when animals awoke from sleep or went under anesthesia.[12] Perhaps most intriguingly, heightened attention seemed to have a similar effect. "In some instances, it was evident that the thought or expectation of food caused the movement of the needle," Caton wrote. "If I showed the monkey the raisin but did not give it, a slight negative variation in the current occurred."[13]

In 1890, Polish physiologist Adolf Beck became the second researcher to report oscillating currents in the cortexes of animals. Beck found similar decreases in brain potentials in response to certain sensory stimuli, but he also found something new: these changes often occurred not only in the brain area associated with the stimulus but across the entire cortex. Beck's study, published in a leading German physiological journal, drew the attention of scientists across Europe.[14] Many hoped it would someday become possible to read human mental states by decoding these so-called brain waves.[15]

Yet by 1902, when Berger conducted his first "PE" experiments (his acronym for "psychophysical energy"), research on spontaneous cortical currents had lain dormant for more than a decade. Meanwhile, numerous scientists had followed in the footsteps of Fritsch and Hitzig, using electrical *stimulation* to investigate the relationship between specific cortical regions and physical or mental functions. There were two reasons for the disparity. First, the practical value of identifying localized brain functions was more obvious than that of studying the electrical rhythms of the cortex as a whole. And second, the era's technology—primitive galvanometers and electrometers—made it difficult to record those rhythms, especially through the thickness of a skull.[16]

Berger began his quest by trying to detect currents in exposed dog brains. He used a capillary electrometer—essentially, a tube of mercury and sulfuric acid fitted with copper wires. When a pulse of electricity traveled through the instrument, it triggered a bounce in the fluid level, which was captured on a photographic plate. In five trials, however, he observed such oscillations only once.[17]

Discouraged, he turned to more conventional methods of measuring brain activity, tracing cerebral blood flow and temperature during different mental states in patients who were missing parts of their skulls due to accidents, surgery, or other causes. He published two books on his findings. Yet these investigations brought him little closer to his ultimate goal of quantifying PE.[18] In 1907, he made another attempt to observe cortical currents in animal brains. This effort, too, proved futile.[19]

In 1910, Berger swapped his old electrometer for a high-tech upgrade: a string galvanometer. Developed by the Dutch physiologist Willem Einthoven (who won a Nobel Prize for it in 1924), this device had recently been adapted to create the first practical EKG machines. More sensitive and accurate than earlier galvanometers, the instrument used a silvered-quartz or platinum fiber stretched between two magnets to measure low-voltage electrical signals. When the string carried a pulse of current, the magnetic field made it vibrate; this movement was projected onto a running film sheet, which recorded it as a series of waves.[20] Yet this technique, too, failed to satisfy Berger's hopes. Out of nine experiments on a dog brain, only one seemed to show changes in current related to tactile, visual, or auditory stimuli—and that one was ambiguous. "Eight years!" he wrote in his diary. "Trying always, time and again."[21]

Soon afterward, Berger married one of his lab technicians, Baroness Ursula von Bülow, and grew too busy with familial and professional responsibilities to continue his clandestine experiments.[22] Meanwhile, a few other researchers began harnessing the new galvanometer technology to gain deeper insights on cortical currents. In 1913, Ukrainian physiologist Vladimir Pravdich-Neminsky published the first paper to include photographic evidence of those currents. Recording from dogs' intact skulls as well as their exposed brains, he'd detected spontaneous fluctuations ranging from twelve to thirty-five per second. They seemed to take two main forms: high-amplitude "waves of the first order" and lower-amplitude "waves of the second order." Pravdich-Neminsky called his recordings *electrocerebrograms*—a term that Berger dismissed as a "linguistic barbarism," due to its mixture of Greek and Latin roots, and that he would eventually replace with a more euphonious coinage. For now, though, he was stuck on the sidelines as others encroached on territory that he had come to think of as his own.[23]

The next incursion came in 1914, when Napoleon Cybulski—the physiologist who'd mentored Adolf Beck—published the world's second set of electrocerebrograms. These came from monkeys and dogs, and they included further revelations. Cybulski found that when he stimulated animals' peripheral nerves, the frequency of their brain waves increased. In another coup, he recorded the cortical currents of dogs during experimentally induced epileptic seizures, capturing wildly jagged peaks and valleys.[24]

Later that year, World War I broke out, and Berger left Jena to serve as an army neuropsychiatrist on the western front. Still, he never forgot his self-appointed mission; instead, perhaps inspired by the sufferings of the brain-damaged and shell-shocked soldiers he treated, he began to ponder its possible benefits to humanity. For the first time, his diary entries on PE suggest concerns beyond the purely scientific or metaphysical. "Physiological questions mark my path!" he wrote. "Psychophysiology of mental diseases... Psychic energy and the body, but only from a practical viewpoint of establishing cortical measurements for psychiatry and neurology."[25]

After the armistice, Berger was appointed full professor and director of the Jena psychiatric clinic. His workload was heavier than ever, but in 1920, he resumed his animal experiments with the string galvanometer on evenings and weekends. Again, he told no one but a small circle of assistants—and again, he failed to get usable readings.[26]

In 1924, a new idea occurred to him: To get around the problem of detecting current through bone, why not try working with patients who had skull defects as he'd done in his studies of cerebral blood flow and temperature? He began with a seventeen-year-old university student named Zedel, who'd undergone two craniotomies for the removal of a brain tumor. After several weeks of tinkering—changing the positions of the clay electrodes, using different types of galvanometer threads, fiddling with the magnifier settings—Berger achieved a breakthrough. On July 6, after placing two electrodes four centimeters apart over a scar on the boy's scalp, he was able to obtain "continuous oscillation of the galvanometer string." For the first time in history, he had captured images of human brain waves.[27]

Berger came up with a new word for these recordings: *electroencephalograms*, or EEGs. The earliest ones, however, revealed little about the brain's workings; they were too faint and blurry to be analyzed in detail and showed no obvious change in wave patterns when Zedel performed mental tasks. When Berger tried to improve the machine's performance by increasing the current in the electromagnets, the vibrations often shattered the expensive filament. So he ordered a larger, more powerful string galvanometer. During the months it took to arrive, he wrote prayerful entries in his diary: "Is it possible," one read, "that I might fulfill the plan I have cherished for over 20 years and...create a kind of brain mirror[?]"[28]

But the new device brought more disappointment. The machine was highly susceptible to electrical interference, whether from the clinic's power supply or from a subject's own body—currents generated by the heart, the muscles of the eyes and face, even by blood flowing through the cerebral vessels. Squinting at the foggy photographic readouts, it was difficult for Berger to distinguish actual brain waves from irrelevant artifacts. Nonetheless, he continued to test patients using protocols like those he'd employed in his earlier studies, searching for electroencephalographic signatures of desire, repulsion, attentiveness, concentration, and alarm.[29] At the same time, he worked furiously to refine his techniques. He tried pad electrodes, foil electrodes, and funnel electrodes, fashioned from copper, silver, platinum, or other metals; he secured them with adhesive tape, rubber bands, or swim caps—or, in the case of needle electrodes, inserted them in the subject's scalp, along with novocaine shots to numb the pain.[30] Perhaps unsurprisingly, given all these variables, the results were wildly inconsistent. He often found it impossible to obtain successful readings, and rejected many EEGs on suspicion of contaminated data.[31]

A turning point came in the spring of 1927, when Berger obtained a Siemens moving-coil galvanometer, the most advanced such device on the market. Instead of a fragile string, this machine used a rotating spool of wire to register currents. It was several times more sensitive than the instrument it replaced; equally crucial, its images were printed in crisp black on white, instead of fuzzy white on gray.[32] These qualities enabled Berger to achieve more consistent results, and to better tell the difference between signal and noise. They also helped give him the confidence to try recording from subjects with intact skulls—beginning with his fifteen-year-old-son, Klaus. In October, after several successful runs, Berger wrote a jubilant entry in his diary: "Eureka! I have indeed found the *Elektren-kephalogramm!*" Just to be sure, he began running EEGs on a broader selection of patients, on employees of the clinic, and often on himself, eventually amassing more than one thousand recordings.[33]

By late 1928, Berger was able to identify two basic waveforms, as Pravdich-Neminsky had—either of which, he realized, might hold the answer to the puzzle he was trying to solve. The first, which he later labeled *alpha*, appeared when the brain was at rest; it had a frequency of about ten waves per second and an amplitude of seven to fifteen micro-volts. The second rhythm, *beta*, emerged when the brain was focused on a task; it was faster and smaller, with a frequency of about twenty waves per second and an amplitude of two to three microvolts. Eventually, he began to focus on alpha waves, which could be produced fairly reliably by having a subject lie with eyes closed, thinking of nothing in particular. This rhythmic pattern seemed to represent the brain's baseline state; examining it closely, he reasoned, might offer clues not only to the "psychophysiology of mental diseases" but to differences in cortical function among individuals and groups.[34]

In 1929, Berger unveiled his new imaging technique in the journal *Archiv für Psychiatrie und Nervenkrankheiten*. The paper, "On the Encephalogram of Man: First Report," was written in an impenetrable style and a tentative tone; it went mostly unnoticed. The following year, he published a second report, in which he introduced the concept of alpha and beta waves. This article—and a simplified version that he wrote for a popular science magazine—earned a flurry of attention from the German press, with headlines touting THE ELECTRIC SCRIPT OF THE BRAIN and THE MACHINE THAT READS THOUGHTS. A team of physiologists at Berlin's cutting-edge Kaiser Wilhelm Institute for Brain Research were less impressed; they failed to replicate Berger's results.[35]

During the five years after his initial report on the EEG, Berger published fourteen more papers in the same obscure psychiatric journal. These articles examined the effects of age, brain damage, epilepsy, dementia, mental illness, drugs, and other variables on alpha brain waves. Still, they convinced few of his peers.[36] Instead, what enabled Berger's research to revolutionize neuroscience—and, eventually, sleep science—were the efforts of some well-connected scientists to debunk it.

———

In 1933, Edgar Douglas Adrian was looking for new challenges. The previous year, at the relatively youthful age of forty-three, the British physiologist had won a Nobel Prize for his research on the peripheral nervous system. That honor came in recognition of several seminal discoveries, including the fact that sensory information is encoded in the firing rates of individual neurons and the principle that impulses shoot down a nerve fiber at full strength, even when they're triggered by a weak stimulus. Recently, he had begun investigating the central nervous system, where he'd encountered a phenomenon that didn't fit the rapid-fire model common in peripheral nerves: large clusters of neurons firing in continuous, synchronized oscillations. Such currents had created interference in his lab's electrophysiology studies at the University of Cambridge, disrupting observations of eel retinas, goldfish brain stems, and the abdominal ganglion of the great diving beetle. Adrian and his twenty-seven-year-old lab partner—physiologist Bryan Matthews—decided to dig deeper.[37]

Reviewing the literature, Adrian and Matthews came across papers by the Kaiser Wilhelm Institute group that mentioned Berger's EEG research. Although the KWI physiologists disputed his findings, the Cantabrigians were intrigued: The alpha waves that Berger described seemed to resemble the oscillations they'd encountered in their animal studies. After reading Berger's own reports, they were also skeptical. As the pair later wrote: "We found it difficult to accept the view that such uniform activity could occur throughout the brain in a conscious subject, and as this seems to be Berger's conclusion we decided to repeat his experiments."[38]

In this endeavor, Adrian and Matthews had certain advantages over the EEG's inventor. For more than a decade, Adrian had been one of the leading innovators in electrophysiology, pioneering the use of cathode ray tubes, thermionic valves, and other amplification technologies that had vastly enhanced scientists' ability to monitor faint nerve impulses.[39] (Berger was unable to afford such luxuries at Jena until 1931 when a funding windfall enabled him to order them from Siemens.)[40] Matthews was a brilliant engineer who'd developed an innovative recording system for neurophysiological research. More sensitive, precise, and reliable than even the best galvanometers, this setup coupled specially designed amplifiers to a moving-iron oscillograph—a device that used an iron "tongue" instead of a string or coil to register weak currents.[41]

In the spring of 1934, the pair deployed a battery of advanced equipment to see if they could record alpha waves from each other's heads. In addition to Matthews's specialized system, which recorded data on photographic paper, they used an ink-writing oscillograph that he'd originally designed as an EKG machine; this provided real-time readings. To call attention to any wave patterns their eyes might miss, they connected their devices to a loudspeaker as well. For electrodes, they used small squares of copper gauze soaked in saline solution.[42]

"We found Berger's alpha rhythms almost at once, but only in one of us," Adrian would recall. "When Matthews was subject...the record showed no regular waves, but when I was the subject the waves at 10 a second appeared whenever I closed my eyes. We were surprised at first to find such great differences in the records from two human subjects but it did not take long to examine other inhabitants of the Physiological Laboratory and we found that a few were like Matthews in showing no sign of the alpha rhythm." Not only had they confirmed the validity of the EEG, but they'd discovered one of its oddities: for reasons that remain unknown, some people find it harder than others to produce alpha waves at will.[43]

That May, the two gave a demonstration of the EEG at the convention of the Physiological Society in Cambridge. Then Adrian embarked on a lecture tour of the United States, in which he shared his excitement about the "Berger rhythm" (Adrian's preferred term).[44] In December, he and Matthews published their findings in the British journal *Brain*. Although they questioned Berger's theory that the beat arose from the entire cortex—suggesting, correctly, an origin in the occipital lobes—they were convinced that it was real. It also seemed to occur in other species. "There is, in fact," they wrote, "a surprisingly close resemblance in the records from man and from the water-beetle." As evidence, they added a chart comparing the alpha patterns of the bug and the Nobel winner.[45]

At year's end, when the British press reported on the experiment as if Adrian had discovered the EEG, he penned an effusive letter of apology to Berger—as well as a correction that was published in several prominent journals. Soon, the shy Bavarian professor was receiving distinguished scientific visitors from around the world. He and Adrian first met in 1937, when both men were invited to speak at the International Congress of Physiology in Paris. Afterward, Berger expressed his gratitude in a note to the Englishman: "I know that I am indebted to you alone that the scientific world became aware of my analyses."[46]

If Edgar Adrian lent Berger's brainchild an air of intellectual respectability, it was Hallowell Davis who ensured that the EEG would be widely adopted by researchers and clinicians. A young physiology professor at Harvard Medical School, Davis had trained under Alexander Forbes (developer of the first vacuum-tube amplifier for string galvanometers), who'd worked with Adrian on peripheral-nerve research in the early 1920s. On Forbes's recommendation, Davis had done his postdoctoral fellowship with Adrian at Cambridge. He'd gone on to become a specialist in the auditory system, studying how sound is transmitted to the brain via the cochlear nerve.[47]

Davis, too, started out as an EEG skeptic. Like Adrian and Matthews, he first learned of the technology in 1933; the news came from his PhD student, A. J. "Bill" Derbyshire, and Derbyshire's medical-student friend, Howard Simpson. After explaining that alpha waves were likely artifacts of the machine itself, Davis encouraged the pair to test Berger's claims.[48] To do so, they employed a new device that Davis had been using to measure neural responses to sensory stimuli: a cathode-ray-tube oscilloscope. This instrument had only one moving part—an electron beam, which could trace even the faintest signals with near-perfect accuracy across a fluorescent screen.[49]

The two students began testing each other for Berger rhythms in January 1934. After three weeks, they told Davis that they'd failed to detect the ten-cycles-per-second pattern, but they asked him to stop by the lab to check their work. There, he watched Derbyshire stick needle electrodes in Simpson's scalp; as the latter sat in a soundproofed room with his eyes closed, the dot representing his brain waves wandered aimlessly across the oscilloscope display next door. Remarking that "three heads are better than two," Davis asked the pair to place the electrodes on him. "I sat in the room and closed my eyes," he later recalled. "Immediately, there were shouts outside. 'There it is! There it is!'" Davis, it turned out, was a strong Berger-rhythm producer; if he hadn't been tested, he and his alpha-challenged students would have gone on thinking there was no such thing. Flabbergasted, he tested other lab members. They were divided almost evenly, he found, into "'Bergers' and 'non-Bergers.' . . . We were convinced Berger was right."[50*]

Davis quickly recognized the clinical and experimental promise of EEGs. For the first time, it would be possible to see how the brain's firing patterns changed during normal and pathological states—and to harness that knowledge for diagnosing and treating disease. He also realized that the photographic recording process used for most electrophysiological research was too slow and expensive to maximize this technology's revolutionary potential. He asked his department's electrical engineer, E. Lovett Garceau, to build an ink-writer (similar to Matthews's EKG device) that could visualize EEG readings instantly and cheaply. Garceau customized a Western Union "undulator," normally used to record telegraph signals from transatlantic submarine cables. His design, incorporating an automatic pen and five-eighths-inch ticker tape, would be adapted for the first commercially produced EEG machines.[51]

Soon after his conversion experience, Davis and his wife, Pauline—a psychiatrist whose zeal for the new technology matched his own—began presenting public EEG demonstrations at Harvard. These events caused a stir comparable to the unveiling of x-ray machines four decades earlier. The *New York Times* ran a front-page story on one of Davis's lectures, headlined ELECTRICITY IN THE BRAIN RECORDS A PICTURE OF ACTION OF THOUGHT. The reporter, who served as guinea pig, was asked to solve a math problem, during which the EEG machine showed his brain waves passing twice from alpha to beta. He was astounded when Davis, seeing

---

* Around the same time, Brown University psychologist Herbert H. Jasper and colleagues were experimenting with EEGs at the Emma Pendleton Bradley Home, a neuropsychiatric hospital for children in Providence, Rhode Island. Jasper went on to publish the first North American study of the new technology, in January 1935, beating Davis to the punch. Herbert H. Jasper and Leonard Carmichael, "Electrical Potentials from the Intact Human Brain," *Science* 81, no. 2089 (January 11, 1935): 51–53.

the second set of beta waves, guessed correctly that the subject "is now checking the answer to see if he was right."[52]

That December, three Harvard neurologists joined Davis's lab: William Lennox, who had a Rockefeller Foundation grant to study epilepsy, and the husband-and-wife team of Frederic and Erna Gibbs. They began running EEGs of two patients with the petit mal form of the disease, which causes lapses of awareness rather than violent spasms that could create movement artifacts in the signal. The group was astonished to see distinctive, three-per-second "spike-and-wave" complexes in both patients' readouts. These findings—reported in a paper published a year later, along with similar results for twelve children—established the first objective diagnostic standard for epilepsy and proved once and for all that it was a brain disorder rather than a psychiatric illness.[53] The study also set electroencephalography on the road to becoming a diagnostic tool for a wide range of neurological conditions, including head injuries, strokes, brain tumors, and brain death.[54]

By the end of 1936, five EEG research centers were operating in the United States alone; more had sprung up across Europe. But it took a charismatic millionaire polymath to make sleep a focus of their investigations.[55]

———————

Before the advent of electroencephalography, researchers had only crude tools for gauging slumber's effects on the nervous system. In fact, there was no way to tell for sure whether a person or animal was sleeping, drifting *near* sleep, or just pretending. Scientists could test a sleeping or sleep-deprived subject's reflexes, but this provided limited information.[56] They could remove or stimulate part of an animal's brain to try to locate centers that turned sleep on and off—but in humans, this approach was not feasible. Or they could measure electrical currents in the peripheral nerves, as one prominent researcher had recently begun to do.[57]

Edmund Jacobson—a former colleague of Nathaniel Kleitman's at the University of Chicago, and author of the best-selling *You Can Sleep Well*—had developed a technique he called "progressive relaxation" to fight the nervousness associated with modern life, which he blamed for a host of physical and psychological ills. Jacobson preached that dreams, along with insomnia, were a by-product of "residual tension," and would dwindle in number and vividness as a patient learned to fully unwind. At his private clinic, he installed bedrooms with homey curtains and furnishings, as well as copper-lined walls to block electrical interference from distorting his tests. Sleepers were hooked up to galvanometers that measured nerve impulses—and thus muscle tension—from face to feet. At the time, Jacobson's approach was widely seen as representing sleep science's technological cutting edge. But his instruments revealed nothing of what was happening in the brain itself, which governed all the phenomena he and other researchers observed.[58]

The EEG offered the first opportunity to peer inside that black box. And no one peered more thoroughly than Alfred Lee Loomis, a middle-aged tycoon who was also the most celebrated amateur scientist of his time. After making a fortune as a Wall Street lawyer and investor in the 1920s, Loomis—a math prodigy and a prolific inventor since his teens—bought a vast stone mansion in Tuxedo Park, New York, and transformed it into what Albert Einstein called a "palace of science." In his basement laboratory, Loomis and his collaborators pioneered a dazzling array of new technologies. They developed the earliest ultrasound devices, created a "microscope-centrifuge" to analyze the impacts of strong gravitational forces on cells, rigged up high-speed cameras to study the physics of lightning, and demonstrated the moon's effects on pendulum clocks. During Prohibition, they distilled their own gin, serving it at banquets for the eminent visitors who stayed in the sixteen bedrooms upstairs. Loomis's houseguests often included Albert Einstein, Enrico Fermi, Niels Bohr, and other top scientists of the era.[59]

In 1934, Loomis set out to improve the electrocardiograph, which was still based on the finicky Einthoven string galvanometer. Tinkering with the placement of electrodes, he stuck one to a subject's head and detected rhythmic oscillations that didn't match those of the heart. Suspecting that they might result from electrical activity in the brain, he asked a physiologist friend—Alexander Forbes—for advice. Forbes introduced him to Hallowell Davis.[60]

Soon afterward, Loomis invited both Davis and his wife, Pauline, to collaborate with him at Tuxedo Park. The millionaire scientist was particularly curious about the EEG patterns of sleep, which other researchers had examined only in passing. Berger and others had observed episodes of alpha activity during brief tests of slumbering subjects. But no one had taken EEG readings of sleepers throughout the night, due largely to the technical challenges of charting brain waves for extended periods. Loomis—who'd long been fascinated with the questions raised by encephalitis lethargica, as well as by Pavlov's theories about sleep and Freud's about dreaming—aimed to go further.[61]

Among the custom-made gadgets in Loomis's lab was the world's largest kymograph—eight feet long and nearly four feet in circumference—which he'd designed to record changes in physiological functions over long stretches of time. Using colored pens that traveled once every minute around the drum's circumference, the device allowed an experimenter to inspect hours of data on a mural-size sheet of paper instead of sifting through half a mile of ticker tape. Loomis and the Davises hooked up the kymograph to a multichannel, oscilloscope-based EEG machine, and built a copper-shielded "sleeping room" sixty-six feet away to prevent electrical interference or other disturbances. Sensors monitored respiration, heart rate, body motility, and facial movements while microphones, infrared cameras, and one-way mirrors enabled more comprehensive surveillance.[62]

When everything was ready, in early 1935, Loomis began inviting his guests—physicists, chemists, biologists, psychiatrists, their spouses and children—downstairs for an afternoon nap. (Later, he and his team began monitoring full nights of sleep.) After each subject was fitted with electrodes and tucked into bed, a phonograph would play a recording of Ravel's *Bolero* as standardized background stimulus to establish his or her normal alpha patterns. Sometimes, Loomis and his colleagues left the sleeper undisturbed; at other times, the scientists watched how the EEG responded when they made noises of different kinds (low conversation, coughing, slamming a door) or flashed lights of varying intensity. These tests were meant to provide more precise measures of sleep depth than older methods, which could determine only what it took to wake someone up.[63]

On one occasion, Einstein himself served as a guinea pig. "They put him to sleep, and at first he showed the typical slow waves of sleep," a witness later recalled. "Then the EEG changed to the rapid waves of arousal. He awoke suddenly, asking for a telephone. He called his laboratories in Princeton to tell his colleagues there that he had been reviewing his calculations of the day before and discovered an error which should be corrected. This done, he was able to go back to sleep again."[64]

Einstein's EEG matched what the Tuxedo Park team was finding with other subjects: brain waves during sleep varied far more than anyone had suspected. Besides the familiar Berger rhythms, Loomis's instruments revealed short bursts of low-voltage activity that the team called "spindles." There were also stretches of slower, high-voltage "saw-toothed" waves and periods when the peaks and valleys seemed entirely random. Strikingly, these new waveforms did not appear when a subject was in a merely sleep*like* state, such as hypnosis or an alcoholic stupor.[65]

It took more than a year for the researchers to group the patterns into five distinct rhythms that appeared in succession as a person fell asleep—a discovery they reported in 1938 in the *Journal of Experimental Psychology*. They cataloged their taxonomy alphabetically, from what they identified as the shallowest stage of slumber to the deepest:[66]

A—*Alpha*. Alpha rhythm appearing in trains of various lengths. The eyes may be slowly rolling, under closed eyelids, as indicated by an electrode over the eyebrow.

B—*Low voltage*. A quite straight record, with no alpha rhythm and only low voltage changes of potential. Rolling of the eyes may occur.

C—*Spindles*. Line slightly irregular with 14 per second spindles of 20–40 microvolts every few seconds.

D—*Spindles plus random.* The spindles continue together with large random potentials 0.5 to 3 per second. The random voltages may be as high as 300 microvolts.

E—*Random.* The spindles become inconspicuous, but the large random potentials persist and come from all parts of the cortex.[67]

This list represented the first attempt scientists had ever made to impose a structure on sleep based on brain activity rather than behavioral and physiological signs. Yet the significance of these findings remained unclear. Loomis and his colleagues made only tentative efforts to correlate EEG patterns with phenomena such as dreaming or body movements, for example. Nor did they establish whether the shifts between patterns followed any kind of order as the night progressed. Exploring such questions would be up to others.[68]

The dozen papers Loomis published on sleep (and a conference on "The Electrical Potentials of the Brain" that he hosted at Tuxedo Park in November 1935, attended by fifty pioneering electrophysiologists) inspired a handful of scientists to follow his lead.[69] At the University of Chicago, two colleagues of Kleitman's—physiologist Ralph Waldo Gerard and his PhD student Helen Blake—probed the connection between brain waves and sleep depth by whispering, "Are you awake?" to sleepers who were wired to an EEG machine, and counting the seconds until the subjects opened their eyes. They found light sleep to be associated with alpha and faster "feeble and irregular" waves while deep sleep was marked by slow, high-voltage waves lasting one-half to three seconds. The new technology confirmed one of the earliest discoveries in sleep science: these slow waves peaked about an hour after sleep onset, corresponding with the period of maximum sleep depth noted by researchers as far back as Ernst Kohlschütter in the 1860s.[70]

In 1938, Kleitman joined Gerard and Blake for another EEG study; the results convinced the group that uninterrupted sleep generated *six* EEG stages, which cycled predictably throughout the night. They also gave the slow waves a name: *delta.* Among the study's thirty-three subjects, eleven suffered from narcolepsy; these individuals exhibited more delta activity than normal during slumber, suggesting that electroencephalography could be useful in diagnosing this and other sleep disorders. The team also woke subjects during different stages and asked if they'd been dreaming. While no correlation was evident between dreaming and any particular stage, the ability to recall a dream's contents seemed greatest when alpha waves had been absent for two to sixteen seconds and no delta waves were present.[71]

All these studies contained tantalizing hints about the mysterious activities of the sleeping brain, suggesting that further digging could yield revelations with real-world impact. Astonishingly, however, none of the researchers went on to explore those clues. Loomis, ever restless, turned his attention to radar—a technology that he would revolutionize during World War II, playing a key role in the Allied victory.[72] Davis focused his electrophysiological investigations on waking life, developing new techniques for diagnosing and treating deafness.[73] Gerard went on to spearhead innovations ranging from the intracellular recording microelectrode to the peer-review system for government research grants.[74] Blake married a young physicist, George Raymond Carlson; after taking his surname, she vanished from the historical record.[75]

Berger's research career ended in 1938, when he reached the mandatory retirement age of sixty-five. Historians once believed he was forced out by the Nazis for opposing the regime, but more recent scholarship paints a different picture. Although he never joined the party (unlike most faculty members at the University of Jena), he'd been a dues-paying "supporting member" of the paramilitary Schutzstaffel, or SS, since 1934. He also served on a Nazi eugenics court that heard appeals from people who'd been condemned to sterilization for "feeblemindedness"; of the six cases he handled, none were granted a reprieve. It is possible that Berger, known to colleagues for assiduously avoiding conflict, was only trying to stay out of trouble, like millions of other "good Germans." If so, his efforts did not quite succeed. After becoming director of a sanatorium in

Bad Blankenburg, he fell into a profound depression, exacerbated by heart troubles and a painful skin condition. He hanged himself in 1941.[76]

Kleitman, for his part, remained more interested in sleep's physiological patterns than in their neurological correlates. After reporting on his own and others' brain wave experiments in *Sleep and Wakefulness*, he abandoned the EEG. The machine he'd used was consigned to a storage room. There it remained for more than a decade—until a troublesome student asked to borrow it.[77]

A New Continent

Eugene Aserinsky would always remember the first time he knocked on Nathaniel Kleitman's office door. It opened slightly, he later wrote, and "a man with a grey head, a grey complexion and a grey smock peered through the crack and enquired abruptly, 'Yes?'" Each had reason to be wary of the other: Aserinsky, because the prospect of studying sleep filled him with anticipatory boredom; Kleitman, because the PhD student seeking his mentorship lacked an undergraduate degree.[1] The professor may also have detected something worrisome in his supplicant's demeanor: a hint, perhaps, of sullen resentment. Kleitman, then fifty-four, was known as the father of his field, and Aserinsky—a slight, black-haired twenty-seven-year-old[2] with aquiline features and the bearing of a bantamweight fighter—had a problem with patriarchs.[3]

The man who would transform sleep science never intended to do anything of the sort. He may have been the least likely candidate imaginable for the role. Yet his combative brilliance, cussed obstinacy, and prickly self-reliance proved well suited for his mythic task. If the EEG was sleep's magic key, he was the apprentice who rescued it from the cave, discovered its secret powers, and unlocked the portal to a hidden world.

Aserinsky grew up in Brooklyn, in a household (like Kleitman's in Kishinev) where Yiddish and Russian were the primary languages.[4] His mother died when he was twelve, shortly after his older sister left home.[5] Eugene was marooned with his father, Boris—an immigrant from Ukraine who'd scrapped his way to a DDS at New York University, launched a low-rent

Aserinsky with his family in 1952, the year he made his great discovery—with the help of his son, Armond. *Courtesy of Jill Buckley.*

dental practice, and developed a gambling habit that kept him perpetually on the edge of poverty. Boris presided over nightly pinochle games in their walk-up apartment, where players paired off in remorseless competition. He often recruited his sharp-witted son as a partner, and the two developed a signaling system that enabled them to cheat without detection. Beyond this, Boris showed little interest in parenting. Around 10:00 p.m., when the boy asked about dinner, the usual response was "I'm not hungry, how can you be hungry?" Boris possessed a cutting tongue and a quick temper; when the family's small dog annoyed him, he would kick it across the room. This hard-knocks background would color not only Aserinsky's relationships with authority figures but many of his other interactions as well.[6]

Although his nocturnal tasks often left him too exhausted to attend school, Depression-era teachers tended to forgive absences, and his academic prowess enabled him to skip two grades. After graduating from high school at fifteen, Aserinsky spent more than a decade searching for his calling. He attended Brooklyn College, majoring successively in social science, Spanish, and premedical studies, but never earned a diploma. He switched to dental school at the University of Maryland, supporting himself as a bookkeeper for an ice company and a social worker for the state unemployment agency.[7] During his studies there, he married Sylvia Simon, a smart and sensitive young woman from a family as destitute as his own, who was drawn by his restless energy and omnivorous intellect. But a physical disability blocked Aserinsky's path: he was blind in his right eye, probably due to a neglected case of amblyopia (or lazy eye), leaving him with impaired depth perception. Unable to handle the surgical component of his studies, he dropped out in late 1942.[8]

Sylvia gave birth to a son, Armond, the following March; shortly afterward, Aserinsky was drafted into the army and shipped to England. He later told Armond that he'd spent his hitch unloading munitions trucks and interrogating the occasional German prisoner. (His vision problem spared him from combat, he explained.)[9] Though he would claim publicly that he'd been an explosives handler for the army or a machine gunner for the Office of Strategic Services—the precursor of the CIA[10]—records list his rank as private first class, with a specialty in administrative duties.[11] Sylvia took a filing job at an aircraft plant. When the war ended, her husband returned to work at the Baltimore unemployment office.[12]

In his late twenties, friends convinced Aserinsky that he was wasting his talents. Backed by GI Bill tuition aid, he decided to pursue a doctorate in a field that had long fascinated him: organ physiology. He chose the graduate program at the University of Chicago, which was known for admitting gifted students from unconventional backgrounds, and aced the entrance exam.[13] After enrolling in September 1948,[14] he was dismayed to learn that the only available adviser specialized in a discipline he considered a literal snooze. Nonetheless, he wrote, "neither Kleitman nor I could operate independently of the labor market." The shortage of students eager to engage in sleep studies meant "a major criterion for the selection was that the candidate have a heartbeat." Having passed that test, he sat down with the stern-faced sleep maven to discuss potential projects.[15]

The meeting that launched Aserinsky's voyage of discovery took place in early 1950.[16] At this point in Kleitman's career, he may have been a little bored himself. Although *Sleep and Wakefulness* had inspired a modest surge of sleep research over the past decade, the field had made only incremental progress. A few months earlier, however, there had been a major breakthrough—at a rival institution across town. Northwestern University neurophysiologists Horace Magoun and Giuseppe Moruzzi identified a waking center in the brain, confirming Constantin von Economo's hypothesis that such a thing existed. It was located in the ascending reticular formation, a weblike network of nerve centers in the brain stem. In one of the few sleep-related studies since the 1930s to rely on EEG readings, the researchers found that by electrically stimulating part of the formation, they could produce brain waves indicating alertness. These patterns persisted after the stimulation ended; the stronger the current, the longer they lasted. For slumber to occur, the pair concluded, the waking center must receive signals from elsewhere in the brain to suspend operations.[17]*

Kleitman could not have welcomed this discovery, which demolished his theory that sleep was a passive phenomenon that occurred when the brain was deprived of sensory stimuli.[18]† As for his own work, he had reached an uninspiring plateau. He was still the world's only full-time sleep scientist and the leading authority on his subject. The newspapers quoted him regularly on the health benefits of a consistent sleep schedule, the folly of forcing a "night owl" child to turn in early, and the wisdom

---

* See diagram, page 289.

† Magoun and Moruzzi showed that the real reason cats fell into permanent sleep in Frédéric Bremer's 1935 *cerveau isolé* experiment (see Chapter 3) was not that the forebrain was cut off from sensory input but that it stopped receiving waking signals from the reticular formation.

of taking one's own temperature throughout the day to determine one's ideal bedtime.[19] After a quarter century at the University of Chicago—"probably the longest period of 'probation'" in the institution's history, he remarked half jokingly to Paulena—he'd just been promoted to full professor.[20] But he hadn't authored an important paper in ages. His biggest recent success was scoring a $10,000 grant (worth about $100,000 today) from the Swift meatpacking company for a study of the sleep habits of babies. "The scientists will seek to determine whether a 25 per cent increase in protein content of the infants' diet will induce a more restful slumber," the *New York Times* reported. To Kleitman, such a gig must have seemed, to use a Yiddish-inflected phrase of the time, strictly from hunger.[21]

Still, he was always capable of genuine curiosity, and an item in *Nature* had whetted his investigative appetite. He pushed a copy of the journal across his desk toward Aserinsky. It was opened to a letter by the British physicist Robert Lawson, who reported that during a train trip, he had observed an unvarying rate of blinking reflexes in two passengers whether their eyes were open or closed in rest; these twitches ceased only when the subjects fell asleep. "What disturbed Kleitman," Aserinsky later recalled, "was that the blinking stopped abruptly with sleep onset rather than stopping gradually." The apprentice physiologist's next assignment would be to test Lawson's constant-rate hypothesis. Before beginning, Kleitman admonished, Aserinsky was to bury himself in "all the literature on blinking, thereby becoming the premiere savant in that narrow field."[22]

Although Kleitman's orders reflected his own growing interest in eye behavior during the sleep-wake cycle (he published a study on that topic the following year),[23] Aserinsky regarded the effort as pointless busywork. Nonetheless, he dug in as instructed. He spent the next several weeks in the stone-walled lab, trying to devise a way to record blinks using an old string galvanometer while avoiding contact with the gray eminence in the next room. Perhaps due to his own childhood conditioning, Aserinsky perceived Kleitman's social awkwardness and old-world formality as coldness and hauteur—an impression that did not alter with time. "On

occasion," Aserinsky wrote, "when it was absolutely essential to communicate with Kleitman, I would knock on the door and wait for a slightly
irritated 'Yes?'" And now, he had to step over the threshold and admit
failure.[24]

Kleitman then suggested that his advisee try observing eyelid movements in infants, using subjects from the ongoing baby-food trial. Aserinsky began visiting the children in their homes, peering at their eyes
as they lay in cribs outfitted with Kleitman's motility recorders. But he
found himself unable to distinguish true blinks from the quivering of the
infants' partially closed eyelids. Once again, he was forced to knock on
the "dreaded door."[25]

As the younger man described his dilemma, Kleitman's face registered
deepening gloom. This time, however, Aserinsky had prepared an alternative proposal: Why not record the presence or absence of *all* eyelid
movements, even those caused by eyeball motion? To his relief, Kleitman assented. Aserinsky was now confronted with the task of designing
an experiment that he considered "about as exciting as warm milk" and
whose theoretical objectives were unclear.[26]

After a few months, illustrating what he came to identify as the "golden
manure" principle (by which painstaking exploration of unpromising
material brings a rewarding result), he found something relatively interesting: in each hourly motility cycle, there was a twenty-minute period in
which the eyes stopped moving, after which subjects typically awakened.
This discovery brought two immediate benefits. First, Aserinsky could
take a much-needed nap himself, knowing he would not miss anything.
And second, he could tell mothers when their babies would wake up. They
were "invariably amazed at the accuracy of my prediction," he recalled,
"and equally pleased by my impending departure."[27]

In early 1951, when Aserinsky completed the infant eye-movement
project, Kleitman suggested that he skip a master's thesis—having already
done the equivalent amount of research—and go straight for his doctorate. Although the doctoral committee protested that it was highly
irregular to leap from a high school diploma to a PhD, they ultimately

relented.[28] For his part, Aserinsky was grateful for a shortcut toward a paying job. The family was jammed into a tiny, plywood-walled apartment heated by a kerosene stove; in winter, they slept under piles of coats and blankets. (Built as temporary housing for military personnel and their families during the war, these "barracks" flats served married graduate students at the University of Chicago in the early '50s.) Sylvia found work as a department store salesclerk and a nightclub cigarette girl, but her meager wages, combined with Aserinsky's GI Bill stipend, barely covered basic expenses.[29] Even the rent on his typewriter was a stretch. Once, their daughter Jill later told a reporter, he stole potatoes for dinner. "I think he saw himself as a kind of Don Quixote," she said, recalling his investigative persistence in the face of these privations. "Ninety percent of what drove him was curiosity—wanting to know. We had a set of Collier's Encyclopedias, and my father read every volume." Clearly, he and Kleitman had more in common than either would acknowledge.[30] But in his own writings, Aserinsky insisted that what drove him was desperation to become a breadwinner.[31]

To avoid remaining a perpetual student, Aserinsky needed a successful research project. He decided on a "gamble," as he later put it, betting that if he watched the eyes of adults throughout the night—as no one else had ever done—he would find something worth reporting on. His plan was to scrutinize eye movements for amplitude, frequency, and overall pattern relative to the EEG sleep stages postulated by earlier researchers. He figured he might identify an adult equivalent of the twenty-minute "no eye movement" period in babies. And given the large amount of space taken up by the visual and eye-muscle areas of the cerebral cortex, he hoped that examining brain waves and eye movements in tandem "would yield some unrevealed aspect of brain function."[32]

No funds were available to buy a new EEG machine, so Kleitman permitted Aserinsky to use the antique stored in the basement of Abbott Hall (as the physiology building was now called).[33] Since the 1930s, as the devices came into widespread use for diagnosing and researching epilepsy, brain injuries, and other neurological conditions, a half dozen manufacturers had gone into the EEG business. In the United States, the two leading companies were Grass and Offner, both founded by engineers who'd worked in university laboratories with some of the pioneers of electroencephalography. Franklin Offner had designed amplifiers and oscillographs for Ralph Waldo Gerard at the University of Chicago before striking out on his own. He and Albert Grass—who'd custom-built devices for the Gibbses at MIT and Harvard—now mass-produced sophisticated machines. The latest models boasted state-of-the-art tube electronics, with up to sixteen channels and relatively robust safeguards against interference; they were known as *polygraphs*, because they could simultaneously record brain waves and electrical activity elsewhere in the body. Instead of ticker tape, they used broad sheets of folding graph paper, which were far easier to handle and to read.[34]

The machine that Aserinsky hauled up to the second-floor lab was not one of those newfangled EEGs. It was a prototype Offner Dynograph, hand-built by Frank Offner in 1936 or '37. This relic came in two parts, a control unit and an ink-writer, each the size of a large suitcase. It appears to have had only three channels—enough to record brain waves and eye movements, but not to correlate them with motility, heartbeat, or other sleep behaviors. It spewed out inch-wide paper tape.[35] And as Aserinsky began teaching himself how to run the device, its performance proved to be wonky in the extreme. The pens would sometimes begin scratching out signals before he'd attached electrodes to a subject, or barely budge when they should have been scribbling.[36]

Aserinsky asked a fellow grad student named Shirley Bryant, who was known as an electronics whiz, for help getting the contraption into working order. Bryant agreed, on the condition that his classmate provide a schematic diagram. When Aserinsky phoned Offner, however, the inventor expressed "utter amazement...that this particular machine was still in existence" and told him that no such document existed. Nonetheless, Bryant managed to adjust the primitive polygraph so that in calibration mode—in which the operator tests for accuracy using a standard signal—its readings corresponded to reality.[37]

Taking readings from a sleeping subject was a different matter. Aserinsky hoped to run simultaneous EEGs and EOGs (or electrooculograms), which track eye movements by measuring changes in electrical potentials between the front and back of the eyeball.[38] At Kleitman's suggestion, he sought out Edmund Jacobson (the former University of Chicago physiologist and author of *You Can Sleep Well*) for pointers in the latter technique.[39] To hone his skills, Aserinsky enlisted eight-year-old Armond—recruiting his son for his gamble as his own father had once done.[40]

The lab where the pair worked had changed little since Kleitman established it in the 1920s. The main room, which held the Dynograph, was furnished with marble-topped tables, sinks, and shelves full of brass-plated equipment; in the smaller room, an army cot sat beneath an unused fume hood.[41] Each practice run required arduous preparation. First, Aserinsky used a razor blade to scrape away patches of the boy's hair, then scrubbed the scalp and the skin around the eyes with sharp-smelling acetone.[42] After squirting blobs of conductive goo known as *collodion* onto Armond's head and face, the researcher placed electrodes on those spots, securing them with adhesive tape.[43] He plugged the leads into a jack box, whose cable snaked beneath the door to the polygraph outside.[44] Then, as Armond lay on the cot, Aserinsky ordered him to close his eyes and move them in different directions—often cursing as he fiddled with the finicky controls or searched the device's innards for a burned-out tube.[45] To avoid interruptions for bathroom trips, a bottle sat on the floor by the bed.[46]

Despite the tedium and discomfort, Armond cherished these hours with his father, who tended to be testy and hypercritical at home. "I knew my dad needed help," he recalled years later, "and I was flattered that he talked over his findings with me and always took what I said very seriously." Eventually, he began helping to set up the equipment himself.[47]

By December, Aserinsky felt ready to try Armond on his first sleep run. But when he scrutinized the half mile of tape in the morning, he realized that more bugs needed working out. Aside from artifacts that could have come from currents in his son's skin or the lab's electrical wiring, he saw signals that looked oddly like those generated by the quick, coordinated eye movements Armond had made during their calibration exercises. Another sleep session brought similar results. How could he be sure these patterns weren't just glitches?[48]

Aserinsky consulted with some "very bright" colleagues, as he described them, who came up empty. He then turned to one of the most experienced electroencephalographers in the country—Frederic Gibbs, codiscoverer of the spike-and-wave EEG pattern in epileptics. The expert's response was crushing: Gibbs advised abandoning EOGs, which were notoriously hard to decipher, in favor of mechanical recording of eye movements. "This was virtually a death sentence to the project," Aserinsky later wrote. He had invested too much time developing his technique to drop it and start learning another one.[49] He was thirty years old. His wife was pregnant with their second child.[50] They were living in a wooden icebox. He couldn't afford to start over.

For weeks, Aserinsky shuttled between panic and despair. Then inspiration struck: If he compared the signals from the left and right eye channels while simultaneously comparing the frontal EEG with the EOG on the same side, he could more easily distinguish between actual eye movements and mere noise. "Of course, it was idiotically simple," Aserinsky wrote. (Nonspecialist readers may disagree.) He immediately used the new technique in a sleep run with a student volunteer—and again, the recording showed the same eye movements.[51]

Aserinsky was beginning to believe that the phenomenon was real. To make sure, he went into the sleeping room one night when the polygraph pens showed a burst of activity and saw that the volunteer's eyes were moving vigorously. He called the young man's name and got no response. Even stranger: when Aserinsky checked the tape, the EEG signal—fast, jittery, and low-voltage—looked as if the subject were wide-awake.[52]

This was not the first time an observer had seen a sleeper's eyes moving; people had noted such wanderings for centuries, without investigating further. More recently, a few scientists had reported slow, rolling motions, visible at sleep onset through closed lids. But none had detected this type of darting eye movement, let alone recorded it electronically—or linked it with brain wave patterns that had long been thought to indicate that a sleeper was surfacing briefly into consciousness.[53]

In early 1952, Aserinsky reported these bizarre findings to Kleitman, who advised him to replicate them before speculating further. He began running tests on more volunteers, recording a few minutes of sleep per hour to save expensive paper. The "rapid eye movement" period, as Aserinsky began to call it, proved to be no fluke. (He also considered the term *jerky eye movement* but rejected it for fear that the slang meaning of *jerk* would inspire dumb jokes.) Even these brief samplings showed that REM sleep, as it came to be abbreviated, occurred several times a night in most subjects.[54] Indeed, further investigation would show, it occurred in *all* normal humans.[55]

But why? One possibility seemed obvious. Remembering a line from Edgar Allan Poe (who described his raven's eyes as having "all the seeming of a demon's that is dreaming"), Aserinsky began waking his subjects and asking them, "Did you dream?" The answer during REM was yes far more often than in other periods of sleep, and recall of dream content far better. One subject generated a "REM hurricane" that almost unhinged the polygraph's eye-movement pens. Rushing to the man's bedside, Aserinsky found him muttering "slurred but occasionally intelligible" words. As suspected, he'd been in the grip of a terrifying nightmare.[56]

This was strange indeed. According to the prevailing wisdom (including Kleitman's), dreams happened when the shutdown cerebral cortex was unable to provide a rational interpretation of sensory impulses; supposedly, those impulses instead triggered unfettered associations in lower brain regions. Yet the EEG indicated that the cortex was highly active during REM sleep.[57]

Aserinsky wasn't sure what to make of all this, but he was eager to know more. For example, was there a connection between the eye movements of REM and the imagery of dreams? To explore this question, he recruited a blind undergraduate, who arrived on the appointed night with his seeing-eye dog. For several hours, the EOG registered only small, erratic movements. But when they grew more active, the researcher decided to go have a look at the sleeper's eyes. He opened the door to the next room and heard a menacing growl, followed by a "general commotion"— and realized he'd forgotten about the young man's dog. "By this time," Aserinsky wrote, "the animal took on the proportions of a wolf, and I immediately terminated the session, foreclosing any further exploration along this avenue."[58]

Scientists would eventually learn that blind people, too, show the physiological signs of REM sleep, though their eyes may move less than those of their sighted counterparts.[59]

―――――――――

As the evidence for REM mounted, Kleitman grew increasingly engaged. In October 1952, he provided Aserinsky with a tech upgrade—a desk-size Grass Model III, which had been donated to the physiology department by the American Heart Association.[60] This polygraph probably dated to the mid-1940s (it had four channels rather than the eight found in newer models), but it was far more reliable than the ancient Offner. It was also capable of recording more kinds of data, and its broad folds of graph paper made that information easier to visualize and interpret.[61] Aserinsky began sampling more frequently throughout the night; to get a more detailed picture of the REM period, he incorporated respiration, heart rate, and motility into his measurements.[62]

Kleitman soon showed other signs of enthusiasm for the project. In December, he asked the National Institute of Mental Health (NIMH) to renew Aserinsky's minuscule fellowship, writing that the young scientist had "discovered a distinct type of eye activity which is found only in sleep and may be associated with dreaming." He also assigned his mentee an assistant: a second-year medical student named William Dement.[63]

For Aserinsky, who preferred to work alone, Dement's arrival was a mixed blessing. On the one hand, he would be able to get some shut-eye at home while his amanuensis supervised a few sleep sessions.[64] This must have been especially welcome in light of Aserinsky's growing domestic troubles. After giving birth to their daughter the previous April, Sylvia had fallen into a severe depression. For Aserinsky, the stress of this situation—combined with the family's financial straits and the pressures of completing a PhD—would have been daunting even without having to pull all-nighters in the lab.[65]

On the downside, he would have to spend time training the newcomer and then trust him to run the show solo. Dement turned out to be a quick study. But one night, after leaving him in charge, Aserinsky awoke around three in the morning with a premonition that he should return to the lab. There, he found his helper asleep in a chair next to the polygraph while the pens scratched wildly and graph paper spilled across the floor. "On a mantel nearby," Aserinsky wrote, "was a gargoyle-like lamp with a demoniacal visage which was connected to the subject so that the latter, in case of emergency, could signal to the investigator to come immediately into the sleeping room. The eyes of the demon lamp were aglow, indicating that the subject had turned the lamp on some time ago." Aserinsky flung open the sleeping-room door to check on its occupant, enabling the subject to observe the dressing-down that doubtless followed. "It was inevitable," Aserinsky wrote tartly, "that many years later, Dement would title a book, *Some Must Watch While Some Must Sleep*."[66]

Aserinsky, of course, had experienced his own misadventures over the course of his learning curve; once, he'd forgotten to ground the mattress springs, leading to the near electrocution of a wired-up subject who accidentally touched them. With Dement on board, however, the experiments quickly gained momentum. Patterns began to emerge in the data, indicating that REM sleep was a complex phenomenon involving a variety of physiological systems.[67] In addition to jerky eye movements, unexpected brain waves, and vivid dreams, REM seemed to be associated with elevated heart rate, respiration, and body motility. It could last from a few minutes to nearly an hour. It even appeared to follow some kind of schedule, though its rhythms were not yet clear.[68]

To Aserinsky, these results seemed eminently worthy of a doctoral degree. But Kleitman—always a stickler for thoroughness—refused to fully endorse his student's findings until he'd scrutinized them from every angle. The scientist lay down for two sleep sessions himself but exhibited no REM during the first run, probably due to a technical problem; he had three REM periods during the second session but recalled dreaming only once. After this ambiguous test, he lent Aserinsky a small home-movie camera to document the eye movements. To avoid awakening the subjects, the researchers increased the lighting gradually using a rheostat; nothing could be done, however, about the camera's loud whir. Despite the noise, the experimentees slumbered on. This was another surprise: although people's brains and bodies seemed more active during REM, they were as hard to rouse as in the profoundest depths of slow-wave sleep.[69]

At this point, Kleitman suggested that Aserinsky prepare an abstract and present an oral report on his discovery at the upcoming conference of the Federation of American Societies for Experimental Biology (FASEB) in Chicago. Before the meeting, however, Kleitman asked his advisee to arrange a typical sleep session for his benefit; he wanted to see the experimental procedures firsthand, he said, so that he could answer any questions thrown at him during the meeting. "I was elated at the prospect of finally showing him the fruits of my labor," Aserinsky recalled, "and I suggested to him that I already had a subject readily available for such a session." To Aserinsky's puzzlement, Kleitman insisted that the subject be his own daughter Esther, then twenty-two.[70]

The demonstration must have been satisfactory, as Kleitman voiced no further reservations. Yet the incident hinted at conflicts soon to come. It was only years later—while conducting a televised REM demonstration with Armond as the sleeper—that Aserinsky guessed at his professor's probable motivation: "to eliminate even the remotest possibility that there might have been collusion between myself and the subjects."[71]

As he wrote the abstract of his study, Aserinsky should have been in a state of happy anticipation. He'd uncovered a phenomenon that suggested the sleeping brain was far more active than anyone had suspected—one that also challenged longstanding beliefs about dreaming. He was about to deliver the kind of presentation that can launch a fabulous career. Yet he was dogged by a growing conviction that Kleitman aimed to steal his thunder, exploit his labor, and block his progress. The result, tragically, was his premature exit from sleep science.

It is hard to judge the validity of Aserinsky's grievances, which he published in a neuroscience-history journal in 1996—when Kleitman, at 101, was incapable of responding.[72] Kleitman's own accounts of the discovery of REM shed no light on the personal dynamics involved, and he never commented publicly on his dealings with Aserinsky. (A conspicuous omission hints at his private feelings: Discussing REM in the second edition of *Sleep and Wakefulness*, Kleitman mentions the discoverer's name only in the footnotes.)[73] What's clear is that by March 1953, when the FASEB conference was imminent, Aserinsky's antipathy toward his adviser had become almost unbearable.[74]

In his telling, several factors contributed to the rift. First, Kleitman was now suggesting that the REM study didn't merit a doctorate; to earn a degree, Aserinsky should expand his research—perhaps into the developmental changes of eye movements throughout childhood. Then there was the matter of who would be credited in the abstract. When Aserinsky broached the question, he recalled, Kleitman listed four options: "I could be the sole or the senior author, or he could be the sole or the senior author." Judging from Kleitman's tone and body language, Aserinsky surmised that he was "not apathetic" about the matter. He deemed it wise to include Kleitman's name—though as junior author, since (by Aserinsky's calculation) he'd had only a "tangential" role in the work.[75]

That concession, however, soon led to the decisive break. As coauthor, Kleitman told Aserinsky, he would discuss plans for announcing the REM discovery with a representative from the university's public relations office. Aserinsky would not participate; instead, he was to monitor the conversation from behind Kleitman's office door, "which on this occasion was to be kept slightly ajar." He was instructed to listen for any statements he judged incorrect and inform the professor afterward.[76]

Outraged by this "unconscionable" slight, Aserinsky resolved to end his peonage as quickly as possible. He held his fire until he'd spoken at the conference, earning brief mentions by the Associated Press and the *New York Times*. (The AP's take: "Your eyes apparently dance when you dream.") Soon afterward, he asked physiology department chair John Hutchens—who'd succeeded Anton Carlson in 1941 and had been impressed by Aserinsky's work—to intercede with Kleitman. The three men hashed out a deal whereby Aserinsky would be granted his PhD as soon as he had an employment offer at another institution. Still, Kleitman tried to keep him, offering a loan of $200 and—drawing on his own experience as a penniless grad student—"some unforgettable paternal advice on how to economize by eating chicken necks, which were not only economic but very tasty too." Aserinsky took the cash but sent out his CV.[77]

Meanwhile, he'd grudgingly coauthored a journal paper with Kleitman, presenting their preliminary results. After submitting it in April, he raced to finish his remaining REM research and complete his dissertation. By June, he'd left Chicago for a research position at the University of Seattle, studying whether electrical currents could prod migrating salmon to climb fish ladders around dams.[78]

Aserinsky and Kleitman's two-page article, "Regularly Occurring Periods of Eye Motility, and Concomitant Phenomena, During Sleep," was published in *Science* that September. It reported on trials involving twenty adults, who exhibited a new type of eye movement—"rapid, jerky, and binocularly symmetrical"—that seemed connected with dreaming. The authors described their tentative findings on the physiological changes that accompanied REM, along with an estimate of its frequency over an average night (three or four episodes, judging from periodic sampling). "The fact that these eye movements, EEG pattern, and autonomic nervous system activity are significantly related and do not occur randomly," they wrote, "suggests that these physiological phenomena, and probably dreaming, are very likely all manifestations of a particular level of cortical activity which is encountered normally during sleep."[79]

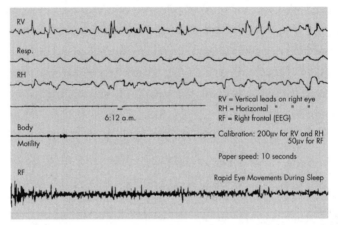

The first published chart of REM sleep, from Aserinsky and Kleitman's landmark 1953 paper. Key: RV = Right eye, vertical movements. RH = Right eye, horizontal movements. RF = Right frontal EEG (or brain waves). The other readings are for respiration and body movements. ©*AAAS, 1953*.

Within a few years, this paper would be recognized as revolutionary—not only for sleep science but for our basic understanding of neurophysiology. Once researchers began digging into REM, they came to a startling realization: Sleep wasn't a mere "complement of the waking state" as Kleitman had asserted in *Sleep and Wakefulness*; it was *two* states, which were as different from each other as sleep was from wakefulness.[80] Yet the report made barely a ripple at first. Indeed, when Aserinsky submitted an expanded version to the *Journal of Electroencephalography and Clinical Neurophysiology*, its editor politely turned him down.[81*]

Aserinsky and his erstwhile mentor did land a more comprehensive paper on REM in the *Journal of Applied Physiology* in 1955.[82] By then, he'd become a faculty member at Jefferson Medical College in Philadelphia,

---

* In an ironic twist, that editor was Herbert Jasper, who two decades earlier had authored the first EEG study to be published in the United States.

where he specialized in animal respiration.[83] His wife, meanwhile, spiraled into mania and alcoholism; in 1957, Sylvia died of suicide at a mental hospital in Pennsylvania. Aserinsky struggled on as a single parent, devoting just enough time to his work to avoid getting fired. Trying to be a better father than his own had been, he cooked dinner for his kids each evening; after putting five-year-old Jill to bed, he sat on the couch with Armond, fourteen, expounding on his new research, decrying the idiocy of his new boss, and watching hours of old movies on TV. After two years, he remarried, to a widow named Rita Goldman. He had long since divorced himself from his former field.[84]

The French sleep scientist Michel Jouvet later wrote that REM represented "a new continent" in the brain; before long, whole fleets of explorers would set out to chart its contours. Yet Aserinsky himself initially had little sense of its importance. Even Kleitman failed to grasp how fundamentally the map of sleep was about to change.[85]

But Bill Dement got it. He recognized the headlands when they were just a dot on the horizon.

7

Chasing Dreams

For all their differences, William Charles Dement had one thing in common with Eugene Aserinsky: a problematic progenitor. When Dement was born, in July 1928, his father, Charles, was forty-eight years old and working as an accountant for the Internal Revenue Service. The age difference might have been partially responsible for the distance between them. The Depression likely contributed as well: Charles, an accountant, was unemployed through much of the '30s, and his withdrawal may have reflected a sense of humiliation. In any case, he held himself at a physical as well as emotional remove—spending long hours in the garage, where he kept his tools and his whiskey bottle, or out playing cards with his cronies. Like Aserinsky's father, Dement's was a habitual gambler, betting on everything from slot machines to baseball games to presidential elections, and generally losing. "He wasn't much of a father," Dement said later. "Practically all the parenting was done by my mother."[1]

Kathryn Severyns Dement, thirty-seven when her only child arrived, was as industrious as her husband was feckless. Born to Belgian immigrant farmers in a sod cabin in Nebraska, she'd migrated westward with her family, eventually landing in Washington State. Shrewd and energetic, she'd worked as a schoolteacher, attended business college, and run successfully for auditor of Benton County—the first woman to hold elected office there. Though Kathryn left the paid workforce after becoming a mother, she kept the household afloat when the hard times hit. She shepherded their savings, took in sewing, babysat for neighbors. As Charles chased fleeting jobs, the family relocated from Wenatchee, Dement's birthplace, to the larger city of Spokane, then back to Wenatchee. Still, thanks largely to Kathryn, Dement remembered his childhood as fundamentally stable and secure. He rode a secondhand bike and seldom wore new clothes, but (unlike some of his classmates) he never went hungry or found himself homeless.[2]

From an early age, Dement stood out as a student; in his spare time, when he wasn't playing capture the flag on the street or listening to *The Green Hornet* and *Ma Perkins* on the radio, he devoured science fiction and articles about the technological marvels of the day: atom smashers, electron microscopes, jet engines.[3] He joined a Boy Scout chapter that met at an evangelical church; at one revival meeting, entranced by the fire-and-brimstone preacher, he marched to the immersion tank to be baptized, followed by his thirty troop mates—earning the nickname "Trailblazer Dement."[4] Although his religious impulse faded quickly, his capacity for infectious enthusiasm would be one of the keys to his later career.

Another would be the work ethic he inherited from the maternal line. When he was thirteen, Dement and his parents moved in with his paternal grandfather in Walla Walla. The old man, who owned a nearby wheat ranch, had been disabled in an automobile accident, and Kathryn took over his care from a small staff of paid helpers. She nursed him, took in boarders, and cooked and cleaned for all the occupants. Dement got his driver's license in eighth grade (the laws were laxer then) so that he could ferry Grandpa Frank on errands. Soon, he was contributing to the household finances, laboring after school as an insulation installer and a field hand. For a spirited, gregarious adolescent, the responsibility took its toll. But such burdens were common for kids of his generation, and they were lightened by his mother—a warm and nurturing woman with a dry wit and keen eye for human foibles.[5] "I couldn't have a conversation with my father," Dement said, "but I could talk to her."[6]

It's striking that each of the three founding figures of modern sleep science had serious father issues. Kleitman's was dead; Aserinsky's was pathologically self-centered; Dement's was barely there. That dad-shaped bruise may have been a major source of their drivenness—the wound they sought to heal by earning recognition and acclaim. It's also noteworthy that Kleitman and

Dement, who went on to build vast professional networks and lasting institutions, had a resource Aserinsky lacked: a close relationship with a strong,
supportive mother. The degree to which such factors influenced the trio's individual trajectories is, of course, impossible to say for certain. One characteristic
that set Dement apart from the others, however, was that he liked to ponder
such questions: He came to sleep research out of an interest in what makes people tick. His emotional intelligence, along with his talent for turning objective
data into compelling human stories, would also be crucial to his later success.

Those qualities began to shape the course of his life in 1946 when Dement
joined the army after briefly attending Whitman College in Walla Walla. His
principal goal was to return home with a GI Bill scholarship. Stationed on the
island of Kyushu in occupied Japan, he took a training course in journalism
and was chosen to establish a regimental newspaper. Dement found that he
loved interviewing fellow soldiers and writing about camp life. In 1948, after
finishing his tour of duty, he landed at the University of Washington. Unable to
find a dorm room amid the postwar rush of enrollments, he and some friends
pitched in and bought a ramshackle houseboat on Lake Union. He'd planned
to continue with journalism, but all the classes were full, so he signed up for
Introduction to Psychology—and discovered the teachings of Sigmund Freud.[7]

Freud's central idea was that the human psyche is ruled largely by unconscious drives and desires, many unacceptable to civilized society; neuroses
arise when individuals repress shameful or dangerous thoughts and feelings
to keep them from consciousness, tying their own "libidinal energy"—
or life force—in knots. These tangles, he asserted, have their origins in
painful childhood experiences, ranging from overly strict toilet training to
sexual trauma. Freud's main tool for cutting through such blockages was
"free association," in which patients said whatever came into their heads
under the direction of analysts trained to help decode its symbolic content.
This technique—often using dreams as raw material—ostensibly enabled
one to unearth deeply buried wishes and conflicts, achieving both a liberat-
ing catharsis and a better-integrated sense of self.[8]

By the late '40s, when Dement took his first psychology classes at the University of Washington, the analyst's couch may have been the most ubiquitous

piece of furniture in popular culture.[9] As he later wrote: "Every other book or movie, it seemed, had some element of Freudian psychology or some character being psychoanalyzed. Cartoons and jokes featuring a goateed psychiatrist popped up in everything from *The New Yorker* to *Reader's Digest*... There was a belief that Freudian psychoanalysis could explain every aspect of our problems: fears, anxieties, mental illnesses, and perhaps even physical illness."[10]

One factor that drew Dement to this territory was a desire to pursue his own liberation. In high school, he'd begun to cast off the conventions of his farm-town upbringing, identifying as a New Deal Democrat against his parents' rock-ribbed Republicanism and taking up jazz bass.[11] (He also played football, though without notable distinction.) By the time he got to college, he'd become something of a bohemian. He formed a combo that played Seattle nightclubs for tips; on Fridays, he hosted jam sessions on his houseboat that attracted local musicians—including Ray Charles and Quincy Jones, who were just beginning their stellar careers—and members of touring bands.[12] When the president of his fraternity objected to these interracial gatherings, Dement quit in disgust.[13]

Dement (left) with his band, late 1940s. His experience as a jazz musician would help shape the course of his scientific career. *Courtesy of Catherine Dement Roos.*

Tall and rawboned, with an equine face and a shock of dark hair, he projected the confidence of a front man. Yet for all his apparent boldness, he was shy and inhibited around women. "I had anxieties about sex," he later told an interviewer. In light of his family dynamics, Dement may have considered the Oedipus complex a plausible explanation for what ailed him. He tried psychoanalysis, but it was a flop; he found he couldn't free associate. "The analyst said, 'I don't think this is working, and it's costing you a lot of money,' which I thought was kind of ethical," he recalled.[14]

Nevertheless, by his senior year, Dement had set his sights on becoming a psychiatrist. After reading an interview in *Time* with University of Chicago chancellor Robert Maynard Hutchins—a champion of individualized education and academic freedom—he applied to medical school there. He received his acceptance letter in the spring of 1951 and moved to Chicago that fall.[15]

Dement was in his second year, grinding through a required neurophysiology course, when Nathaniel Kleitman appeared as a lecturer on the topic of "higher brain function." A small, aging man in a charcoal suit, Kleitman had a reputation as a dull speaker. But when he began talking about his work on sleep, Dement snapped wide-awake. "Like Sigmund Freud," he later wrote, "I had become fascinated by 'the problem of consciousness.' What is it, and how does it work?...During that lecture it occurred to me that the way to understand consciousness is to understand what must be given up to enter unconsciousness."[16] The realization, he recalled, "was like an arrow hitting a bulls-eye."[17] After class, he sat gathering his thoughts for a few minutes. Then he marched down the hall and knocked on Kleitman's door.[18]

There was no answer. Taking a deep breath, he knocked harder. Suddenly, the portal opened just a crack, and a gray head peered out.

"Professor Kleitman, I would like to work in your lab," Dement blurted.

"Do you know anything about sleep?" demanded the professor.

"Well—no."

"Read my book." The door swung shut.[19]

Dement bought a copy of *Sleep and Wakefulness* and found it riveting. After reading it cover to cover, he knocked on Kleitman's door again. "Okay," the scientist told him. "You can help Gene Aserinsky, who's trying to record eye movements during sleep."[20]

———————

When Dement arrived at the lab for his first training session, in late 1952, it struck him that Aserinsky "didn't seem terribly excited by what he was documenting." The all-night observations seemed to be exhausting the thirty-one-year-old researcher and putting strain on his marriage. Meanwhile, Dement, twenty-four, was "unmarried, energetic, and incredibly excited...to have the chance to be documenting something that had never been observed before." He grew more so when Aserinsky explained the theory he was testing: "Dr. Kleitman and I think that these eye movements might be related to dreaming." For Dement, who believed ardently in Freud's credo that dreams were the "royal road" to the unconscious, it was as if he'd been handed a treasure map. He was eager to follow wherever it might lead.[21]

In his books and other writings, Dement would take credit for two important developments in Aserinsky's research. One was the decision to include women in the study. As Dement remembered it, he suggested that a student named Pamela Vickers—his then girlfriend—take a turn on the lab cot. Kleitman, leery of scandal, agreed only on condition that Aserinsky act as chaperone, so the latter spent the night sleeping on the office sofa.[22] (Three other female classmates subsequently participated, including the future Hollywood actress and screenwriter Elaine May.)[23] Dement also recalled that it was he who'd shot the movies of sleepers' eye movements that helped convince Kleitman that REM was real.[24] In later years, Dement complained that he should have gotten a mention in the earliest REM papers. "Today," he told a journalist in 2004, "a student who did as much work as I did would have been listed among the authors."[25]

Aserinsky, for his part, resented Dement's claim that early rapid eye movement research was a team effort. "If anything is characteristic about the REM discovery," he wrote, "it was that there was no teamwork at all. In the first place, Kleitman was reserved, almost reclusive, and had little contact with me. Secondly, I myself am extremely stubborn and have never taken kindly to working with others." There had been fifty-one reportable sleep sessions between October 1952 and May 1953; of these, Aserinsky calculated, Dement had run only five. His assistant, he insisted, "had nothing to do with the REM discovery in my laboratory, although he subsequently became prominent on his own in clinically oriented studies."[26]

In science, of course, a discovery is only as good as the data gathered to prove its validity. Whether Aserinsky could have completed that buttressing without help is unknown. In any case, Dement's leadership in exploring REM's implications—eventually igniting an explosion in sleep science and harnessing its energy to form a new branch of medicine—is beyond dispute.

After Aserinsky's unhappy departure from the University of Chicago in June 1953, Dement continued to study the phenomenon. His curiosity was sparked by an observation that Aserinsky and Kleitman had omitted from their paper on REM: that this newfound sleep state contradicted the accepted explanation of how dreams arose—as a result of the mostly deactivated "higher mental centers" misinterpreting random sensory input from the sleeping body. Instead, according to the polygraph evidence, dreams appeared mainly during the rapid eye movement period, accompanied by brain waves indicating arousal. There must be some *purpose* for these intervals of high activity. But what was it?[27]

Dement thought the answer might lie in a theory first proposed by Freud: that dreams provided a "safety valve" for libidinal energy, which would otherwise erupt in the waking state as psychosis.[28] REM, the young researcher speculated, might be key to maintaining sanity. When he scanned the literature, he found few studies exploring the relationship between dreaming and mental illness. One author, however, claimed that schizophrenics did not dream, bolstering Freud's (and Dement's) hypothesis. To test it, he arranged to study patients at Manteno State Hospital, about forty miles south of the university.[29]

Opened in 1930, Manteno was designed as a cutting-edge mental institution, with spacious brick "cottages" set on one thousand wooded acres; to aid in their rehabilitation, occupants tended crops on an on-site farm. But by the autumn of 1953, when Dement arrived, the hospital was severely overcrowded and understaffed, with more than 8,000 patients and 450 attendants—only 21 of whom were trained nurses. Along with psychoanalysis, psychosurgery (also known as lobotomy) and crude electroshock therapy were favored forms of treatment. Manteno's polygraph department, however, was among the most advanced in the country; it was run by EEG pioneer Frederic Gibbs, who was researching brain wave patterns in epilepsy and mental illness. Dement spent several months there, conducting sleep tests on schizophrenic volunteers and medical students who worked at the facility.[30]

The readings, it turned out, showed virtually no difference between schizophrenic subjects and controls. "In fact," Dement later wrote, "they all had regular periods of REM sleep with voluminous eye movement, and all reported dream content to the best of their ability."[31] The only difference, he found, was that the patients tended to dream of "isolated, inanimate objects"—such as hats and ripped coats—"with no overt action whatsoever." This seemed to shed more light on the effects of schizophrenia (or perhaps lobotomies) than on the purpose of REM.[32]

Disappointed but undaunted, Dement returned to campus. He began probing REM more deeply, running polygraph tests two nights a week in Kleitman's lab while keeping up a full schedule of med school courses. To ensure that he never missed a rapid eye movement period, he implemented a new protocol: recording continuously throughout the night instead of at random intervals as both he and Aserinsky had previously done to save money on paper.[33] "In blatant violation of principles I now routinely and loudly proclaim, I became extraordinarily sleep-deprived," Dement wrote. "If I sat in the back of any classroom, I would fall asleep and snore, thereby attracting the attention of the professor, who would promptly order me out of class." At one point, he was summoned to the dean's office to discuss the problem. But he was too jazzed by his research to modify his schedule. After receiving his MD in 1955, he decided to stay and pursue a PhD based on his sleep studies before moving on to a medical internship.[34]

Dement and Kleitman go over sleep test results, mid-1950s. © *Elsevier, 2003*.

That year, Dement coauthored the first of four papers with Kleitman that fleshed out the relationship between EEG patterns, eye movements, body movements, and dreaming.[35] He also began dating twenty-five-year-old Eleanor "Pat" Weber,[36] who was working as a secretary for University of Chicago psychologists Carl Rogers and Rosalind Cartwright. (Rogers is remembered as a founder of the humanistic school of psychology. Cartwright—inspired partly by Pat's enthusiastic descriptions of Dement's work—would later become a leading dream researcher.)[37] A native of Charleston, West Virginia, Pat came from a background at least as challenging as Dement's. Her father had never regained his footing after being forced to drop out of medical school during the Depression, and the family had fallen into poverty. Unable to afford college, she left for Florida after graduating high school; at eighteen, she married a writer for *TV Guide* but was divorced soon afterward and decamped to Chicago.[38]

Despite these rough beginnings, Pat was vibrant, fun-loving, and intellectually aware—passionate about books, music, sports, and causes such as civil rights and nuclear disarmament. Petite and stylish, with her dark hair in a pixie cut, she brought out Dement's frolicsome side, helping to keep his ambition from outrunning his capacity for joy.[39] They lived together for a few months (a highly unconventional arrangement at the time) before marrying at city hall in March 1956.[40] After that, Kleitman allowed the couple to remain in the lab overnight whenever they pleased—without a chaperone. She eventually served as a subject in dozens of sleep runs.[41]

Pat helped deepen Dement's bond with his mentor as well. From the start, that partnership had been mutually respectful, with none of the tension that marred Aserinsky's interactions with Kleitman. Still, there was a gap in age and eminence that had seemed impossible to overcome. Dement was taken aback when Kleitman invited him to bring his new bride to dinner, asking incidentally, "Do you play bridge?" He stammered, "Yes," then fell into a panic at the prospect of proving the lie. Pat, however, was unfazed by his predicament. "Fortunately," he later wrote, "my wife was a very good player and, doing almost nothing else in the [next] several days…she managed to teach me the rudiments of the game." Arriving at Kleitman's modest town house, the couple were "awestruck" by the original Picasso in their entry hall—likely bought cheap on a long-ago trip to Europe—and charmed by the "gracious and welcoming" Paulena. "After a lovely dinner," Dement recalled, "the cards were dealt. Lady Luck really took pity on me that night—my bridge hands were so good it was difficult to make a mistake."[42]

Over the next four decades, there would be many more such games, whether at the Kleitmans' home or the Dements', accompanied by lively and wide-ranging conversation.[43] Kleitman would refer to Dement as "my star pupil"[44] while their relationship came to transcend the academic. For Dement, the older man proved to be something of a surrogate father, much as Anton Carlson had once been for Kleitman. "He was conservative and economical in his personal life, but I was delighted as I gradually learned that he was a genuine populist, humanitarian, and liberal," Dement recalled. "He always asked what was going on in my personal life. He sent wedding presents, gifts when my grandchildren were born, and a never-ending supply of Christmas gifts of food or candy. When tragedy struck our lives, he was more than sympathetic; he was grief stricken for us."[45]

———————

Dement's early papers with Kleitman got national press coverage, in part because they illuminated the experience of dreaming in ways that had never previously been possible. For example, one study challenged the widely held belief that dreams lasted only for a few seconds, even when the action—a journey through a landscape, a chase by a monster—seemed to go on much longer. In fact, Dement discovered, dreams matched the duration of the REM period in which they occurred, from several minutes to over an hour.[46]

Another finding that intrigued reporters (though other scientists failed to replicate it) was that eye movements during REM bore some relation to the action a dreamer was visualizing. As the Associated Press explained in April 1956, under the headline IN DREAMS YOU'RE STAR AND AUDIENCE: "In dreaming, your behavior in bed is much the same as though you were at a Broadway show, two University of Chicago scientists said yesterday. That is, just before your dream 'show' begins, you move and turn restlessly in bed—just as people twist and turn in their seats in a theater before the curtain goes up. Once the dream starts, your body movements stop, except for minor ones, such as finger twitching… [U]p-and-down eye movements are connected with dreams involving climbing and similar activities, presumably including the awful type where you're falling from a height. On the other hand, side-to-side movements characterize dreams concerned with 'watching' activities on a horizontal plane."[47]

But the paper that made Dement's name as a scientist—and transformed REM into a paradigm-disrupting force—was his fourth collaboration with Kleitman, "Cyclic Variations in EEG During Sleep and Their Relation to Eye Movements, Body Motility, and Dreaming," published in November 1957 in *Electroencephalography and Clinical Neurophysiology*. This study scuttled forever the notion of sleep as a period of suspended animation, in which the brain rolled through deeper and shallower phases of quiescence before surfacing back into wakefulness. Instead, it showed that sleep was a highly structured process built around distinct cycles of neural activity.[48]

The paper was based on 126 uninterrupted, all-night sleep tests on twenty-six men and seven women, conducted by Dement over a two-year period. Along with EEG readings, the polygraph recorded eye and body movements; in seven of the tests, the experimenter sat at the subject's side, watching by dim light for motions too subtle for the machine to register. To analyze the results, Dement and his coauthor divided sleep into four stages, based on the proportions of various EEG patterns during different periods that recurred each night. In Stage 1, brain waves displayed fast, irregular, low-voltage patterns, including some often seen during waking. In Stage 2, bursts of the high-voltage activity known as "spindles"— absent during waking—began to appear. In Stage 3, spindles alternated with slow, high-voltage waves (also never seen during waking). In Stage 4, slow waves dominated.[49]

When Dement and Kleitman combed through the data, they found that sleep unfolded in an unvarying order. In the first hour after onset, the EEG descended rapidly from Stage 1 to Stage 4, where it remained for about thirty minutes before returning to Stage 1. At this point, the eyes began jerking back and forth, signaling the arrival of REM. After a few minutes, the eyes stilled and the EEG drifted back down through the lower stages. Then came another bounce to Stage 1 and rapid eye movement. Each cycle lasted about ninety minutes, with four or five bouts of REM per night (representing 20 percent of total sleep time). With each round, the REM phase lasted longer and the descents grew shallower, until wakefulness returned.[50]

Although "Cyclic Variations" made only a small stir when it appeared, it was a monumental achievement—and it became one of the most-cited scientific papers of all time.[51] As the first detailed description of the brain's behavior during sleep, including its interactions with the neuromuscular system, the study laid the foundation for a new era of research. It established the link between REM and dreaming far more precisely than earlier efforts. (When awakened at the end of a rapid eye movement period, Dement and Kleitman found, subjects reported dreams 80 percent of the time—versus only 7 percent for other sleep stages—and recounted complex narratives rather than the simple imagery associated with non-REM sleep.) In addition, the paper helped pave the way for scientists to explore the question that had first drawn Dement to sleep research, and one that philosophers had pondered for millennia: the nature of consciousness.[52]

As the authors noted, the presence of alertness-associated brain waves during REM showed that people could be conscious while asleep— just as people with some kinds of brain damage could be unconscious while awake. Could polygraphs reveal the commonalities and differences between dreams, hallucinations, and waking awareness, and explain why all felt equally real?* Probing the secrets of sleep, it now appeared, could enable researchers to crack one of the stubbornest mysteries of human existence.[53]

Yet for all the eventual impact of Dement's work, his eligibility for a PhD remained uncertain as his research fellowship drew to a close. Although he'd submitted and successfully defended his thesis, he needed to demonstrate proficiency in two foreign languages (an area in which he possessed little talent) before he could receive his degree. Dement had

---

* As it turned out, more sophisticated technologies (such as PET scans and fMRIs) were required to answer those questions, which are still being investigated today. But the "Cyclic Variations" paper helped jump-start the search.

chosen French and German but had made only minimal efforts to learn them. Two tests were required for each language—a written exam given by the biology department and an oral exam administered individually by a faculty member. With some cramming, Dement was able to squeak by on the written tests, which were fairly elementary. But since no one had asked him about the oral exams, he'd pushed them to the back of his mind.[54]

By June 1957, Dement had been accepted for an internship at Mount Sinai Hospital, in New York City.[55] On his last morning in Chicago, after he and Pat packed their goods into a U-Haul trailer, he stopped by Kleitman's office to say goodbye. After some small talk, the older scientist remarked, "By the way, you haven't had your French exam." Kleitman handed him a foreign research article to translate, but it soon became apparent that his student was flailing. An awkward silence followed. When Kleitman inquired about the German exam, Dement admitted he'd neglected that one, too.[56]

At Kleitman's urging, Dement called the elderly professor in charge of the German test, Arno B. Luckhardt, and drove to his home with Pat. And then, as it would so often over the next several decades, Dement's gift for connecting with other humans helped decide the course of his career. "It happened that I was a fan of Professor Luckhardt's favorite German poet," he recalled. "He sat and read poetry to me for some time while my wife waited outside in the summer heat, but as long as he wanted to read, I wanted to listen. He finally told me I was a 'nice young fellow' and he passed me."[57]*

Dement and his spouse headed initially to Charleston, where he met his in-laws for the first time and shadowed a physician at a local hospital to refresh his clinical skills. (Pat's family members were "somewhat dismayed" when he described his sleep research, he wrote, "since it was not obvious to them how I could earn a living.") The couple then drove to New York, where they stayed with one of Dement's med school

---

* Luckhardt had also been supportive of the young Nathaniel Kleitman; he's mentioned in Chapter 2 as the physiologist who visited him in Paris in 1924.

friends—Elliot Weitzman, who would himself become a pivotal figure in sleep science—before finding an apartment of their own.[58] Later that summer, Kleitman sent his protégé another French article to translate, with access to a dictionary and all the time he needed, enabling him to land his doctorate at last.[59]

Dement began his internship on July 1, five months before "Cyclic Variations" rolled off the press.[60] The opening of the new continent was getting underway.

8

Pathway to Paradox

The beacon that drew Dement to New York City radiated from a stately brick-and-limestone hospital wing, half a block from Central Park.[1] There, a psychiatrist named Charles Fisher was testing Freud's theories about dreaming with an empirical rigor that no other researcher had attempted.[2] Small, dapper, and fiercely brilliant, the forty-nine-year-old Fisher had been primed for his quest by a harrowing childhood. Raised in Los Angeles by immigrants from eastern Europe, he lost his mother to suicide when he was eight; he'd watched her vomiting in the kitchen after taking poison. "She did it because she had a bad dream," his father explained before depositing him in a Jewish orphanage where he remained until he turned eighteen.[3]

Although Fisher earned outstanding grades at the public school down the street, he turned down a scholarship to a rabbinical seminary; he had no interest in religion. Instead, after aging out of the orphanage in 1926, he became what was then known as a hobo, riding freight trains and working low-wage jobs from Wyoming to Manhattan. In Greenwich Village, he cohabited with (and eventually married) a freethinking young woman named Betty Krowech, who urged him to apply to college. Supported by her wages as a secretary and his own as a part-time stenographer, x-ray technician, and night watchman, Fisher earned a BS in psychology from the University of Chicago, followed by a PhD in neurobiology and an MD from Northwestern. As a graduate student, he made an important physiological discovery, pinpointing the lesion in the hypothalamus that causes a rare form of diabetes. But the pain and bewilderment of his early years drove him to explore the brain's less tangible mysteries. He went on to a psychiatric internship at St. Elizabeths Hospital in Washington, DC, and earned his certificate in psychoanalysis at a local training institute.[4]

During the war, Fisher served at a US Public Health Service clinic on Ellis Island, treating merchant seamen and coast guard sailors with combat neuroses. Afterward, he joined the faculty at Mount Sinai, built a successful private practice, and developed a research focus—laser-like in its intensity—on the role of visual input in the making of dream content.[5]

Fisher's obsession grew out of a footnote in the 1919 edition of *The Interpretation of Dreams*, in which Freud discussed a recent paper by the Viennese sensory physiologist Otto Pötzl. In a series of experiments, Pötzl had shown subjects pictures using a tachistoscope—a device designed to flash images for a fraction of a second—and instructed them to sketch whatever they remembered immediately afterward. The next day, when he asked the subjects to draw what they'd dreamed the previous night, details from the tachistoscope pictures turned up. Strikingly, however, the drawings contained only elements that the subjects had *not* recalled after the initial viewing. To Freud, this observation carried "a wealth of implications," but the questions it raised went "far beyond the sphere of dream-interpretation as dealt with in the present volume."[6]

Fisher saw Pötzl's findings as evidence for an assertion that Freud made elsewhere in the book: that the dreaming mind uses "subsidiary and unnoticed" impressions from recent days as construction material. And he added his own theory of how the process works. Such images, Fisher suggested, are registered by a part of the mind that Freud called the "preconscious"—a sort of holding area from which memories can be accessed at will. Because they have not been laden with fixed meaning, he reasoned, they are ideal for building the coded narratives that help neutralize our psychic conflicts as we slumber.[7]

In the early 1950s, Fisher set out to pursue this notion further than Freud could have dreamed possible. In a small room near Mount Sinai's twenty-bed psychiatric ward, he carried out Pötzl-style experiments on patients, doctors, medical students, his friends, and his friends' children. Fisher used a slide projector with a mechanical shutter to flash images on a screen—photos or drawings of people, animals, objects, and landscapes, as well as printed words or numbers—for ten milliseconds. But instead of simply having subjects draw what they remembered, he used their sketches as a jumping-off point for psychoanalytic sessions.[8]

Fisher determined that what Freud called "dream-work"—the process of repurposing dimly recollected images to construct cryptic allegories—began within minutes of glimpsing a slide. One patient, shown a photo of two cats and a parakeet, reported seeing two dogs; when she tried to draw one, it came out looking like a bird, which she linked to a traumatic childhood incident involving a vicious watchdog and a pigeon. Fisher also found that subliminal imagery, transmuted through the filter of dream recall, could inspire patients to astounding flights of free association. The day after a physician was shown a six-pointed star with a missing tip, he sketched a dream image of "two spread thighs with a vague hermaphroditic genital," which he connected to a boyhood memory of watching a girl urinate and wondering why she had no penis. He also drew an M shape, which conjured up his "dominating, phallic mother." Whether this insight helped the physician resolve his neurosis is not recorded.[9]

Fisher's first paper on the "Pötzl phenomenon" (as it had come to be known) was published in 1954.[10] Dement read it the following year and was so smitten that he struck up a correspondence with the older scientist. Fisher, in turn, was fascinated by Dement's research on REM, which seemed to bolster the Freudian tenet that dreaming served a vital neurological function. The news that it occurred at regular intervals each night, as part of an inborn cycle, was "a revelation," Fisher later told an interviewer. "I said, 'Jesus Christ, what a thing!'" When Dement mentioned that he was looking for an internship, Fisher invited him to Mount Sinai.[11]

The timing of Dement's arrival was serendipitous for his new mentor, coinciding with a scandal that gave the word *subliminal* the sinister association that it has today. For a century, tachistoscopes had been used for essentially benign purposes—not only to treat neuroses but also to study visual perception, boost reading skills, or (during World War II) train pilots to quickly recognize enemy planes.[12] In April 1957, however, Vance Packard's bestseller *The Hidden Persuaders* revealed a less savory use for images projected at split-second speed: to surreptitiously influence viewers' decisions. Packard cited a recent report in the *London Sunday Times* that "certain United States advertisers were experimenting with 'subthreshold effects' in seeking to insinuate sales messages to people past their conscious guard." According to the article, a cinema in New Jersey had flashed ice cream ads onto the screen during film showings; the result was "a clear and otherwise unaccountable boost in ice-cream sales." The success of this gambit, Packard wrote, suggested that "political indoctrination might be possible without the subject being conscious of any influence being brought to bear on him."[13]

The Hidden Persuaders was a wide-ranging exposé of how companies and politicians used marketing research—particularly, the fledgling discipline of "motivational research," which borrowed techniques from clinical psychology—to manipulate consumers and citizens. The book's excursion into "subthreshold effects" took up less than two pages. But it primed the public for outrage when a New York motivational researcher boasted that he had been responsible for the refreshment-stand experiment.[14]

The researcher, James M. Vicary, held a news conference that September, where he declared that he'd developed a new secret weapon for advertisers—the "invisible commercial." Its first test, he explained, had run for six weeks at a theater in Fort Lee, where 45,699 customers served as unwitting subjects. Two messages were projected repeatedly at 1/3000 of a second, neither of them for ice cream. One urged the audience to eat popcorn; the other commanded, "Drink Coca-Cola." Vicary claimed that the experiment increased Coke sales by 18.1 percent and popcorn sales by a whopping 57.5 percent. After demonstrating the technique to reporters, he announced that he and two partners were starting a firm called the Subliminal Projection Company. They had applied for a patent on the process, he said, and planned to adapt it for the burgeoning medium of television.[15]

Vicary's announcement sparked a firestorm of indignation. YOU'LL BE BRAINWASHED AND NOT EVEN KNOW IT, the *Lincoln Evening Journal* warned.[16] "Welcome to 1984," editor Norman Cousins wrote in the *Saturday Review*, referring to George Orwell's novel of totalitarian dystopia.[17] Even *Advertising Age* signaled its disapproval, headlining its story on the press conference "PERSUADERS" GET DEEPLY "HIDDEN" TOOL. The National Association of Radio and Television Broadcasters asked its members to report any uses of such techniques. Congress and the Federal Communications Commission (FCC) launched investigations of the threat.[18]

Within a few months, Vicary's claims began to fall apart. *Motion Picture Daily* quoted the theater owner as denying that the original experiment had any effect on refreshment sales; a retest at a radio station in Washington, DC, with members of Congress and the FCC in the audience, failed miserably.[19] Years later, Vicary confessed that he'd fabricated his initial results.[20] By 1958, however, the imbroglio had sullied even legitimate subliminal perception research like Fisher's. Scientists who valued their reputations were stowing their tachistoscopes and striking out for other fields.[21]

"The subliminal stuff was blowing up in smoke," Fisher recalled. "I was out of ideas. I didn't know quite where to go."[22]

Dement's sojourn at Mount Sinai would give both men an opportunity to pursue new avenues of exploration. But first, he had to complete the clinical portion of his internship. As the program got underway, in the summer of 1957, he had little time for anything but his assigned duties. "I carried a lump of anxiety every day for several months until I learned the ropes," he later said. There was endless scut work—handling admissions, histories, and physical exams; administering medications; starting IV infusions; performing spinal taps. Over the following year, he rotated through orthopedics, cardiac surgery, abdominal surgery, neurosurgery, pediatrics, and the emergency room, with new skills to master in each department. He was on duty twelve hours a day, on call around the clock.

When he was too exhausted to keep his eyes open, he sometimes caught a few winks on a break room pool table.[23]

That December, Fisher helped Dement apply for a fellowship from the Foundations Fund for Psychiatric Research; it was duly approved, providing an annual stipend of $6,000 (worth about $58,000 today)—double his income from his Chicago days, and enough to sustain his family in relative comfort while he pursued his further studies. The following month, Pat gave birth to a daughter, Cathy, whose arrival added to the household's level of sleep deprivation. When Dement's internship ended, in June 1958, he was eager to embark on the next stage of his career.[24]

Once again, however, he had neglected to prepare for his final exams. He managed to squeak by on Parts 1 and 2 of the National Board test but failed the pediatrics portion of Part 3. "The examining pediatrician asked me to tell him honestly if I ever planned to practice pediatrics," he recalled. "I told him, 'Not only will I never practice pediatrics, I will probably never practice medicine, because I want to do research.'" This was not the whole truth: Dement wanted to have the option of going into practice if circumstances (professional or economic) called for it. Nonetheless, as with his foreign-language exams at the University of Chicago, the proctor graciously gave him a passing mark. Dement was able to move forward where another aspirant—less charming or charismatic, perhaps—might have lost his chance.[25]

Soon afterward, Dement and Fisher converted the tachistoscope room into a sleep lab, with a cot and a polygraph machine. And here, the brief romance between Freudian dream theory and EEG-driven sleep research planted a seed for an important medical discovery.[26]

One of the first tasks Fisher assigned his research fellow was a sleep study of a patient with narcolepsy,[27] a disorder that was only a little better understood than when French neurologist Jean-Baptiste-Édouard Gélineau described it in 1880. It was now known that narcolepsy's core symptoms were daytime drowsiness and sleep attacks, as well as nighttime sleep disruptions—often including hallucinations during sleep onset, violent nightmares, and sleep paralysis, in which the patient awakens from a dream but cannot move. Most patients also experienced spells of daytime paralysis known as *cataplexy*, typically triggered by strong emotions; this had come to be recognized as a separate, though usually co-occurring, disorder. Treatments, too, had advanced somewhat since Gélineau's time, when dubious patent medicines were the only remedies available; doctors now prescribed stimulants such as ephedrine and amphetamines, which controlled the disease at least partially in most cases. Yet the causes of narcolepsy remained a mystery. Some experts suspected that it stemmed from a defect in the sleep center that von Economo identified in 1930 (in the hypothalamus) or in the waking center that Magoun and Moruzzi discovered in the '40s (in the reticular formation).[28] Others insisted that it was a psychosomatic disease.[29]

The latter explanation had become increasingly popular for illnesses whose origins were unclear—including ailments now known to have complex biological causes, such as asthma, peptic ulcers, and inflammatory bowel diseases. Since the 1930s, in fact, a whole school of "psychosomatic medicine" had emerged to study this class of disorders.[30] In some cases, patients with such diagnoses wound up in psych wards, including the one at Mount Sinai.[31]

Fisher, like many Freudians, believed that narcolepsy arose from unresolved libidinal issues and could potentially be cured by psychoanalysis.[32] (A smattering of case studies had reported successes, though their methodology—based on isolated anecdotes—left much to be desired.)[33] His rationale for recording his new patient's REM patterns, however, can only be guessed at. Perhaps it was pure scientific curiosity. Or he may have hoped the polygraph, like the tachistoscope, would become a tool for spurring dream-based free association.

On the appointed night, Dement crowned the narcoleptic subject with a tiara of electrodes. Minutes later, at the control board in the next room, he was startled to see the polygraph needles scratching out the rhythms of REM. Although he had watched infants plunge directly into that sleep stage, adults typically took more than an hour to reach it. But was the man's quickness a unique feature of his neural wiring, he wondered, or was it a characteristic of narcolepsy itself?[34]

In the coming months, Dement and Fisher scoured New York hospitals for more people with narcolepsy to test—but in a city of eight million,[35] they found only four. Shepherded into the sleep lab, they all showed the same tendency to topple straight from wakefulness to REM. To Fisher, this suggested that the patients' emotional disturbances were severe enough to have disordered their dream-production patterns. (Whether he actually psychoanalyzed these individuals is unknown.) Dement thought the onrush of REM pointed to an organic brain malfunction, but he reserved judgment pending further investigation.[36]

Unbeknownst to either scientist, another researcher was on the same trail—at an institution both men knew well.

———

In the fall of 1959, Nathaniel Kleitman, facing mandatory retirement at sixty-five, cashed in a year's worth of unused vacation time and left the University of Chicago. After more than three decades as sleep science's lonely prophet, he felt ready to pass the torch to a new generation of apostles. Kleitman was also working on a revised and expanded edition of *Sleep and Wakefulness*, likely prompted by the interest that REM had begun to generate among scientists, funders, and the general public.[37] He would remain an active presence in the field into his nineties.

Kleitman's successor as director of the sleep lab was Allan Rechtschaffen—a Bronx-bred psychologist who'd been a Golden Gloves boxer in his youth.[38] The lanky thirty-one-year-old was known for his penetrating intellect, exacting scientific standards, and demanding leadership style. ("Al was a tough son of a bitch," a former student said, recalling his merciless critiques.)[39] Like Dement, Rechtschaffen had planned to become a journalist before becoming obsessed with the enigmas of the mind and brain. As a PhD student at Northwestern, he'd been drawn to sleep science after reading about the discovery of REM. Now he was running the facility where that breakthrough took place. When Rechtschaffen took over, he recalled, the old man gave him just one piece of advice: "Clean up in the morning."[40]

Among Rechtschaffen's apprentices was a psychiatry resident named Gerald Vogel—then a devout Freudian, though he would later renounce the faith and become a prominent sleep researcher. And in October 1960, Vogel published the first paper to describe the phenomenon that Dement and Fisher were also exploring: sleep-onset REM periods in narcolepsy.[41]

Vogel conducted his landmark experiment to test the idea that narcolepsy was a psychogenic disorder—more specifically, to test the notion that "the attack of sleep constitutes a defense against the development of painful affects (anxiety and guilt) or a defense against the emergence of unconscious, forbidden, fantasies and impulses." His subject was a forty-two-year-old "married Negro shipping clerk," who sought treatment for both his medical condition and a sense of dissatisfaction with his career.[42]

The paper, published in *Archives of General Psychiatry*, began with a detailed case history. The patient had been diagnosed with narcolepsy at thirty-three, two decades after his symptoms first appeared. A doctor had prescribed amphetamines, which changed the man's life dramatically. "From a fellow who came home, read the paper and went to sleep," he told Vogel, "I became alert, active, and interested." He had taken up drawing, with a skillfulness that astonished his wife and friends. "I quit my long-standing job because security was the last thing I was thinking of," he said. "I wanted the opportunity for advancement."[43]

The subject had found a better job and begun moonlighting as a commercial artist. Yet he felt that he had not risen to a position matching his potential. "Though he does his work well," Vogel wrote, "the patient is unable to assert himself and insist on adequate compensation for this work. For example, on the job, his title and pay is that of shipping clerk. But in reality he supervises the shipping department of a moderately large manufacturing concern. Off the job, the patient does commercial or technical drawing. His fee for this work is often so small that his clients voluntarily pay him more than he asks. This difficulty, i.e., the patient's neurotic inability to insist on adequate compensation for his work, is the second symptom for which he seeks psychotherapy."[44]

Vogel—true to his time—did not consider the possibility that the man's reticence arose from the stigma of his illness or (perhaps more so) from the social constraints facing a Black American in an era of segregation. Instead, he delved into matters of family history and psychosexual development. The patient's parents had separated when he was a toddler; he'd lived with his mother until age five or six, at which point his father had won custody. Vogel speculated that the boy had felt rejected by his mother, and had been dominated by his paternal grandmother, who "ruled the roost." As an adult, the patient displayed difficulty expressing anger and reaching orgasm, as well as a tendency to fall asleep during intercourse. He painted semiabstract nudes whose imagery hinted at "an expectation of hostile unresponsiveness and exploitation by women." A fear of abandonment by maternal figures, Vogel wrote, led the patient to defend himself by "strong repression of both libidinal and hostile affects and fantasies."[45]

Vogel hypothesized that one function of a narcoleptic attack was to produce a dream that provides "hallucinatory gratification" of an unacceptable fantasy. If so, he conjectured, a patient would begin dreaming soon after the attack began. To investigate this proposition, he employed what he referred to as "the EEG technique of Kleitman, Aserinsky, and Dement."[46]

One afternoon, after suspending the patient's narcolepsy medication for forty-eight hours, Vogel ushered him into the sleep lab and fired up the machine. "The patient fell asleep one minute after the procedure began," the researcher reported. "Simultaneously with the onset of sleep he was almost certainly dreaming and within three minutes after the onset of sleep he was definitely dreaming." Vogel awakened the patient after eight minutes and asked him to describe his dream. A few minutes later, the procedure was repeated, with similar results. On the third and final run, the patient fell asleep immediately, but his need to dream seemed to have been satisfied; sixty minutes passed before REM began. Vogel compared these times with polygraph records from two hundred full nights' sleep and five daytime naps in normal individuals, compiled by Rechtschaffen and his PhD student Edward Wolpert: in all but one of those runs, sleepers had taken at least thirty minutes (and usually far longer) to reach their first REM period.[47]

Vogel's paper included a description of the first dream of the session, which began with a doctor giving the patient a urine and saliva test. The doctor then led the man down a corridor lined with rooms whose windows had "frilly and feminine" curtains and left to "find out what was wrong." The dream ended with the patient wandering the halls, utterly lost. When Vogel asked for his interpretation, the patient said the dream reminded him of his feeling that "there is something physical as well as mental about narcolepsy." Vogel—again in keeping with his times— dismissed this analysis. Instead, citing the frilly curtains, he suggested that the dream represented a "wish for care and support by a maternal figure and the anger at abandonment." The experiment, he concluded, "supports the proposition that the narcoleptic patient makes use of his sleep for the projection of fantasy which is gratified in a dream in a way unacceptable during waking life."[48]

Vogel had beaten Dement and Fisher to the punch in reporting sleep-onset REM in narcolepsy. But he would not have the last word on the symptom's significance. Its true implications began to emerge a few years later.

Soon after Dement started his fellowship with Fisher, he began pursuing a basic question about REM itself: What would happen if sleepers were deprived of it? The answer, he thought, was key to understanding what REM—and dreams—were really for. He made his first observations in this area in January 1959, when a New York disc jockey named Peter Tripp decided to stay awake for 200 hours to raise money for the March of Dimes.[49] (The previous record was 187.)[50] The thirty-two-year-old Tripp broadcast from a glass booth in Times Square, keeping up a lively patter as passersby stopped to gawk. Periodically, a team of medical researchers—Dement among them—escorted him to a nearby hotel room to check his vital signs, take lab samples, and test his alertness and cognition.[51]

After two days and two nights, Tripp began to have occasional hallucinations, a common effect of prolonged sleep deprivation. Paint specks on a table morphed into insects; he saw cobwebs in his shoes. Past the 100-hour mark, simple tests of mental agility were torture, and the visions grew more disturbing. A scientist's tweed jacket became a mass of furry worms. Tripp opened a bureau drawer in his hotel room and screamed when it burst into imaginary flames. In the booth, a wall clock assumed the face of an actor friend; he began to wonder whether he was himself or the friend. By 170 hours, he'd spiraled into paranoid delusions. He accused the researchers of trying to poison him and of conspiring to send him to prison for an unspecified crime. On the ninth and final morning, a physician asked Tripp to lie down for an examination. Gazing up at him, the deejay became convinced that the dark-suited man was an undertaker about to bury him alive. Tripp leaped for the door and fled down the hall, with several doctors in pursuit.[52]

To Dement, Tripp's psychosis was strong evidence for the Freudian notion that dreaming was essential to sanity. Yet many experts still thought of the REM period—when most dreams occurred—as merely a stage of light sleep with no unique importance. One way to test the latter theory was to find out how the brain responded after a period of sleep deprivation. When Tripp finally staggered into bed, Dement hooked him up to a polygraph machine. If REM was nothing special, he reasoned, it would probably disappear for a few hours while his subject enjoyed a prolonged bout of deep, mostly dreamless, slow-wave sleep. If REM periods were in some way necessary, however, they would appear on schedule or earlier, and might increase in duration.[53]

The hint of an answer arrived early on: Just thirty minutes after falling asleep, Tripp had one of the longest REM periods Dement had ever recorded. Over the course of the DJ's thirteen-hour slumber, he experienced significantly more REM than normal. And he awoke perfectly sane.[54] "It was as if the loss of REM sleep during the marathon had led to a REM sleep pressure that was spilling over into wakefulness as the psychotic symptomatology," Dement later wrote. That pressure, he thought at the time, was "right in line with Freud's theories."[55]

Dement set out to study "REM pressure," as he came to call it, by depriving people *specifically* of REM sleep. Night after night, he would escort volunteers (usually graduate students or unemployed actors) into the room near the psych ward, settle them onto the cot, and bedeck them with electrodes. At the first sign of REM, he would shake them by the shoulder and call their name. He then let them go back to sleep until the telltale tracings reappeared on the polygraph chart, at which point he woke them again. The subjects endured as many as five consecutive nights of this annoyance, after which their sleep patterns were measured during several undisturbed "recovery nights." In the morning, Dement or Fisher would assess each volunteer's psychological and physical state.[56] They repeated the process with non-REM sleep, waking volunteers for brief periods several times a night, to differentiate the effects of REM deprivation from those of sleep deprivation in general.[57]

Early results confirmed that REM pressure was real. As Dement reported in *Science* in 1960, subjects deprived of the sleep stage fell into it more quickly each consecutive night and had to be awakened more frequently to ward it off—up to thirty times a night after prolonged deprivation. On recovery nights, they experienced "REM rebound," spending an average of 29 percent of total sleep in REM versus the 20 percent seen in normal sleepers. During the day, many REM-deprived subjects experienced heightened anxiety, irritation, and difficulty concentrating. One subject quit the study after a panic attack. Some volunteers reported a marked increase in appetite, accompanied by weight gain of three to five pounds. Significantly, these effects did not occur in subjects awakened during non-REM sleep stages.[58]

Yet to the researchers' frustration, none of the REM-deprived subjects exhibited any of the psychotic symptoms that Tripp had shown. (Only years later did it occur to experts that he might have been suffering side effects from the stimulant methylphenidate, or Ritalin, which he took during the last sixty-six hours of his marathon to help him stay awake.)[59] Hoping that further trials would bring more conclusive results, Dement secured a grant from the National Institutes of Health to rent a large apartment on Riverside Drive and turn part of it into a sleep lab. This enabled him to devote himself fully to his project while spending more time with his family. Sometimes he served as a subject himself, with Pat or a colleague running the polygraph. Kleitman, who stayed with the couple whenever he was in town, occasionally submitted as well. More often, Dement recruited paid volunteers through help-wanted ads.[60]

One Barnard grad student who answered the call, eager to earn some extra cash by sleeping, was a member of the Rockettes, the famous dance troupe that performed at Radio City Music Hall. Other members soon followed her example. "Thus began the following routine," Dement later wrote. "A lovely woman, still in theatrical makeup, would arrive at the apartment building and ask the doorman for my room. In the morning she would reappear, sometimes with one of my unshaven and exhausted male colleagues who had spent the night monitoring the EEG. One day, the doorman could finally stand it no longer. 'Dr. Dement,' he demanded, 'exactly what goes on in your apartment?'" The young doctor just smiled.[61]

Dement also conducted other types of experiments in his apartment lab. One involved testing sleep depth in children by playing an electronic tone at increasing volume; this trial was curtailed by a neighbor pounding furiously on the door and informing the researcher that he had awakened half the building. On another night, Dement attempted to measure the effect of a new drug called LSD (a powerful hallucinogen that was just beginning to interest scientists) on a subject's sleep cycles. Although the man complained of feeling "very uncomfortable," the dose was evidently too low to have any measurable impact.[62] Meanwhile, over several months of trials, evidence that REM deprivation led to insanity had not emerged.[63]

A new revelation about REM arrived during this time, however. It came from an explorer who was searching for something else entirely.

A native of France's mountainous Jura region, Michel Jouvet was dark-haired and dashing, with a taste for adventure that was not limited to his area of study. Jouvet had dreamed of becoming a naval officer or an anthropologist before the Nazis invaded his country in 1940. He fought for the Resistance as a teenager, then joined an Allied ski patrol unit in the Alps, but succumbed to family pressure after the war, and followed his father into medicine. Jouvet chose neurosurgery for its difficulty and complexity; growing fascinated by the mechanics of consciousness, he went on to pursue a doctorate in neurophysiology at the University of Lyon. He learned to run an EEG machine and to surgically isolate portions of the brain.[64]

Awarded a Fulbright fellowship in 1954, Jouvet spent a year doing research in the Long Beach, California, lab of Horace Magoun—the former Northwestern University neuroscientist, since relocated to UCLA, who'd codiscovered the waking center in the reticular formation of the brain stem. He returned home by way of Hawaii, Fiji, New Zealand, Australia, the Philippines, Japan, Hong Kong, Macao, Vietnam, Singapore, Malaysia, Thailand, Cambodia, Burma, India, Pakistan, Lebanon, Turkey, Greece, and Italy, carefully limiting his expenses to the equivalent of ten dollars a day. After a brief pause in his travels, he headed off to complete his fellowship at the University of Brussels. His adviser was Frédéric Bremer, whose *cerveau isolé* experiment (see Chapter 3) had influenced Kleitman's thinking before Magoun and Moruzzi's discovery upended the theory that sleep onset was a passive process.[65]

In 1957, Jouvet won a grant from the US Air Force—which was supporting European neurophysiology research as part of a postwar aid program—to start his own small lab in Lyon. For his first project, he set out to study how Pavlovian conditioning worked in the brain. He was particularly interested in habituation, the process by which repetition of a conditioned stimulus (such as a buzzer) diminishes response (say, salivation) when it isn't accompanied by the associated reward (a bowl of food).

Pavlov had theorized that this process, during which animals often fell asleep, was triggered by an inhibitory signal from the cerebral cortex.[66]

Jouvet wanted to see whether habituation was really governed by the cortex or some other brain region. He began by surgically implanting electrodes in cats' brains, then training the animals to associate an electronic tone with a flash of light. Once the conditioning was complete, EEG readings showed that the light alone triggered a response in the auditory cortex, as if the tone had also played. After prolonged repetition, the strength of the response decreased—just as Jouvet had expected it would. At this point, however, an obstacle arose: the animals kept dozing off, disrupting the brain wave measurements.[67]

To get around this problem, Jouvet decided to experiment on cats that were already asleep. Because their eyes were closed, he played the tone instead of flashing the light. The first beep, he found, produced a cortical response that lasted two to three minutes. By the fifth or sixth repetition, there was no reaction at all.[68]

Now Jouvet was ready to search for the origin of the signal that inhibited the response. This required removing different parts of cats' brains, waiting until the animals were asleep, and then observing how the remaining parts of their brains responded to the repeated tone. But another problem arose: Cats without a cortex never showed the telltale EEG curves of slow-wave sleep. How could Jouvet tell when to administer the stimulus? He needed to look for a sleep-related EEG pattern in another brain region. Fortunately, he found one: In normal cats, the slow-wave phase was also marked by high-voltage spikes in the hippocampus (a seahorse-shaped structure in the midbrain). When those appeared, it would be time to play the tone.[69]

Besides monitoring the brain's response to the stimulus, Jouvet wanted to see how that response affected physical behavior. So he began the next stage of the experiment by attaching electrodes to a cat's neck to record neuromuscular (EMG) activity on the polygraph. This raised yet another difficulty: One effect of removing an animal's cortex is heightened muscular rigidity. Unlike in a normal cat, Jouvet found, there was no startle reaction when he played the tone for the first time. And there was no sign of relaxation when he played it repeatedly.[70]

Jouvet noticed something else, however—something so bizarre that it made him set aside the original purpose of his study. Every thirty to forty minutes, EMG activity in the sleeping cat flatlined. This meant that the animal's muscles had gone completely slack—a condition known as *atonia*. These episodes, which lasted about six minutes, were connected with a spindle-shaped EEG pattern in an offshoot of the brain stem called the *pons*.[71]*

What on earth was going on? Baffled, Jouvet ran EMG tests on sleeping cats with intact brains—something no one else had ever thought to do. Astonishingly, he found the same cycle of muscle atonia. Stranger still, it was accompanied by brain wave patterns in the cortex resembling those during waking. Yet the creatures were harder to rouse during these periods than during slow-wave sleep. "This was," he later recalled, "a paradoxical finding."[72]

The French researcher now realized he'd stumbled upon the same phenomenon reported by Dement in a recently published paper: rapid eye movement sleep in cats. But because Jouvet had found it in decorticated animals, whose eyes couldn't move in the first days after surgery, he proposed a name that evoked its broader weirdness: *paradoxical* sleep.[73] (The two terms eventually came to be used interchangeably, with *REM* the standard nomenclature in English-speaking countries.)

Jouvet didn't just invent a new label for REM sleep; his findings cast it in a whole new light. In 1959, he and his collaborators published their revelations in the journal *Comptes rendus des séances de l'académie de la Société de biologie et de ses filiales*.† First, REM was triggered by the pons—a brain structure common to all mammals, situated in an area that controls breathing, heartbeat, and other essential life functions. (Non-REM sleep seemed to originate higher in the brain, though precisely where remained

---

* See diagram, page 289.
† Translation: *Minutes of the Meetings of the Academy of Biology and Its Subsidiaries*.

a matter of debate.) This suggested that REM wasn't just about dreaming; it must be physiologically necessary, at least for animals in the evolutionary line of cats and humans. Furthermore, atonia—that is, paralysis—was a central feature of REM sleep. The reason animals (including people) lay still during REM wasn't that they were absorbed in "watching" their dreams; it was because the brain made it impossible for them to move, aside from minor twitches.[74] Together, as Jouvet later put it, these qualities marked REM as a "third state of the brain, as different from sleep as sleep is from wakefulness."[75]

Eager to follow these findings wherever they might lead, he threw himself into sleep research full-time. He also began a correspondence with Dement and traveled to visit the American in New York City. The two scientists, Dement wrote, would become "lifelong transatlantic partners."[76]

———

As the new decade began, other researchers were falling under the spell of REM. One was Howard Roffwarg, a psychiatry resident at Columbia who offered to become Dement's assistant after hearing him give a lecture on "dream deprivation" at the New York Academy of Medicine.[77] Roffwarg started helping out at the apartment lab, and the pair soon embarked on a joint study of REM in infants. Observing babies in Presbyterian Hospital's neonatal ward, they determined that newborns spent more than 50 percent of their sleeping time in REM. Here was another clue to REM's importance: its sheer quantity in the first weeks of life suggested it played a part in neurodevelopment. But what was its specific function?[78]

A growing number of scientists were asking similar questions. At the University of Chicago, Rechtschaffen was recruiting a cadre of ambitious young researchers; he'd recently moved into a bigger laboratory, in a brownstone building around the corner from Abbott Hall.[79] Psychologist Joe Kamiya, who'd learned EEG techniques from Dement in Kleitman's sleep lab, was now building one of his own at the University of California, San Francisco.[80] At Downstate Medical Center in Brooklyn, Arthur Shapiro and Donald Goodenough were recording the REM patterns of self-described non-dreamers (who had fewer eye movements than subjects who remembered dreaming) and professed insomniacs (who often claimed they'd been awake when the machine showed otherwise).[81] The pair sometimes collaborated on such work with colleagues at New York University.[82] Polygraph-equipped sleep labs were popping up beyond the major urban centers as well, at institutions like the University of North Carolina and the University of Oklahoma.[83]

Thanks largely to the excitement over REM, such facilities were beginning to proliferate in other countries, too—particularly France, Italy, and Great Britain. In June 1960, the Ciba Foundation held a three-day symposium on "The Nature of Sleep" in London, attended by more than thirty international experts. The first-ever scientific conference on the topic, it was chaired by Australian neurophysiologist Sir John Eccles, who would soon win a Nobel Prize for his work on synapses. Among the speakers were Jouvet, who reported his latest findings on paradoxical sleep, and Kleitman, who reviewed recent research by Dement and others on the nature of dreaming. (Dement himself—whose second child, Elizabeth, was born a few weeks earlier—stayed home.) Frédéric Bremer expounded on neural mechanisms of the waking state; Giuseppe Moruzzi, on the role of lower–brain stem structures in synchronizing brain waves.[84]

REM was still new enough that, during the discussion periods, some of the eminent attendees expressed doubt that it really occurred in *all* human beings—even those who, like many sober-minded scientists, seldom recalled their dreams. Nonetheless, the conference hinted at the future of sleep research in one important way: All the presenters focused on the physiology of sleep rather than the psychological function of dreams. This may have reflected the mostly European makeup of the group; the Freudian strain in sleep research, exemplified by Vogel's paper on narcolepsy and Dement's early REM deprivation studies, was almost entirely an American variant. Even in the United States, however, the tide would soon begin to turn, driven in part by an explosion of knowledge about the workings of the brain. As neuroscience emerged as an independent discipline in the 1960s, sleep science would increasingly fall under its sway.[85]

One branch of that discipline was just beginning its great leap forward: neurochemistry, which studies the molecular compounds that modulate the functioning of the nervous system. Eccles himself had established nine years earlier that most nerve impulses are transmitted chemically rather than electrically, but only a handful of the substances involved had been identified so far. In his closing remarks, he called for a new approach to sleep science that would focus on such research. "[I]t is quite remarkable how little in fact we do know about the subtle levels of biochemical change that can be occurring," he said. "There will certainly be room for many investigations."[86]

Over the next few years, his words would prove prophetic.

———

That same month, another conference pointed toward a very different facet of sleep science's future. Strangely, however, almost no one in the field bothered to attend.

For a quarter century, researchers from around the world had gathered annually at the Cold Spring Harbor Biological Laboratory, on Long Island, to discuss a topic deemed timely by a panel of experts. In 1960, the Cold Spring Harbor Symposium on Quantitative Biology (as the prestigious meeting had come to be known) tackled the theme of "Biological Clocks"—a watershed moment for a discipline that would come to be known as *chronobiology*.[87] Although sleep-wake cycles figured in several of the presentations, only one sleep researcher (Downstate Medical Center's Arthur Shapiro) was present.[88]

Chronobiology did not yet have a formal name; it was referred to simply as the study of biological rhythms. Its practitioners investigated the relationship between an organism's *endogenous* rhythms (those governed by internal clockwork) and the *exogenous* rhythms of its environment.[89] This field had long occupied a kind of parallel universe to that of sleep science, but the two domains had rarely interacted in a meaningful way.[90]

In part, this was a legacy of sleep science's founding father. Nathaniel Kleitman's findings on the diurnal fluctuations of body temperature and alertness, and his experiments with arbitrary sleeping rhythms, were major contributions to both fields.[91] In *Sleep and Wakefulness,* he'd cited the work of some researchers now regarded as chronobiological pioneers.[92] Yet his theory that habit and volition were the main drivers of the human sleep-wake cycle diverged from what those researchers were finding in other life-forms. Following his lead, modern sleep science had never fully grappled with the question of whether hidden clocks might drive the patterns of slumber and alertness in people as they appeared to do in other creatures.[93]

The Cold Spring Harbor Symposium was organized by two of chronobiology's seminal figures, German physiologist Jürgen Aschoff and British-American biologist Colin Pittendrigh, who would later help forge important new paths in sleep research.[94] Eventually, those parallel universes would begin to merge.

But that grand reunion was not yet on anyone's calendar.*

---

* For more background on chronobiology, see Appendix B.

9

A Society of Scientists

In March 1961, William Dement and Allan Rechtschaffen organized their own conference at the University of Chicago's Billings Hospital—the first such meeting for US sleep researchers. The two-day affair drew thirty-six attendees. Kleitman flew in from his new home in Los Angeles, where he and Paulena had moved to be near their daughter Esther. Howard Roffwarg and Charles Fisher came, too, along with Joe Kamiya, Arthur Shapiro, Donald Goodenough, and other pioneers. Conscious of their generally limited financial resources, Rechtschaffen arranged lodging at the Shoreland Hotel for six dollars a night.[1]

In contrast to the physiological emphasis of the London and Cold Spring Harbor gatherings, most of the sixteen papers presented in Chicago focused on the connection between REM and dreaming—often in psychoanalytic terms.[2] David Foulkes, a psychology postdoc in Rechtschaffen's lab, presented a paper reporting that dreams occurred more frequently during non-REM sleep than Dement and others had calculated—a finding that was later replicated in several labs, helping to kick off Foulkes's career as one of the field's foremost contrarians. Dement himself reported on a study of age differences in REM sleep: preliminary results suggested that the proportion was highest in infancy, dipped in early childhood, rose again at puberty, and declined from young adulthood to old age. There were also sessions on technical topics, such as polygraph operation and scoring sleep stages on an EEG. One evening, U of C psychiatrist William Offenkrantz hosted a "beer party" at his home.[3]

The conference's success inspired Dement and Rechtschaffen to launch an organization that would hold one annually—the world's first professional society for sleep researchers. By December, they'd chosen a tentative name: the Association for the Psychophysiological Study of Sleep (APSS). The two appointed Kamiya, a Japanese American who'd grown up on a chicken farm in Northern California and spent his late teens imprisoned with his family in a wartime internment camp, as secretary-treasurer. Besides keeping the minutes of meetings and balancing the books, his job was to recruit as many new members as he could find.[4]

The establishment of the APSS was a quantum leap in sleep science's emergence as an independent discipline—equivalent to the formation of the American Cancer Society or the American Heart Association in earlier decades as a formative step for those areas of research. It supercharged the field's ability to define priorities, set goals, advocate for its interests, and develop a collective sense of self. It served as a seedbed for friendships and social networks that would help propel new enterprises, both investigational and institutional. And it would become a launching pad for the field's leaders—beginning, most spectacularly, with Dement.

His rise was not quite meteoric. Although Dement's REM papers in the mid-'50s had brought him wide recognition, he'd remained in the shadow of Kleitman, his mentor and frequent coauthor. He was now three years into his second apprenticeship, with Fisher, during which the gratifications had been far fewer and less immediate.

Dement published nothing at all in 1961, his first dry spell in half a decade.[5] He was working on just one really exciting project: a study that grew out of a taxicab ride with Rechtschaffen the previous year, when each discovered that the other was investigating sleep-onset REM periods in people with narcolepsy.[6] Unlike Rechtschaffen's former student Gerry Vogel, who'd been the first to report this phenomenon, neither man was convinced that it resulted from psychological factors. The trouble, they agreed, was finding enough patients to provide statistically meaningful data. They'd decided to join forces, running sleep trials of narcoleptic adults under a shared protocol—Dement at Mount Sinai, Rechtschaffen at the University of Chicago.[7] The study was still in its early stages, however, and it wasn't likely to be completed anytime soon.

Meanwhile, the still-officially-nameless APSS was starting to take off. The 1962 conference (again held at Billings Hospital) attracted sixty-nine researchers—nearly twice as many as before. An extra day was required to accommodate all the presentations. Once more, most talks focused on the psychological aspects of dreaming. But one full session, chaired by Arthur Shapiro, was dedicated to physiological studies. Among the lecturers was Dement's med school buddy Elliot Weitzman, now a neurologist at Albert Einstein College of Medicine, who'd recently followed his friend into sleep science. A brilliant researcher who would later play a pivotal role in uniting the field with chronobiology, Weitzman was studying slumber in nonhuman primates. He showed a film of monkey REM—the animal's eyeballs ping-ponging wildly behind half-closed lids—that amazed the crowd with its resemblance to the human variety.[8]

For Dement, the high point of the conference was that afternoon's talk by Michel Jouvet, who'd traveled from France for the occasion. By creating lesions in different parts of the pons, Jouvet reported, he and his team had found the approximate location of the nerve cluster that triggered the loss of muscle tone in REM, as well as a separate cluster that generated the waking-like brain waves associated with the sleep stage. For the first time, scientists were beginning to map the place where dreams came from and to pinpoint the sources of REM's baffling behaviors. "I will never forget the excitement of hearing his presentation at that second Chicago meeting," Dement recalled.[9]

Back in New York, the thrill faded quickly. That year, Dement pumped out three papers with Fisher, Roffwarg, and a small team of other collaborators, but only one was in any way groundbreaking: a study on color in dreams, which found that it was present 83 percent of the time (a much higher proportion than in previous reports). The other two merely elaborated on themes that he'd explored before—the relationship between eye movements and dream content, and the effects of REM deprivation on subjects' mental states.[10] He'd begun to feel he was spinning his wheels. "Here I was, both an MD and a PhD, looking out on an exciting new field in science, and not having much to show for it," he recalled.[11]

Dement was losing patience not only with his work at Mount Sinai but with the psychoanalytic theories that had first drawn him to sleep research—and with his neo-Freudian fellowship supervisor.[12] Fisher is still remembered for an aphorism he coined in 1959: "Dreaming permits each and every one us to be quietly and safely insane every night of our lives."[13] Dement, however, had become more interested in the mechanisms behind the madness than the insanity itself. And he increasingly felt the older scientist's tutelage to be more of a hindrance than a help. "He was too managing of me," Dement said. "He never stayed up [for the experiments], but he would call and keep me on the phone for an hour to tell him what had happened. It was just kind of a worthless waste of time."[14]

Fisher, for his part, thought Dement was simply unwilling to share credit. "I would see the subjects every morning and make sure they were not going nuts," he recalled. "Dement needed me. He wasn't a psychiatrist. But he began to get more and more secretive and wanted to do experiments by himself and not let me know about it." Fisher blamed some of this behavior on Dement's own round-the-clock working hours: "Arguably, he became paranoid because of chronic sleep loss." Although this explanation seems a bit far-fetched, the tension between the two men was clearly wearing on them both. In the summer of 1962, as the end of Dement's fellowship neared, Fisher suggested he might be happier at an institution with a robust academic research program—unlike Mount Sinai, whose medical school had not yet been established.[15]

Dement took the hint. He called an acquaintance from his days in Kleitman's lab—David Hamburg, who'd been a psychiatric researcher at Chicago's Michael Reese Hospital before becoming chief of the adult psychiatry branch of the National Institute of Mental Health, and was now chair of the psychiatry department at Stanford University.[16]

"I said, 'I'm looking for a job,' " Dement later recalled. "He said, 'You're hired.' "

Not long afterward, he and Pat packed up their car again—this time, with two daughters, ages five and two, in the back seat—and drove west.[17] Dement's migration marked a new chapter in the evolution of sleep science. The field's center of gravity was about to change.

10

The Frontier Shifts

When the Dements arrived at Stanford, in January 1963, they must have been struck by the contrast with their previous postings. Compared to the somber Gothic cloisters of the University of Chicago, or the imposing urban fortress of Mount Sinai, the Palo Alto campus was almost giddily bright and expansive. Graceful Spanish-style buildings, with red-tile roofs and sandstone arches, shimmered in the Northern California sun, amid sprawling lawns dotted with palms, oaks, and eucalyptus. Dement had been hired as an associate professor, at a generous salary of $15,000 (about $121,000 today). Faculty members could get low-interest loans to build houses on university-owned lots, and after a few months in an apartment, the family moved into a single-story home constructed by developer Joseph Eichler—the pioneer of "middle-class modernism"—with glass-and-timber walls, skylights, and a landscaped atrium.[1]

David Hamburg, the psychiatry department chair, gave Dement complete freedom in establishing what would come to be known as the Stanford Sleep Research Center. But like many universities at the time, Stanford was experiencing an explosion of scientific inquiry—touched off by the Soviet Union's 1957 launch of the satellite *Sputnik*, which spurred the US government to spend lavishly on R&D—and little lab space was available.[2] Dement received a small workroom and an office at the Stanford Medical Center, with the promise of an upgrade when a new building was completed.[3] Despite the cramped quarters, he felt a sense of emancipation. For the first time in his career, he could follow his curiosity wherever it led him.[4]

To start with, that meant doubling down on the questions that had obsessed him since he first began studying REM: Why was dreaming necessary? What made the brain set aside 20 percent of its sleeping time for it, scheduling sessions at ninety-minute intervals? Why did REM come rushing back after a sleeper was starved of it for a few nights? Dement still suspected that dreams were a safety valve against psychosis, even though his deprivation studies in Fisher's lab had found little evidence for that theory. Perhaps, he thought, the five-night maximum of those trials had been insufficient. Maybe he hadn't deprived sleepers of *enough* dreams to prove their sanity-protecting function.[5]

In his new lab, Dement began running longer trials. Yet he soon ran up against a frustrating barrier: After about a week of deprivation, the only way to stop subjects from REMing (and thus dreaming) was to wake them up two hundred times a night. At that point, it became impossible to distinguish the effects of REM loss from those of near-total sleep deprivation. In two cases, he managed to extend the run to fifteen days by administering small doses of amphetamine; rather than confine the volunteers to the lab, he had an assistant follow them around town and prevent them from dozing long enough to enter REM. And still, the psychosis Dement was seeking failed to appear. One of the two-week volunteers did become slightly paranoid but not alarmingly so. The other, a medical student who was normally quite reserved, had the opposite response.[6] "I felt peculiarly uninhibited and carefree, not the least bit concerned with what others might be thinking," he recalled. "I wanted to go into all the nightclubs, but especially the ones with the loudest, and, better still, sexiest entertainment."[7]

The failure of these experiments to produce anything approaching insanity was one factor that made Dement eager to explore new realms. Another was his worry that longer periods of REM deprivation might actually do some kind of lasting harm—if not to the volunteers' mental health, then in some other area. He began to wonder if he'd been asking the wrong questions. Maybe, as Jouvet's experiments with cats suggested, what made REM so important wasn't dreaming. Maybe it was something else. If so, the right question was: What could that something be?[8]

———————

Dement was not the only American sleep scientist who was drifting away from the psychoanalytically tinged view of REM that had drawn so many to the field just a few years earlier. That trend was unmistakable at the March 1963 meeting of the Association for the Psychophysiological Study of Sleep—the APSS's third conference and the first to be held outside Chicago.[9]

Nearly two hundred researchers gathered at Downstate Medical Center in Brooklyn, New York. Among the crowd was Kleitman, who was about to publish the second edition of *Sleep and Wakefulness*, with studies by Dement and other members of the new generation listed among the citations.[10] Also present was Eugene Aserinsky, who was now teaching physiology at Jefferson Medical College; a decade after discovering REM, he hoped to pick up where he left off. At his son Armond's urging, Aserinsky had written to Kleitman of his plans. "It was good to learn that you have renewed work on rapid eye movements during sleep," the latter had replied. "I believe that you have ability and perseverance but have had... personal hard knocks to contend with. Let us hope that things will be better for you in the future." Kleitman also reminded his ex-student of an outstanding loan of one hundred dollars, which the latter promptly repaid.[11]

At the meeting, Aserinsky's presence caused a small commotion. "People were shocked," Armond told a reporter. "They looked at him and said, 'My God, you're Aserinsky! We thought you were dead!'"[12] Dement recalled that the returning self-exile "seemed a little put off by the excitement over REM deprivation and speculation over REM sleep and psychosis, and perhaps a tiny bit chagrined that he had not been able to exploit his discovery."[13] Aserinsky later set up a sleep lab at the Eastern Pennsylvania Psychiatric Institute, publishing a handful of studies over the next two decades—none of which, sadly, had much impact on the field that he'd revolutionized.[14]

Scientifically speaking, what made the 1963 conference remarkable was that four of the six sessions focused primarily or exclusively on physiological issues; only one centered exclusively on dream content. Dement chaired a panel on the neurophysiology of sleep and dreaming, with panelists including Michel Jouvet and Elliot Weitzman. The discussion focused on the neural pathways of REM sleep rather than its psychological functions. Although some researchers continued to explore slumber's elusive imagery, the spotlight was shifting to its intricate circuitry.[15]

There were multiple reasons for the change. For one, Freudianism was going out of fashion, and with it the concept of dreams as keys to repairing dysfunctional psyches. The rise of neuroscience was another factor; increasingly, researchers were able to identify the physiological phenomena that drove brain functions rather than relying on abstractions such as id, ego, and "libidinal energy." Perhaps most important, sleep scientists were beginning to find clues that might lead to more effective treatments for real-world ills.

———————

Soon after the APSS meeting, Pat Dement gave birth to the couple's third child, John Nicholas (who came to be known as Nick).[16] When her bleary-eyed husband returned to the lab, he began digging deeper into one of the most devastating disorders of the sleep-wake cycle: narcolepsy.[17]

The impetus was the study that Dement and Rechtschaffen had begun after their taxi ride three years earlier. Published that August in the journal *Electroencephalography and Clinical Neurophysiology*, the paper revealed that abnormalities in REM sleep were a central feature of the disease. The pair and their coauthors (Charles Fisher at Mount Sinai and grad students Edward Wolpert and Stephen Mitchell at the University of Chicago) had conducted all-night sleep trials of nine volunteers with narcolepsy, along with a control group of nine healthy subjects. Among those with narcolepsy, seven exhibited sleep-onset REM periods, or SOREMPs; in the healthy group, none did. The researchers also ran EEG tests on two nar-coleptic volunteers during daytime sleep attacks; in both cases, the men's eyes darted as in REM.[18]

Citing Jouvet's discovery that REM sleep was controlled by the pons, Dement and his colleagues suggested that narcoleptic symptoms might indicate a malfunction of that brain stem structure. But the pons wasn't the *only* structure involved in REM, they noted: Researchers had recently found that stimulating an area higher in the brain stem (the mesencephalic reticular nucleus) during sleep could also trigger the rapid eye movement period and that stimulating the hippocampus could interrupt it. Narco-lepsy might result from an injury or defect anywhere in this system. Or it could stem from a biochemical glitch—or from some combination of the two.[19]

To solve the mystery, Dement knew, scientists would have to study many more people with narcolepsy. Again, the first problem was finding them. When calls to local physicians turned up no patients, he posted an advertisement in the *San Francisco Chronicle*, describing the disease without naming it: "If you are sleepy all the time and have attacks of muscular weakness when you laugh or get angry, write Stanford profes-sor at this address." About one hundred people responded. Judging by their symptoms, Dement determined that at least half of the subjects did, in fact, have narcolepsy; hooked up to a polygraph, most of this group showed SOREMPs as well. In almost all cases, doctors had misidentified the ailment[20]—as epilepsy, perhaps, or depression.[21] Other respondents had been correctly diagnosed but were told that narcolepsy was psychoso-matic and could be treated with talk therapy.[22] Those who were receiving amphetamines or Ritalin (a newer stimulant) often found that the drugs interfered with normal sleep, caused nervousness and loss of appetite, and became less effective over time.[23]

Dement was struck by the suffering of his subjects, whose somnolence hobbled them in countless ways, and whose cataplectic collapses often resulted in painful injuries.[24] He was also surprised at their sheer numbers. Judging by the size of the *Chronicle*'s readership, the response suggested that narcolepsy affected 0.07 percent of the population, making it far more common than experts believed. That translated to tens of thousands of patients in the United States alone. (More rigorous studies by Dement and others later found a prevalence of 0.03–0.05 percent, remarkably close to his hunch.[25] Today, experts estimate that 135,000 to 200,000 Americans have the disorder.[26]) If he could nail down the relationship between REM and narcolepsy, he realized, the disease could be diagnosed more accurately and, perhaps, treated more effectively. He could make a real difference in people's lives.[27]

Yet he felt no inclination to focus exclusively on that quest. On the contrary, he wanted to broaden his explorations as far as his abilities would allow.[28]

That September, Dement traveled to France for the International Symposium on the Anatomical and Physiological Basis of Sleep, organized by Jouvet at the University of Lyon. At this three-day conference, twenty-one scientists from nine countries gave talks on their research. Dement, who spoke on REM deprivation, found that he was the sole presenter focusing on the impact of sleep (or its experimental disruption) on human behavior. All the others were trying to trace the neurological processes that governed sleep and waking, mainly using animal models.[29]

In its survey of the latest cutting-edge research, the conference offered new insights on themes that had threaded through the history of sleep science. There were talks on cerebral metabolism, recalling the ancient debate over whether blood flow increased or decreased in the slumbering brain. Investigators had found that both sides were partly right: Circulation in the cerebral cortex diminished during slow-wave sleep but rose during REM.[30] (In other brain regions, the correlation would prove less straightforward.)[31] Several presentations focused on the search for sleep centers in the brain, showing how far that quest had come since their existence was first proposed in the late nineteenth century. It was now becoming evident that there were several such centers—in areas including the ascending reticular formation, the pons, the hypothalamus, the thalamus, and the basal forebrain—which interacted in complex feedback loops.* And at least seven lectures delved into the neurochemistry of sleep, reflecting a surge of excitement similar to the buzz around hypnotoxins five decades earlier.[32]

That development owed much to new findings about neurotransmitters, the chemicals that carried electrical impulses across the synapse—or gap—between nerve cells. First discovered in the early 1900s, these substances came in dozens of varieties, grouped into four major types: acetylcholine, amino acids (such as GABA), peptides (such as oxytocin and vasopressin), and monoamines (noradrenaline, dopamine, and serotonin). Some neurotransmitters were excitatory, meaning that they could provoke the next neuron to fire; others were inhibitory, meaning that they prevented firing; and still others could switch roles, depending on a dizzying array of circumstances. Through a mysterious code, they conveyed thoughts, sensations, regulatory signals, and all the other data that an organism needed to function in the world. These substances were stored in tiny chambers at the tip of each neuron's axon, or "tail." When the neuron fired, its molecular messengers swarmed into the synapse to bind with specialized receptors on the other side.[33]

By the early 1950s, the idea of chemical transmission in the peripheral nervous system was well established. Not until the early 1960s, however, did experts generally agree that neurotransmitters were also present in the brain—thanks largely to advances in electron microscopy and fluorescence spectroscopy, which made it possible to peer into cells on an atomic level. Now, scientists wanted to find out what those chemicals were doing there.[34]

Dement suggested one possibility in his talk at the Lyon conference: that the cyclical buildup of some "hypothetical substance" (whether a neurotransmitter or a metabolic waste product) might be responsible for triggering REM sleep, which would then "use up" the chemical. Other

---

* See diagram, page 289.

researchers reported on experiments designed to ferret out such connections. Swiss physiologist Marcel Monnier, for example, had surgically connected the circulatory systems of two rabbits. When he induced slow-wave sleep in one animal by electrically stimulating its thalamus, the other drifted off as well—apparently due to something in the donor rabbit's blood. A Mexican colleague, Raul Hernández-Peón, had found that cholinergic drugs (which mimic acetylcholine) brought on non-REM sleep, followed soon afterward by REM, when injected into parts of a cat's brain. A Frenchman, Paul Mandel, presented evidence from a range of studies that GABA and monoamine neurotransmitters might also help regulate sleep.[35]

The meeting left Dement feeling as though he'd been rocketed into a new, futuristic era; he later described it as "truly epochal." But he was equally thrilled by a visit to Jouvet's lab, where the scientist and his team—including his former student Danièle Jouvet-Mounier, now his wife—demonstrated their sophisticated techniques for monitoring cats' brain waves and sleep behaviors.[36]

Back in Palo Alto, Dement resolved to focus his REM-deprivation work primarily on those creatures, which would eliminate many of the practical limitations inherent in human research. Cats were easier to handle than larger animals, such as dogs or monkeys, yet their brains resembled those of humans more closely than did those of rats or mice. They could be monitored nonstop without complaint, beyond occasional meowing. They loved to nap. And they could be kept from REMing indefinitely, without the need to be taken to nightclubs.[37]

In his next study, Dement set out to investigate two questions that would be difficult to answer using human volunteers: What would happen if a creature were deprived of REM sleep for several weeks? And did some kind of chemical really trigger REM?[38]

───────────

Dement designed a protocol that in some ways resembled Henri Piéron's early twentieth-century search for hypnotoxins, in which the French researcher starved dogs of sleep for days and then injected their bodily fluids into other dogs to see if slumber ensued. One difference was that the animals in this new study would be deprived of only one stage of sleep; the other was that they would remain alive.[39]

Still, Dement—who was personally fond of cats—knew they would experience significant unpleasantness, beginning with surgery to implant a skull plate with attachments for electrodes and fluid-transfer tubes. So he established rules to minimize the animals' suffering. Whenever possible, he instructed his staff, they were to be taken out of their cages for recreation and snuggling.[40] And after completing a series of experiments, each cat would have its plate removed and be offered to lab workers or local households for adoption.[41*]

Constrained by his tiny lab at the medical center, Dement began with just two subjects—one called Fang and the other known as Yellow Fang. Fang was prevented from achieving REM sleep while Yellow Fang served as control subject, getting as much REM as he wanted. Both spent eight hours a day attached to polygraph machines, with Fang being awakened whenever the readout showed signs of REM. When the cats weren't napping, three lab workers took turns playing with them. One assistant sometimes brought the animals to her house, where they seemed to enjoy watching her dog.[42] To determine whether some sort of "REM juice" (as Dement nicknamed it) was building up in Fang's brain, lab techs periodically syphoned his cerebrospinal fluid into Yellow Fang's cranium.[43]

After more than a month without REM, Fang showed no sign of impairment. Eventually, however, the fluid transfers seemed to increase Yellow Fang's REM sleep. Encouraged by this hint of progress, Dement decided to start a larger version of the study.[44]

For that, more space would be required. Fortunately, Dement knew someone who could provide it—not a university administrator but an

---

* Jouvet, too, was known as a cat lover—keeping several at a time as house pets and lavishing them with affection—despite the undeniable misery his experiments inflicted. Another paradox from the prophet of paradoxical sleep. Barbara E. Jones, "In Memoriam: Michel Jouvet 1925–2017," *Sleep Medicine* 41 (2018): 116–117.

ambitious undergraduate named Peter Henry, who'd worked in Jouvet's lab the previous summer and wanted to test Dement's hypothesis himself. Henry had managed to acquire some rooms in a nearby house that was scheduled to be torn down in several weeks, and the team set up shop there.[45]

In the new location, a medical student on Dement's crew designed an automated system to keep subjects from achieving REM sleep. Dubbed "Big Bertha," it consisted of a treadmill cobbled together from rolling pins and a washing-machine motor, with a tray of water at the rear of the track. A cat could walk to the front and catch twenty seconds of sleep before risking a dunking—insufficient time, at least initially, to progress to REM. The animals spent sixteen hours a day on the contraption, with breaks to wander the lab, eat, undergo various tests, and do polygraph runs.[46] After a few weeks of deprivation, however, subjects would begin REMing the moment they returned to the start of the ride. At that point, the researchers found, the most effective way to ward off REM was to hold the cat in one's lap and poke it gently in the nose whenever its muscles went slack.[47]

This study ran for two and a half months. By the end, one cat—a black male named Othello—had gone without REM for sixty-nine days, and several others had been deprived for nearly that long. The cerebrospinal fluid transfers continued to show somewhat positive results. Yet as before, none of the REM-starved cats exhibited any signs of damage.[48]

One striking change did emerge among the REM-starved group, however: Its members seemed more excitable. A Stanford neurophysiologist who was studying how the brain responds to auditory stimuli—in this case, a click of sound—found that the EEGs of REM-deprived cats showed a stronger reaction. Later, when a postdoc tried using electrical stimulation to awaken those subjects, a shock that would normally be harmless sent one of them into convulsions.[49] That kind of heightened responsiveness turned up in the cats' behavior, as well. After prolonged deprivation, some were quicker to attack a rat that was placed in their cage. Othello, among others, showed an amplified appetite for both food and sex. Far from being weakened by the experiment, he was "a stronger, healthier, and better cat on REM deprivation day 69 than when we started," Dement wrote.[50] The animal's condition mirrored that of the nightclubbing medical student in the human experiment.

Dement began to suspect that one of REM's functions might be to help regulate basic drives. That brought him back to his old idea that REM served as a safety valve that prevented psychosis. Maybe certain mental illnesses—those involving pathological excitability, like mania or paranoid schizophrenia—resulted from an excess of REM juice. If that were the case, he speculated, some kind of drug might help correct the imbalance.[51]

So far, though, Dement had found no clear evidence that prolonged absence of REM was harmful, whether in cats or humans. "The whole thing is very puzzling," he told a journalist. "Here you have a state that's controlled by a delicate mechanism, a state that seems so necessary that it preempts regular sleep. Yet they can get along without it. What can it be for? Can it be some kind of cosmic joke?"[52]

In early January 1964, Dement participated in a human experiment that thrust him back into the headlines—while befuddling him further about the need for REM. Reading the newspaper, he learned that a San Diego high school senior named Randy Gardner had just launched an audacious science fair project: to beat the world record for sleeplessness by staying awake for 264 hours.[53] (A Honolulu deejay named Tom Rounds had surpassed Peter Tripp's mark in 1959, with 260.)[54] Although Gardner was already 80 hours into his stunt, Dement was eager to observe its effects over the remaining 184. He called Gardner's parents, who gratefully accepted his offer of medical supervision.[55]

When Dement arrived in town, he found the seventeen-year-old to be alert and cheerful—a marked contrast with Tripp at that stage. As the days and nights wore on, the scientist and a colleague who later joined him reported that Gardner did experience some "waking dreams." (At one point, the skinny, bespectacled white kid imagined he'd become a famous Black football player.) He grew increasingly grouchy and had bouts of "heightened suspiciousness." He failed to solve simple arithmetic problems. As in previous studies of long-term sleep deprivation, these symptoms worsened between 3:00 and 7:00 a.m. and improved each afternoon, suggesting the influence of an internal clock.[56]

Yet what impressed the researchers was how *well* Gardner performed during his ordeal. The teenager dominated the basketball games they organized to keep him awake. On Night 10, he wandered the city with Dement; at an arcade, the pair engaged in a one-hundred-round pinball contest, and Gardner won every game. Given a psychiatric interview at 262 hours, he was "well oriented as to time, place, and person," the scientists wrote. His thinking was logical and coherent, and showed "no loss of contact with reality." Unlike Tripp, in other words, Gardner remained essentially sane. He gave a telephone press conference on the eleventh and final morning, assuring reporters from across the country that he could have lasted several more days if he didn't have to go back to school. And after 14 hours and 40 minutes of sleep—marked by a whopping ten REM periods—he returned to his classes with no lasting symptoms.[57]

What accounted for the difference between Tripp's and Gardner's responses? Dement suspected several factors. Gardner, unlike the Ritalin-popping deejay, had used no stimulants—not even coffee. When Tripp set his own wakefulness record, he was twenty years older than Gardner and less physically fit. Moreover, previous studies had shown that prolonged sleep deprivation could send subjects with incipient mental illness into full-blown psychosis.[58] Tripp, who would be indicted for payola just weeks after his stunt, was likely carrying hidden emotional burdens; he might have been on the verge of a breakdown all along.[59]

Dement still believed that loss of REM was what made prolonged sleep deprivation psychologically dangerous for some people, but he was less certain than ever about what might make REM necessary for everyone else. He would continue searching for the answer.[60]

That March, Dement hosted the fourth-annual conference of the Association for the Psychophysiological Study of Sleep at Stanford. About three hundred scientists attended the three-day meeting, during which they finally voted to make the organization's name official. The trend away from dream research continued; of the sixty-five papers presented, only seventeen discussed that topic.[61]

One of those was delivered by a new member—psychologist Rosalind Cartwright, Pat Dement's former boss at the University of Chicago.[62] Cartwright, who'd moved on to the University of Illinois College of Medicine, had found her way to sleep science in 1962, after her husband abandoned her and her two young daughters. Beset by insomnia and anxious dreams, she'd decided to do something productive with her nights: after hiring a babysitter, she started a sleep lab in the men's room of an unused psychiatric unit, replacing the bathtubs with beds and using acoustic paneling to wall off the urinals.[63] The APSS had fewer than half a dozen women when Cartwright joined, and she was the first to speak at its yearly meeting.[64] Her presentation concerned a hallucinogenic drug called Ditran, which blocked the neurotransmitter acetylcholine; its effect, she reported, mimicked the dream state, and might help patients work out emotional problems if they took it under the guidance of a psychotherapist.[65]

Dement himself gave five talks, which spanned the burgeoning range of his interests. He discussed the EEG patterns of REM sleep in cats whose brains were cut off from sensory input using the *cerveau isolé* preparation. He described his team's latest findings on sleep and dreaming after prolonged wakefulness; on sleep patterns in schizophrenics; on REM deprivation in humans; and on the role of REM in narcolepsy.[66]

Besides moving Dement further into the limelight as a researcher, the 1964 APSS conference thrust him into a new role: social catalyst. As the host of the meeting, he was in charge not only of the conference program but also of the after-party. It took place at his family's home, with the crowd spilling from the glassed-in living room, across the redwood deck, and down to the broad lawn.[67] *Meet the Beatles* had just come out, and the Dements were early fans. It's likely that the chords of "I Want to Hold Your Hand" jangled from the stereo—and that Bill and Pat, trailing cigarette smoke, led the dancing. Such scenes would become common as the decade unfolded and would help forge Dement's status as something more than a pioneering scientist. He was on his way to becoming his field's first rock star.[68]

————————

Not that he looked like one. At thirty-six, Dement still wore the bristly crew cut, narrow tie, and white coat of a respectable lab director. He was becoming a familiar face in newspaper and magazine stories and TV interviews. But his public appearances gave little hint of the contradictory qualities—including a fierce work ethic paired with playful creativity, a flair for visionary abstraction tied to a gift for human connection, and a capacity for evangelical ardor leavened by a supple flexibility of mind—that would help him become a paradigm-busting figure.

One manifestation of Dement's trailblazing tendencies was the narcolepsy clinic he launched soon after the conference—the first such facility in the world. Although Dement had never planned to practice medicine, his research on the disease made him reconsider that decision. He and psychiatry resident Stephen Mitchell (who'd left Rechtschaffen's lab for Stanford) had already begun treating some of the participants in their narcolepsy studies. By starting a dedicated practice, Dement hoped to attract more such patients and improve their often woefully inadequate care. Soon, he and Mitchell were managing over one hundred cases—taking polygraph readings and exhaustive histories to confirm patients' diagnoses; adjusting their medications to enhance both daytime alertness and nighttime sleep. There was just one problem: Most of the patients "either didn't have jobs or were poorly paid," Dement recalled, "probably resulting from their history of falling asleep on the job." Many failed to pay their bills, and the clinic went broke within a year. But the seed of sleep science's next transformation was sown.[69]

Dement's playfulness was readily apparent to his family and to the small circle of colleagues and friends with whom he felt comfortable enough to show it. At home, he designed elaborate environments for the kids' birthdays and Halloween, transforming the atrium into a haunted house with strobe lights or a fully rigged pirate ship. He made up bed-time stories with a running cast of characters, starring a villainess called Maleficent (modeled on the Disney character, but more mischievous than evil). To the neighbors' horror, he placed psychedelically painted barrels on the roof as a gesture of domestic exuberance. "He needed to entertain himself," his daughter Cathy Dement Roos recalled. "Let's say you're play-ing badminton. He'd say, 'Let's play with our left hands!' Or 'Let's play with one eye blindfolded!'"[70]

Pat was a coconspirator in the high jinks. Carrying on her own natal family's tradition, she was a tireless game player, leading everyone in rounds of euchre, hearts, spit, Monopoly, and Scrabble. A devoted chef, she threw "croquet gourmet" parties, featuring multicourse feasts and flowing wine along with the backyard competition. The guests often included eminent scientists from Stanford and elsewhere, some of them future Nobel Prize winners. (One of the latter, physicist Burton Richter, broke a rib during a touch football game on the lawn.) Although Pat had never been to college, she more than held her own. She was an insatiable reader and a savvy conversationalist—tiny and wry, with a tobacco-roughened voice. "She was one of those people who everybody talks to," Nick Dement said. "Anybody in the sleep field, Stanford students, my dad's assistants— they'd either be in the kitchen while she was cooking or sitting with her on the back porch. That includes my close friends, who should have been opening up to me. They were opening up to her instead."[71]

Dement fostered a similarly antic atmosphere in his lab while insisting on high standards of precision and efficiency. In mid-1964, he relocated his animal operations to more spacious quarters—a Quonset hut at the

Menlo Park Veterans Administration Hospital, about three miles from the Stanford campus. *New Yorker* journalist Calvin Trillin visited the following year and composed a ten-thousand-word portrait of Dement in his new domain. The profile, in one of America's leading cultural organs, described its subject as "the most prominent REM Sleep researcher in the country," noting that annual NIH funding for the field had grown from zero to $2 million (an astounding 5 percent of the agency's grant budget) since he'd published his landmark 1957 paper on the topic. Yet for all his distinction, the article showed a man whose self-presentation was evolving with the times. "A widespread image of scientific research," Trillin wrote, "shows a white-coated scientist staring up at a test tube in his glistening laboratory, carefully protected from germs and worldly concern. If such a vision could survive a trial with Dement, dressed in an old sports coat and work pants and driving a borrowed car from borrowed building to borrowed building, it would surely not be likely to survive the cat laboratory at the Menlo Park hospital."[72]

Dement's hut had a rounded ceiling, a large room at one end, and a narrow hall flanked by several smaller chambers. One of the small rooms served as a machine shop for building and repairing equipment. Another was crammed with cages—the domiciles of Gray Fang, Othello, Plattock, and Buster facing those of Clyde, Fred, Seymour, and Buddy. A sign on the door said, "Dangerous Cats Inside." A third room contained a Big Bertha treadmill as well as cages holding Lilith, Calico, Natasha, Nina, Nicholas, Proteus, and James Bond. Dement had adopted a technique developed by Jouvet for keeping cats from REMing: placing each one on a brick surrounded by shallow water. The animal was able to enjoy non-REM sleep while sitting up—but when REM hit, its muscles began to relax, forcing it to wake up or topple into the drink. The cats spent twenty hours a day on their bricks, an hour on the treadmill or prowling around, and three hours sleeping in the main room (undisturbed, if they were control cats; poked periodically in the nose if not) while connected to the polygraph machine.[73]

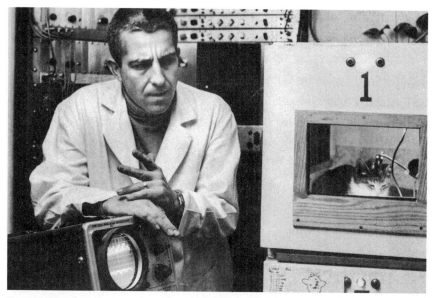

Dement in the Quonset hut lab, mid-1960s. © *Richard Meek/The LIFE Picture Collection/Shutterstock.com.*

One morning, Trillin hung out in a storeroom where cats were taken during the observation period. A sign on its door read, "Sexual Testing Room. Authorized Personnel Only. No Unqualified Females Allowed." This was only partially a joke as Trillin learned when one of Dement's lab assistants—a young woman named Margarita—carried in Gray Fang, Othello, Proteus, and James Bond. Proteus, whose REM deprivation had ramped up his libido, immediately tried to mount James Bond. Trillin noticed other signs tacked to the walls among the sacks of kibble: instructions from Dement, with lists of cats' names under headings such as "Double Feeding" and "Testosterone" (which was given to some REM-deprived cats to see if it would trigger hypersexuality). One sign read, "T.L.C."—a reminder that tender loving care was necessary to ensure that any changes in the cats' behavior didn't stem from a lack of affection.[74]

The conversations in the lab "sometimes sounded rather gossipy," Trillin observed. "It was not unusual to hear remarks like 'Buster didn't seem interested in Plattock today,' and 'Othello's back to his old self.'" On one occasion when a cat had been taken to another building, the writer overheard his subject speaking on the phone: "This is Dr. Dement. Is James Bond over there? Is he being fed?"[75]

Dement's combination of goofy humor, boyish impulsiveness, and professorial absentmindedness earned him a nickname among his staff: "the Big Kid." Although his aptitude for organizing scientific data was extraordinary, he relied on an administrative assistant to keep his office at the Quonset hut from resembling the one at his home, with teetering stacks of paper covering every surface.[76] A former lab technician remembered him yelling to his secretary, "Tibby, have you seen my cigarettes?" Dement was leaving on a trip and (as usual) running late. "He's carrying a crumpled-up backpack, five notebooks, and a suitcase and a briefcase and a sport coat," the tech recalled. His cigarettes, as it happened, were clutched in his hand.[77]

It wasn't that he was scatterbrained. He had his eye on bigger things.[78]

11

A New Vision

The Menlo Park VA Hospital sat at the junction of several tectonic plates whose shifts would contribute to the upheavals of the coming decade. It was here that a Stanford grad student named Ken Kesey had first tried LSD a few years earlier as part of a CIA-financed mind-control experiment, while working nights as an orderly on the psych ward; those experiences inspired him to write the 1962 novel *One Flew Over the Cuckoo's Nest*, whose depiction of heroic madmen battling their soulless keepers helped trigger a backlash against conventional psychiatry and the "Establishment" values it represented. Kesey would become one of the first evangelists of acid, spreading the gospel on a cross-country pilgrimage with a gang of disciples known as the Merry Pranksters.[1]

Shortly before *Cuckoo's Nest* was published, nineteen-year-old Jerry Garcia was riding past the hospital with a carload of buddies when the driver slammed into a guardrail; one friend was killed and two others badly injured. The accident spurred Garcia, a talented guitarist, to get serious about his musical career. He formed a long-haired combo called the Warlocks, which eventually changed its name to the Grateful Dead and became the house band for Kesey's Pranksters. By the late '60s, the Dead were a driving force in the period's musical insurgency—and in the hippie counterculture that was sweeping the world.[2]

In the neighborhood around Dement's lab, other disruptive movements were stirring. Next door to the hospital was the headquarters of the Stanford Research Institute (SRI), where the computer mouse had recently been invented and the internet would soon be born. A few blocks away was the garage where a pair of digitally obsessed hippies, Steve Jobs and Steve Wozniak, would hold meetings of their Homebrew Computer Club in the 1970s—as well as the school auditorium where they unveiled the first Apple computer, launching the personal-tech revolution.[3]

During Dement's stint in the Quonset hut, Stanford itself was rocked by the convulsions of the time. The turmoil began in 1966—the year he and his family attended the Beatles' last public concert, at San Francisco's Candlestick Park. That May, students opposing the Vietnam War occupied the university president's office for three days.[4] In 1967, they walked out on a speech by US Vice President Hubert Humphrey, who had come to promote President Lyndon Johnson's foreign policy.[5] In 1968, protesters burned down the naval ROTC building, then returned to the university president's office to trash it.[6] Later that year, Black Panther leader Eldridge Cleaver drew cheers at Memorial Auditorium when he called Governor Ronald Reagan "a punk, a sissy, and a coward" and challenged him to a duel. (Pat and ten-year-old Cathy were in the audience.)[7] In 1969, activists enraged by Stanford's military research occupied the Applied Electronics Lab, ransacked an administration building, and blockaded SRI, before police dispersed them with tear gas and batons. That October, eight thousand students rallied on campus for the nationwide Vietnam Moratorium.[8]

Although Dement was distressed by the violence on both sides, he sympathized with the protesters, whose ranks included much of his research team.[9] "There was a supportive environment in the lab about being free to take a stance on things," recalled Bill Gonda, the son of a Stanford psychiatry professor, who juggled work as a technician with draft resistance organizing. "He wanted people to be able to speak their minds, while maintaining professional decorum."[10]

Dement also acted as adviser to the Stanford Black Student Union. That role grew partly out of his love for jazz (he occasionally sat in on bass with student bands)[11] and partly from his connection with the Stanford Cardinal football team. In the Dements' first years in Palo Alto, Pat had compensated for the loss of New York City's parks and museums by taking the kids to every college sporting event on the calendar. Dement, a former high school football player and lifelong fan, sometimes tagged along. When Cathy developed a schoolgirl crush on a Cardinal defensive back, Thom Massey, her parents invited him to the house. Massey and his girlfriend, Grace Carroll, soon became regular guests; after dinner, Dement and Carroll would nod off by the fire as Pat and Massey played gin rummy late into the night. Stanford had few Black faculty members, and when Massey and Carroll helped found the BSU in 1967, Dement provided aid and counsel. After the assassination of Martin Luther King Jr. the following year, the group seized the provost's microphone during an assembly and presented the administration with a list of demands—including the hiring of more Black instructors, the establishment of an African American studies program, and the formation of a Black-themed dorm. When the latter request was granted, the group asked Dement and his wife to serve as the house's faculty residents.[12]

Yet throughout this insurrectionary era, Dement was less engaged with politics than with the data that he and his colleagues were unearthing.[13] Every few months, it seemed, there was a new revelation about the mechanisms of sleep—each of which raised new questions and controversies.

Jouvet unveiled one such discovery during his presentation at the 1965 APSS meeting in Bethesda, Maryland. When he destroyed the part of the pons responsible for causing muscle atonia during REM, he reported, sleeping cats acted out their dreams—stalking, attacking, crouching defensively—with their eyes open but blinded by sleep. Jouvet showed a film of this behavior, which left the conference-goers dumbstruck.[14] His finding seemed to solve the mystery of why the brain normally kept sleepers paralyzed during *le sommeil paradoxal*: to keep them from hurting themselves or others. It also suggested that cats dreamed mostly about activities that, in a state of nature, would help keep them alive. Humans, the scientist realized—with their dreams of danger, sex, and challenging social situations—might be doing much the same thing.[15]

This led Jouvet to a new theory of REM's purpose, centered on its possible evolutionary advantages. Like Dement, he suggested that the sleep stage helped regulate instinctual drives. But rather than serve as an outlet to prevent psychosis, Jouvet argued, REM's main function was to tune up neural circuits involved in survival behaviors. (The amped-up cats in Dement's lab might not have fared so well in the wild.)[16] Charles Fisher declared this notion to be of "great interest," citing his own recent finding—using a penile gauge he dubbed a "plethysmographic bagel"—that human males, no matter the dream content, and even as infants, almost always experienced erections during REM.[17]

Dement, however, had begun to doubt that REM served any important purpose in adult animals. (By this time, it had been detected not only in humans and cats but in dogs, monkeys, rats, opossums, sheep, and birds.) He offered an alternative theory in 1966, in a paper he published with his frequent collaborators Howard Roffwarg and Joseph Muzio, both at Columbia. Citing their own observation that newborns spend half their sleeping time in REM, the researchers argued that its function *after* infancy—whatever that might be—was secondary. REM's principal job, they proposed, was to build and strengthen connections throughout the developing nervous system. By spontaneously stimulating babies' sensory and motor pathways, Dement and his colleagues wrote, REM might serve "to bring higher brain centers to an operational capacity requisite to handling the rush of external stimulation in waking experience." An added advantage to this hypothesis, he later joked, was that it was "extraordinarily difficult to test."[18]

By 1969, Dement was investigating another new theory proposed by Jouvet: that sleep was governed primarily by the neurotransmitter serotonin. (Research eventually determined that the chemistry was more complex.) Intrigued by Jouvet's notion that a drop in serotonin levels in some brain areas triggered REM and dreaming, Dement wondered if a persistent lack of serotonin might bring on psychosis. To test this idea, he began running experiments on cats with a drug called para-chlorophenylalanine, or PCPA, which blocked serotonin production. Cats under its influence stopped sleeping, chased imaginary mice, slashed viciously at their handlers, and tried to mount any other cat they saw.[19]

Dement documented the mayhem in a home movie, then flew to Canada to attend a conference. Rummaging through his bags, a customs agent discovered a film canister labeled "Sex and PCPA." Although Dement tried to explain the production's scientific value, the inspectors insisted on seeing for themselves. "This meant spending about half a day at the Montreal airport while they tried to locate a 16-mm movie projector," he recalled. "I was not exactly proved innocent—they just didn't know what to make of five cats having an orgy."[20]

Truth be told, neither did Dement. He was growing weary of pursuing hypotheticals. He missed the feeling he'd experienced while running his short-lived narcolepsy clinic—the rush of having an impact beyond the lab, the journal, or the conference hall. In the era of revolutions, he'd begun to envision one of his own: a new field he called "sleep medicine."[21]

In the near term, Dement simply wanted to set up a practice devoted to sleep disorders—itself an almost unprecedented enterprise. But he also envisioned something far more ambitious: creating a medical subspecialty focused on those disorders, and a system of clinics where it could thrive.[22]

At the time, the notion would have struck many experts as absurd. To begin with, only a handful of such ailments had been recognized: In the latest edition of *Sleep and Wakefulness* (as in the first), these included insomnia; narcolepsy; parasomnias such as sleepwalking, bedwetting, and night terrors; and hyper- and hyposomnias associated with physical or mental illness.[23] In Germany, neurologists Richard Jung and Wolfgang Kuhlo had recently shown that patients with Pickwickian syndrome—marked by obesity and daytime drowsiness—experienced periodic apneas, or short lapses in breathing, during sleep. (The malady was named for a corpulent servant boy in Charles Dickens's novel *The Pickwick Papers*, who falls asleep at every opportunity.) Neurologists Henri Gastaut, in France, and Elio Lugaresi, in Italy, theorized that the Pickwickians' somnolence—which persisted even if they lost weight—resulted from their sleep apneas. But whether the apneas were caused by a malfunction of the patient's brain or their airway remained unclear. In any case, sleep-related respiratory troubles were thought to be quite rare, a misconception that lingered for more than two decades.[24]

Few treatments existed for any of these ills, and those that did exist were administered by general practitioners, neurologists, psychiatrists, or specialists in other established fields. The idea of a "sleep doctor," skeptics later argued, simply made no sense. All sorts of disorders occurred during the daylight hours; who would think of becoming a "waking doctor"?[25]

And yet, several signs were pointing toward the possibility that a branch of medicine devoted to sleep disorders might be not only viable but necessary. One of them was a sea change in public attitudes toward sleep medications that had been gaining strength since 1960, when a sedative called thalidomide (marketed for insomnia, anxiety, and morning sickness) was found to have caused severe birth defects in thousands of children worldwide. In the United States, this disaster led Congress to pass the Kefauver-Harris Amendment of 1962, requiring drug companies to demonstrate the safety and efficacy of their products through "adequate and well-controlled investigations" using "appropriate statistical methodologies." Also in 1962, President John F. Kennedy established a committee to investigate drug dependence—spurred largely by the growing numbers of Americans addicted to barbiturates, long used as tranquilizers and sleep aids. The replacements for those drugs—so-called minor tranquilizers such as meprobromate, or Miltown, introduced in the 1950s, and benzodiazepines such as Librium and Valium, released in the early 1960s—would themselves prove to be potentially habit-forming and mind-addling. By the late '60s, under pressure from regulators and politicians, sleeping-pill makers were turning to academic sleep researchers to provide independent tests of their wares.[26]

Sleep researchers, in response, were beginning to refocus their attention on the disorder that helped give rise to the field in the nineteenth century. Scientists had largely neglected insomnia since the 1920s, when Kleitman made the study of normal sleep his mission. As they probed the effects of sleep aids, however, the mysteries of sleeplessness could no longer be ignored.[27]

Studies were beginning to show that even modern sleep medications produced distorted slumber, posing possible health dangers with long-term use. Sleeping pills of all types—including benzodiazepines—altered the proportions of REM and non-REM sleep and often dulled a patient's alertness the next day. They could leave users with disrupted EEG readings for weeks after they gave up the habit. One of the most prominent critics of these drugs was Anthony Kales, a young psychiatrist at UCLA, who with Rechtschaffen had recently coedited the first manual for standardizing techniques and terminology in sleep research.[28] (In the process, the pair had simplified Dement's sleep stages, numbering them from 1 to 4 plus REM.)[29] Kales, who'd launched an insomnia clinic at Cedars-Sinai Hospital with his wife, psychiatrist Joyce D. Kales, held a symposium on sleep physiology and pathology in 1968, where they and other researchers—including Dement, Kleitman, and Rechtschaffen—spoke. In the preface to the published papers, Kales decried the irrationality of Americans "spending millions of dollars annually for various drugs, sleeping aids, and other remedies in a frantic search for the 'good night's sleep,'" despite "a dearth of scientific and objective data" on their safety and effectiveness. Ironically, however, the conference was sponsored by the Hoffmann–La Roche pharmaceutical corporation, which would also fund much of Kales's future research (as such companies would do for many of the other participants). The deluge of drugs was not about to subside.[30]

Such developments, Dement believed, called for a medical discipline that could translate the latest findings into better care. Patients with daytime drowsiness, for example, should be able to see a physician trained to distinguish between narcolepsy, thyroid-related hypersomnia, and the aftereffects of insomnia—someone who could run an all-night polygraph test rather than simply recommend warm milk at bedtime. Those who complained of poor sleep should see someone who could pinpoint the problem and prescribe an evidence-based therapy—not a GP who would scribble out a prescription for the pill of the moment.[31]

At the end of the '60s, the science of slumber was still in its adolescence; for many people with sleep troubles, no diagnosis was available, let alone a reliable treatment. Yet new discoveries were laying the groundwork for advances that could benefit countless sufferers if only the infrastructure existed to harness such findings.

————————

One particularly fertile area was chronobiology—a term adopted by biologist Franz Halberg in 1969 to describe the study of biological rhythms. Since the 1960 Cold Spring Harbor symposium that established the field, many of its practitioners had come to believe that organisms were governed by the coordinated action of multiple internal clocks. Some also suggested that a master clock in the brain orchestrated animals' circadian rhythms (the patterns of activity and rest that follow a roughly twenty-four-hour cycle). At the Johns Hopkins University School of Medicine in Baltimore, psychobiologist Curt Richter—who'd introduced the term *biological clock* four decades earlier—began searching for the master clock by snipping bits of tissue from rats' brains and observing the results.[32]

In 1965, he announced its probable location: the hypothalamus, which Constantin von Economo had identified as a key control center for sleep and waking in the 1920s. After Richter sliced into an area at the front of this almond-size structure, he reported that rats continued their normal activities—running, eating, drinking, and sleeping—but at entirely random times.[33]* The master clock's precise position within the hypothalamus, and its mode of operation, would not be found until a few years later. But scientists were closing in on a discovery crucial to the rise of sleep medicine.[34]

At Montefiore Medical Center in New York City, the teaching hospital of the Albert Einstein College of Medicine, Dement's old classmate Elliot Weitzman was pioneering a chronobiological approach to sleep research. In 1966, he published the first detailed study of the relationship between

————————

* This experiment was reminiscent of Kleitman's 1932 study (see Chapter 3) in which he found that dogs woke and slept at random times after their cerebral cortex was removed. But those results were never replicated, and their findings contradict what is now known about the neural pathways of circadian rhythms.

the rhythms of the endocrine system (which governs hormone production) and those of slumber. Weitzman focused on cortisol, a hormone secreted by the adrenal glands, whose functions include stress response and blood-sugar regulation. He took twenty-six blood samples from six young men over a twenty-four-hour period, using implanted catheters to avoid waking them during sleep. Before bed, he also hooked them up to a polygraph machine. It was already known that cortisol production followed a circadian rhythm, peaking in the early morning and reaching its low point between evening and midnight. Weitzman, however, found that cortisol levels spiked during each REM period in the second half of a night's sleep—suggesting that the structure of sleep phases could affect biological clockwork basic to human health. Disrupting that structure might throw other physiological rhythms out of whack.[35]

Weitzman's paper caught the attention of NASA, which commissioned a study on shift work—specifically, what would happen to sleep structure and hormone levels if the hours of sleep and waking were reversed for two weeks.[36] The answer: The body's rhythms drifted toward chaos. Subjects fell asleep faster but awakened frequently toward the end of the sleep period, getting less shut-eye overall. Their proportion of REM and Stage 2 sleep decreased sharply, and the timing of all stages was disturbed.[37] Levels of cortisol and creatinine no longer followed their normal patterns.[38] With further studies that nudged sleep science in the direction of chronobiology, Weitzman, too, would help midwife the new medical field that Dement envisioned.[39]

Across the Atlantic, another set of experiments was nudging chronobiology in the direction of sleep science. In the Bavarian village of Andechs, near the castle where he lived with his wife and six children, physiologist Jürgen Aschoff built a unique laboratory known as "the bunker." The underground facility held two apartments, each with a bedroom, a small kitchen, and a bathroom—but with no clocks, windows, or other means of knowing what time it was in the outside world.[40] There, Aschoff and his collaborator, Rütger Wever, conducted some of the first-ever studies of free-running circadian rhythms in humans.[41]

In some ways, Aschoff's initial studies at the facility echoed Kleitman's Mammoth Cave expedition. But there were two important differences: instead of following artificial sleeping schedules, subjects went to bed and got up whenever they chose. And their biological processes were measured far more thoroughly. Throughout their weeks-long stay in the bunker, subjects wore an electric rectal temperature probe, attached to a wall socket by a long cable. They took their own urine samples, which they placed in the corridor for collection several times a day. Their movements were tracked continually by motion sensors in the floors and beds.[42]

The bunker studies confirmed that, in the absence of environmental "time givers"—*zeitgebers*, as Aschoff called them—humans, as a species, are programmed to follow a sleep-wake rhythm that runs slightly longer than a terrestrial day. But these experiments also showed that individuals could respond very differently when such cues were disrupted. It was becoming clear that neither normal sleep nor sleep disorders could be fully understood without considering the vagaries of each person's inner clock. One person might have much more trouble than another in adapting to shift work, for example. Another might become unmoored from twenty-four-hour time after a few all-night study sessions.[43]

———

All these advances helped set the stage for the emergence of sleep medicine. Whether Dement sensed their significance at the time, however, is unknown.

What spurred him to change his own course was a restaurant conversation with Rechtschaffen and Vogel during a visit to Chicago in 1969. "As we dined," he later wrote, "I talked on and on about our REM-deprivation studies when Allan finally erupted, 'Dement, you'd REM-deprive your own mother.' Almost in a heartbeat, Gerry Vogel said, 'That's all right… it wouldn't hurt her.' And then they both laughed uproariously for what seemed about an hour."[44]

At that moment, Dement realized he was fed up with the question that had obsessed him for a decade. Within a few years, Vogel himself would publish a series of studies showing that intermittent REM deprivation could improve symptoms of depression in many patients (a possible explanation for the efficacy of some antidepressant medications).[45] In the '80s, it would be Rechtschaffen who demonstrated that continual REM deprivation—in rats, at least—could be fatal if kept up long enough.[46] But even if Dement had been able to see into the future, he might not have cared.

He was ready to move on.[47]

PART III

AWAKENING

12

The Birth(s) of Sleep Medicine

Historians often date the birth of sleep medicine to the summer of 1970 when Dement launched the Stanford Sleep Disorders Clinic—typically described as the first facility of its kind—at the university medical center.[1] (In reality, it was the world's *second* full-service sleep clinic. The first opened in Prague, Czechoslovakia, in 1949, but its influence was constrained by its position behind the Iron Curtain and the limitations of sleep science at the time.)[2] The Stanford clinic's mission was initially uncertain, however, and its prospects seemed far from bright. In late August, Dement held a press conference to kick off his new venture. Although he had invited a dozen media outlets, only the *San Francisco Chronicle* bothered to send a reporter. And the resulting story raised a storm of ridicule that threatened to scuttle the fragile enterprise.[3]

As he inspected the exam rooms and sleep-study chambers, with their four state-of-the-art polygraph machines, the reporter "asked all the right questions," Dement later wrote. "What did we plan to do? Who would be our clientele? What did we hope to learn?" Toward the end, the visitor posed one more query, ostensibly off the record, just to satisfy his own curiosity: "Should young children sleep with their parents?"[4]

Although no one had yet researched the matter scientifically, Dement, feeling expansive, offered his own opinion. "It made no sense to me, I said, to isolate an infant or small child just for the convenience of the parents. The image of a little one awakening from a nightmare and being forced to suffer its terrors alone in a darkened room simply seems wrong—always has and always will."[5]

"What about sex?" the reporter asked.[6]

"I don't believe that parents should ever have sexual intercourse in the presence of small children," Dement replied, "but it wouldn't take too much creativity to think about having sex somewhere else, or even having one room for sex and another room for sleep." It might even be better, he continued, warming to his subject, "to have sex much more often in the daytime, when the children are at school or napping!" There was evidence, he explained, that the circadian sexual rhythm might actually peak during the daylight hours.[7]

About two weeks later, Dement was awakened before dawn by a ringing telephone. "This is CBS Radio, New York City," said the voice in the receiver. "Can we talk to you a few minutes?"[8]

Dement declined, noting that the time in California was 5:00 a.m. But as he drifted back to sleep, he realized, with a rush of horror, what the call had been about. He ran outside and grabbed the morning paper, whose page 2 feature—under the innocent headline SLEEPLESS NIGHTS FOR SCIENCE—detailed his views on optimal bed-sharing habits, familial and connubial. Turning on the radio, he heard a commentator quip (using Stanford's nickname): "That's the way they do it down on the Farm, ladies and gentlemen." The story, he discovered, had gone out over the wire services nationwide. "It was very embarrassing," he wrote, "especially the part about Professor Dement enjoying sex in the afternoon."[9]

Thankfully, the Stanford administration let him off with a mild scolding.[10] In truth, however, there was little reason for anyone to take Dement's clinic seriously. While its small staff possessed exceptional expertise, as well as the cutting-edge diagnostic tool of all-night polygraphy, they could offer no breakthrough therapies, no track record of extraordinary results.

What truly set the facility apart, he later explained, was its guiding principle: "If patients complained about their sleep, it was their *sleep* that should be examined, *not* their wakefulness. *This was the conceptual breakthrough* that established sleep disorders medicine." But that wasn't enough to keep the facility afloat. In its first year, referrals averaged just five per week. Most of these were people with narcolepsy, drawn from across the country to one of the few centers qualified to treat the condition. The rest suffered mainly from insomnia or a smattering of other sleep disorders. Because insurers refused to cover the majority of sleep-related procedures, the clinic scraped by almost wholly on research grants.[11]

Dement believed there was a huge potential audience for what he was offering. But he had no idea how to tap into it.

––––––––––––

More successful was a new educational venture, which grew out of Dement's advisership to the Stanford Black Student Union. In response to the organization's demand for a Black-themed dormitory, the administration had designated Cedro House—a boxy stucco structure that primarily accommodated premeds—as a "concentration house" for freshmen of African ancestry.[12] (About 40 percent of Cedro's residents were Black, compared to 4 percent of Stanford's undergraduate population.)[13] In September 1970, Dement and his wife became the house's resident fellows, moving with their children into a small cottage next door.[14]

Over the year the couple spent at Cedro, their lives grew deeply intertwined with those of its young occupants. Bill provided academic and career mentorship; he and Pat helped nurture the students through their daily challenges, crises, and triumphs as well. Cathy, Elizabeth, and Nick ate breakfast in the dining hall each morning and hung out in the common areas after school.[15] The dorm evolved into "a real family, with kids and dogs," recalled Woody Myers, who became a close friend of the Dements and, eventually, health commissioner for the state of Indiana. "Bill was sort of the father figure, and Pat was the mom taking care of all of us."[16]

That winter, Dement began speaking on sleep science several evenings a week in Cedro's lounge. His main motivation, he wrote, was "to keep the students indoors during one of the most disruptive and chaotic times in the University's history." In recent months, antiwar protests at

Stanford had repeatedly devolved into melees, with classes canceled as rock-throwing revolutionaries battled riot police. But many students still appreciated a lively lecture, and growing numbers from other dorms began flocking to Dement's presentations. The next quarter, he offered to teach a formal course on the subject—Psych 157, Sleep and Dreams.[17] To his surprise, 450 students preregistered. Since no classroom had been reserved, he arranged to hold the class in Stanford Memorial Church, preaching the gospel of slumber from its pulpit. Sleep and Dreams filled eight-hundred-seat Dinkelspiel Auditorium the following year, and it remained one of Stanford's most popular courses for the next three decades.[18]

The inherent interest of the subject matter fueled some of the class's acclaim. But the key was Dement's showmanship, which flowered in the "do-your-own-thing" era. At forty-two, he was still tall, slim, and vigorous.* Although he wore rumpled khakis and ink-stained shirts in the lab,[19] he dressed for class. He had exchanged his uptight suits for a groovier look, with oversize collars, flared pants, and wide, flamboyantly patterned ties; his graying hair now flowed into Captain Kangaroo bangs. (In the mid-'70s, he added a walrus mustache, which came and went over the following decades.) He paced the podium like a tent-meeting revivalist, leading his audience in chants to drive home key points. He showed film clips of REMing eyeballs and cataplexing narcolepsy patients. He set up a sleeping section for students who needed to catch up on their shut-eye; when those seated elsewhere dozed off in spite of his exertions, he spritzed them with a water pistol.[20]

His approach, he explained, came from the popular British chronicle *1066 and All That*, whose motto was "History is what you can remember." Dement made sleep science memorable for generations of undergrads—some of whom went on to advance the field as researchers, clinicians, health care executives, or policy-makers.[21] Sleep and Dreams eventually inspired similar courses at colleges across the country.[22]

---

* He had recently given up cigarettes after dreaming he'd been diagnosed with lung cancer and had only a short time to live. William C. Dement, *Some Must Watch While Some Must Sleep* (San Francisco: San Francisco Book Company, 1976), 102.

While Dement's course was taking off, however, his clinic continued to struggle. Then, in June 1971, he led a delegation of Stanford researchers to the first international conference of the APSS in Bruges, Belgium, where a chance encounter changed the equation.[23]

The Bruges conference was notable in many ways. Howard Roffwarg chaired a panel discussion on the effect of sleep on memory and learning, a new area of exploration. Elliot Weitzman led a panel on periodicity in sleep, featuring chronobiologist Franz Halberg—the first representative of that discipline to speak at an APSS meeting. A session on the behavior of individual neurons during sleep, with panels led by Michel Jouvet and Canadian neurophysiologist Barbara Jones, showed the vast distance that the study of brain currents had come in the nineteen years since Aserinsky stumbled on REM in the scribblings of a primitive Offner Dynograph. A symposium on computers in sleep research heralded another new technology that was transforming the field.[24]

What made the conference a pivot point in the history of sleep medicine, however, was the moment when the Stanford clinic's resident psychiatrist, Vincent Zarcone, ran into a young French neurologist named Christian Guilleminault. The latter had spent six months at the clinic the previous year on a research fellowship, and he was frustrated at the unwillingness of his European colleagues to establish similar centers. Over lunch, Zarcone had a long talk with the Marseille native and began to suspect he might make a valuable addition to the staff.[25]

Guilleminault, like his compatriot Jouvet, had started out wanting to be an ethnographer of remote tribes. When that didn't work out, he'd gone to medical school. During his residency, he'd been electrified by Jouvet's early discoveries on the neural circuitry of *le sommeil paradoxal*. It struck him that if brain cells behave differently during waking, slow-wave sleep, and REM, "then the control they exercise over many [bodily] functions—including vital functions—would be different, too," Guilleminault later explained. This revelation awakened him to the need for a new clinical field, which (he insisted) he called "sleep medicine" before anyone else did. The fact that his department head dismissed the idea only heightened its appeal to the proudly iconoclastic Guilleminault, who chafed at the formality and conservatism of French academia. He went on to start a laboratory devoted to sleep disorders at the Pitié-Salpêtrière Hospital in Paris, defying his superiors' resistance.[26] One of his current obsessions was sleep-related breathing problems—in particular, the apneas that European researchers had recently found not only in patients with Pickwickian syndrome but in a few nonobese individuals as well.[27]

Back at Stanford, Zarcone persuaded Dement to recruit Guilleminault, who accepted a position as associate director of the sleep clinic and laboratory, and visiting associate professor at Stanford.[28] He would serve as de facto medical director, freeing Dement to concentrate on research, administration, teaching, fundraising, and public outreach.[29] But Guilleminault's own research proved crucial to shaping the nascent field.

The new hire—an elfin thirty-three-year-old, with oversize square-rimmed glasses and a thicket of dark hair—arrived in January 1972.[30] Soon afterward, his team was running a sleep study on a middle-aged man of normal weight who complained of insomnia. When the polygraph showed periodic heartbeat irregularities, Guilleminault went into the bedroom and found that they marked episodes in which the patient stopped breathing.[31] After fitting the patient with sensors to measure respiratory function, he saw that the man's apneas triggered EEG readings indicating disturbed sleep or wakefulness. They also coincided with spikes in pulmonary artery pressure, suggesting (as neurologist Elio Lugaresi had recently speculated) that sleep apnea could damage a patient's cardiovascular system.[32]

Guilleminault recognized that this kind of life-threatening breathing disorder might be far more prevalent than anyone had imagined. From then on, he made respiratory measures a routine part of the clinic's all-night sleep studies, along with EEG, EOG, EMG, and cardiac tests.[33] One of the clinic's psychiatrists, Jerome Holland, dubbed this combination "polysomnography"—the term still used today for a procedure that has become standard around the world.[34]

The team soon found that sleep apnea was astonishingly common and that corpulent Pickwickians formed only a small minority of those afflicted. Many sufferers came to the clinic to be tested for narcolepsy; they reported extreme daytime sleepiness rather than insomnia. One man had dozed off while driving toward a railroad crossing and crashed into the gate as the train roared through. A session in the sleep lab showed the real cause of these patients' crippling drowsiness: They were experiencing hundreds of apneas each night. Some snored and gasped so loudly that they could be heard in the control room. In the morning, however, they remembered nothing.[35] The Stanford doctors had uncovered what appeared to be a hidden, and potentially deadly, epidemic.

Earlier researchers had found three types of apnea in Pickwickian patients: "obstructive," in which a restriction of the upper airway cuts off air supply while the patient fights to draw breath; "central," caused by a failure of the brain to send signals to muscles involved in breathing; and "mixed," in which both factors played a part. Guilleminault and his colleagues found that these categories also applied to non-Pickwickians. After examining dozens of apnea patients, they concluded that the obstructive type was the most widespread.[36]

The team also found that sleep apnea was not limited to adults. Among the first patients they diagnosed were a ten-year-old boy and a thirteen-year-old girl. The children had tested negative for narcolepsy and the other conditions then known to cause daytime drowsiness—low blood sugar, depression, endocrine troubles, or encephalitis. Both, however, had high blood pressure, and the boy's was progressing toward a perilous level. It took months for Guilleminault and Dement to convince other specialists at Stanford that the children's health woes stemmed from the stress their apneas placed on their cardiovascular systems. Finally, as the boy began to develop kidney and heart failure, the clinic received permission to perform the only treatment then available for sleep apnea—a tracheostomy, in which surgeons cut a hole in the patient's throat that was kept open during slumber and plugged while he was awake and breathing properly. The procedure was brutal, but it usually worked.[37]

After the surgery, the boy's blood pressure returned to normal as did his cognition. "The tremendously debilitating, relentless cloud of sleepiness lifted, and an alert young person emerged," Dement later wrote. For the girl, however, it was too late; although a tracheostomy relieved her drowsiness, years of nighttime oxygen deprivation had caused permanent brain damage. Her IQ had sunk to 75.[38]

Starting in 1973, Guilleminault published a series of papers (coauthored with Dement and others) that helped transform science's conception of sleep apnea—from a rare symptom of Pickwickian syndrome into a widespread disorder of its own. He coined the term "obstructive sleep apnea syndrome," later abbreviated as OSA, to describe the most prevalent variety. He devised a metric known as the *apnea-hypopnea index* to gauge the severity of symptoms and found evidence linking sleep apnea to sudden infant death syndrome, or SIDS.[39]

Another transformation took place during Guilleminault's first year at the Stanford Sleep Disorders Clinic: The facility began to thrive.[40] In part, this reflected the clinic's corner on a potentially limitless market; with each passing month, more desperate patients came seeking solutions to problems that had baffled other doctors. But the turnaround also owed much to the associate director's talent and drive.

Unlike Dement, Guilleminault was an experienced clinician. He'd treated patients at several French medical centers and had briefly run the neurology department at a hospital in Algeria.[41] He was a gifted diagnostician as well. Before undergoing polysomnography, Stanford patients were given a battery of questionnaires, including the Cornell Medical Inventory, the Minnesota Multiphasic Personality Inventory, and a "general sleep and dream inventory." They were tested for daytime sleepiness and sent home with a diary to record their nightly sleep quality.[42] Under Guilleminault's guidance, the team found reliable ways to distinguish OSA from central sleep apnea; Kleine-Levin syndrome (which causes jags of prolonged sleep) from idiopathic central nervous system hypersomnolence (which causes constant sleepiness); psychophysiological insomnia (sleeplessness driven by anxiety) from pseudoinsomnia (anxiety about insomnia in someone who has normal sleep).[43] And they learned that several other obscure sleep ailments were more widespread than previously thought. Among patients who sought treatment for insomnia, for example, a surprising number turned out to have nocturnal myoclonus (now known as *periodic limb movement disorder*), in which involuntary leg jerks woke them every thirty seconds or so. This discovery prompted Guilleminault to add leg-muscle readings to standard polysomnographic protocols.[44]

In a startling 40 percent of patients with insomnia, the Stanford team determined, the cause was long-term use of sleeping pills—a finding that bolstered critics' arguments against such drugs. Most of those sufferers regained normal slumber after being gradually weaned from the medications. Another 20 percent had psychiatric issues and were referred for treatment. About 10 percent suffered from sleep apnea. The remaining 30 percent received a diagnosis of idiopathic insomnia, which meant that physicians could not determine the underlying cause. These individuals were typically prescribed flurazepam, a benzodiazepine thought to be less likely than other sleep meds to create tolerance—that is, require increasing doses to remain effective.* (On the downside, the medication could create severe daytime grogginess.) Patients diagnosed with other disorders were generally treated according to the accepted protocols of the time.[45]

The team also tried out experimental therapies. For some people with narcolepsy, they found, tricyclic antidepressants controlled the symptoms of daytime sleep attacks, cataplexy, and sleep paralysis more effectively than amphetamines or Ritalin.[46] For patients with parasomnias such as sleepwalking or recurrent nightmares—which Canadian neurologist Roger Broughton had recently shown to arise from problems in transitioning out of non-REM sleep—drugs that suppressed slow-wave EEG patterns, such as benzodiazepines, seemed to help.[47]

---

* Experts now consider the development of tolerance to be a risk with all benzodiazepines.

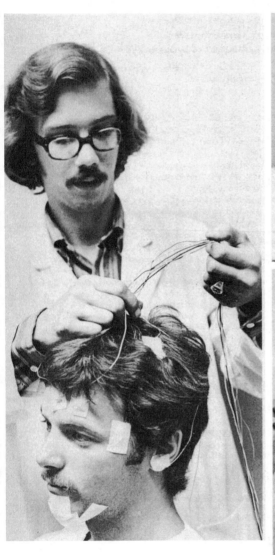

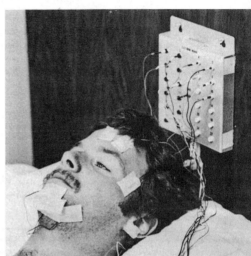

Staffers at the Stanford Sleep Disorders Clinic run a sleep study on a patient, early 1970s. © *Taylor & Francis, 1973.*

By late 1972, Dement, Guilleminault, and Zarcone felt ready to share their know-how. That November, they launched a continuing medical education (CME) course for physicians titled Sleep Disorders: A New Clinical Discipline—another event sometimes cited as marking the birth of sleep medicine. Doctors from around the globe soon began attending the course, which covered normal sleep and its opposite: the growing taxonomy of insomnias, apneas, hypersomnias, and narcolepsies (with or without cataplexy and sleep-onset REM periods). The available remedies were often far from perfect; many patients with obstructive sleep apnea, for instance, preferred to go untreated rather than endure a tracheostomy. Yet the course offered a basic tool kit for doctors hoping to improve their skills and inspired a few to start clinics of their own.[48]

Sleep medicine was beginning to find its feet. Enabling it to soar, however, would require more work in the lab.

13

Decoding the Clockwork

Soon after Dement launched his clinic, two explosive discoveries about biological clocks opened new frontiers in sleep science. The first involved molecular genetics, a field that was just beginning to take off as scientists harnessed new tools to identify genes and their functions by analyzing DNA. In 1971, researchers at the California Institute of Technology reported that they'd found the first "clock" gene—confirming chronobiologists' theory that all cells contain genetically programmed clocks, which form networks to govern the timing of an organism's life processes.[1]

The discoverers were PhD student Ronald Konopka and his professor Seymour Benzer, who'd set out to investigate how genetic factors influence circadian rhythms. As subjects, they chose *Drosophila*, or fruit flies, which normally follow a roughly twenty-four-hour cycle of activity and rest. After dosing the flies with a DNA-disrupting chemical, the scientists observed three types of "clock mutants" among the insects' offspring: one group with a circadian rhythm of nineteen hours; another with a rhythm of twenty-eight hours; and a third with no discernible rhythm at all. Konopka and Benzer traced all three mutations to a single gene, which they named *period* (abbreviated as *per*).[2] Scientists would eventually find over half a dozen other such genes—including one in humans.*

---

* The researchers who later identified *per*'s molecular structure—Jeffrey Hall, Michael Rosbash, and Michael Young—won the 2017 Nobel Prize in Physiology and Medicine. Sadly, neither Konopka nor Benzer lived long enough to share in that honor. See Joseph S. Takahashi, "The 50th Anniversary of the Konopka and Benzer 1971 Paper in PNAS: 'Clock Mutants of *Drosophila melanogaster*,'" *PNAS* 118, no. 39 (2021), https://www.pnas.org/doi/full/10.1073/pnas.2110171118.

The second breakthrough came in 1972 when researchers identified the master clock that—in mammals, at least—coordinates the actions of all those tiny timers. Two teams made the discovery almost simultaneously: Robert Moore and Victor Eichler at the University of Chicago, and postdoc Friedrich Stephan and his professor Irving Zucker at the University of California, Berkeley. Studying rats, both sets of scientists found a neural circuit that ran from the eyes to the anterior hypothalamus (the area that Curt Richter had identified in 1965 as the clock's likely location).* When the researchers destroyed a cluster of cells at the end of the circuit, known as the *suprachiasmatic nucleus* (SCN), many of the animals' bodily and behavioral functions—including sleep, feeding, and body temperature—no longer followed regular patterns in the absence of external cues.[3]

Other researchers soon found that the SCN sent signals to the pineal gland, controlling production of melatonin and other neurotransmitters. Later experiments showed that SCN cells kept up their electrochemical rhythms even when placed in a petri dish. Eerily, when such cells were transplanted into an animal whose SCN had been disabled, the recipient's daily cycles began moving to the donor's beat. The structure's rhythms could be reset by exposing an animal to light at different times—but it seemed to operate using an unidentified set of receptors in the retina rather than the rods and cones that enable sight. The nerves that sent data from the eyes to the SCN were separate from those that fed the visual cortex.[4]

These studies proved that circadian rhythms in animals across the evolutionary spectrum are governed by what Kleitman had (skeptically) referred to as "some general properties of protoplasm."[5] They also provided a physiological explanation for the difficulties humans had in adjusting to unorthodox sleep-wake regimens. And they underscored the warning implicit in Elliott Weitzman's study of sleep stages and hormone secretion on a night-shift schedule: messing with sleep cycles could muddle the intricate coordination among the body's networks of clocks.[6]

---

* See Chapter 11.

But applying the new findings to real-world situations would be impossible without further research. There was clearly some flexibility built into the system as evidenced by people's ability to pull all-nighters without lasting harm (even multiple all-nighters, like Randy Gardner), or to adjust to schedules that strayed moderately from the norm (as Bruce Richardson had done in Mammoth Cave). How much deviation could the inner clocks handle, and how much was too much? What variables determined where those boundaries lay? These questions weren't so different from those Kleitman had attempted to answer, but they could now be explored more precisely and with a better sense of the underlying mechanics.

For Dement, one of the most intriguing enigmas in this realm was the relationship between circadian clockwork and sleep stages. Weitzman had already shown that reversing the hours of sleep and waking disturbed the proportions of the REM and non-REM stages. But what would happen to those stages if you abandoned the twenty-four-hour cycle altogether? Dement turned to the method Kleitman had developed in the 1930s: placing humans on a sleep-wake schedule radically out of sync with the Earth's rotation. The figure he chose was ninety minutes—the average duration of a REM-non-REM cycle.* As in a normal day, subjects would try to sleep for one-third of the time.[7]

Dement dubbed this method the "90-minute day."[8] He'd first tried it in 1971, in a makeshift lab that he set up in the basement of Cedro House. At that point, his main purpose was to engage students' interest in sleep science, and the experiment was too small to produce usable results. In 1973, however, he decided to launch a full-scale study. To oversee it, he tapped a young lab assistant named Mary Carskadon[9]—a choice that

---

* It was also the duration of what Kleitman had hypothesized was a "basic rest-activity cycle" (BRAC), which manifested during waking in fluctuations of alertness and efficiency—a theory for which research has found mixed evidence. See, for example, Aljoscha C. Neubauer and Heribert H. Freudenthaler, "Ultradian Rhythms in Cognitive Performance: No Evidence for a 1.5-h Rhythm," *Biological Psychology* 40, no. 3 (June 1995): 281–298; Mitsuo Hayashi, Kayo Sato, and Tadao Hori, "Ultradian Rhythms in Task Performance, Self-Evaluation, and EEG Activity," *Perceptual and Motor Skills* 79 (1994): 791–800.

would help shape the course of sleep science and, eventually, its impact on public health.

———————

With a round face, dark bangs, and alert gray eyes, Carskadon (pronounced "car-SKAD-en") was not an aspiring scientist; if she hadn't been a relative of Dement's wife, she might never have set foot in his lab. Her father was a civil servant and Air Force Reserve officer, her mother a homemaker. She'd spent her early years in suburban Kettering, Ohio, and rural Elizabethtown, Pennsylvania, where she and her two older brothers roamed the woods and fields. Like her cousin Pat, she came from a game-loving clan, where the kids learned the intricacies of bridge, hearts, and euchre as soon as they could hold the cards. Evenings, the family watched *Ozzie and Harriet* and *Father Knows Best* on TV. Sundays, they attended the local Methodist church and listened to the Mormon Tabernacle Choir on the radio.[10]

As a child, Mary was smart and scrappy, an avid reader drawn to the sci-fi adventures of Tom Swift and the crime-busting exploits of the Hardy Boys. She had a stubborn streak and a wide-ranging curiosity—and in junior high, she developed a yen to learn about the human brain. When she asked if there were any books on the topic, however, the school librarian answered, "No." Would the response have been different if she'd been a boy? At the time, the thought didn't occur to her.[11]

At Elizabethtown Area High School, Carskadon got straight As. A nimble and competitive athlete, she made the field hockey and tennis teams despite what she later described as "a significant weight challenge." But no one suggested that she could accomplish great things. Her mother worried that she might be hurt if she set her sights too high. Her father encouraged her to become a schoolteacher; that was a way, he said, for a girl to "make a difference." The idea left her cold, but she had no better ones. As a senior, she asked her guidance counselor about nearby Gettysburg College, which was known for its teacher training program. He advised her not to apply; its students were mostly male, and she would never get in. Piqued, she applied *only* to Gettysburg and proved him wrong.[12]

There, instead of studying teaching, she followed her curiosity on a tour through the liberal arts. Her favorite subjects were psychology and eighteenth-century literature; she majored in psychology because it seemed more practical. Still, she had no real career plan. After graduating in 1969, she found employment as a social worker with the welfare department in Washington, DC. She rented a house with friends, with whom she marched in that year's mammoth anti-war demonstrations. After three months, however, she quit the job in despair, overwhelmed by the gulf between her indigent clients' needs and her power to help them. Moving back home, she took a series of temporary gigs and was accepted into a graduate program in educational psychology at Penn State. "It was something to do," she said.[13]

In the summer of 1970, at a relative's wedding in West Virginia, Carskadon ran into Dement, for whom she'd served as a sleep-lab subject on a visit to New York City when she was twelve. They had a pleasant talk. A couple of weeks later, she recalled, "the phone rings and it's Bill Dement, asking if I would come work for him because he'd been so taken with me at the wedding." (In reality, she learned long afterward, her father had told Dement she was a little lost and asked if he would hire her.) Two of Carskadon's friends borrowed a station wagon from the dean of students at Gettysburg and drove her to California. "They pushed me out the door," she said, "and I landed on Dement's doorstep."[14]

Her first assignment was to haul the family's trash to the dump in Bill's pickup truck as they prepared to move into Cedro. She rented a room in a faculty member's renovated garage and drove each morning to the animal lab in a VW Beetle she'd bought from Christian Guilleminault.[15] The lab had recently moved from the Menlo Park VA to the basement of Stanford Medical School's anatomy building—a space that Dement's long-haired lab staff had excavated using pickaxes and jackhammers, with the boss manning a backhoe.[16] Carskadon was given a closet-size office in an unfinished corner, down the hall from Dement's more spacious but equally windowless digs. Along the corridor stood workrooms, storage rooms, a surgery room, and an air-conditioned computer room housing a refrigerator-size PDP-11—a popular "minicomputer" of the time.[17]

During her first months at Stanford, Carskadon was miserable with homesickness, and she periodically announced her resignation. But Dement put her off by giving her increasingly challenging tasks to perform. He taught her to apply electrodes for sleep runs; to score sleep stages from polygraph records; to analyze study results; to run drug studies; to run more complex trials. He made her a teaching assistant in his Sleep and Dreams course. He presented her with a copy of *The Elements of Style* (the bible that guided his own prose), then gave her manuscripts to polish. It wasn't just that she was family; he saw qualities in Carskadon that she didn't see herself. "She was an excellent writer," Dement said. "She would edit my grant proposals and really make them sing." He prized her quickness at learning new skills, her organizational acumen, her gift for connecting with students.[18]

He also sometimes made her furious. One such incident occurred during the run-up to the 1971 Bruges conference. When Dement was choosing lab members to accompany him, Carskadon asked if she might come along. "He said he could only support people who were going to stay in the field," she recalled with a rueful laugh. "So he sent a couple of guys who were in the field for maybe another couple of years." At the time, Carskadon interpreted this slight as a sign of residual sexism. Despite his apparent enlightenment, she thought, Dement clung to the norms of an era when paid work was widely seen as a stopover for young women en route to marriage and motherhood. But she later wondered if there might be another explanation.[19] Dement did, after all, include one woman in Stanford's 11-member contingent—a young neurochemist named Pamela Angwin. (Altogether, about 45 of the conference's 305 attendees were female.)[20] Perhaps he was actually trying to goad Carskadon into committing herself to the cause of sleep science.[21]

If so, the snub was partially effective; it strengthened her growing determination to prove herself in the lab. Yet when Dement asked her to run the ninety-minute-day study in 1973, she still wasn't sure this was the work she had been born for.[22]

The study's basic design was simple, but its execution would be grueling: In a lab at the Stanford Medical Center, five undergraduate volunteers would try to sleep for 30 minutes and stay awake for 60, repeating the routine over five consecutive 24-hour periods—a total of 120 hours. Later, another subject would be added to the mix: a forty-seven-year-old man with narcolepsy, chosen to demonstrate how this regimen affected the sleep-onset REM periods associated with the disease. Throughout each trial, Carskadon and her assistants would take shifts running the polygraph and administering tests of mood, sleepiness, and cognitive performance. They would also serve four meals during each 24-hour period, provide 120 milliliters of fluids per person every 90 minutes, and hand out snacks on demand. The goals were to see what happened to REM and non-REM sleep under these conditions and to gauge the effect on subjects' mental and emotional condition during waking. More generally, the aim was to find out if people slept differently when they went to bed at different points in their circadian cycle—a question relevant to shift work, sleep disorders, and other areas in which scheduling might have a significant effect on sleep quality.[23]

The results held several surprises about sleep. For example, in healthy volunteers, REM appeared only in alternating cycles—but when it did occur, it arrived much more quickly than usual. And it *never* occurred during the cycle that began at midnight. The effects during waking were unexpected, too: using a simple test known as the Stanford Sleepiness Scale, the team found that the more REM subjects had, the more alert they felt; the more slow-wave sleep, the drowsier.[24]

Although the details were puzzling, three big takeaways were clear: In healthy people (but not in those with narcolepsy), the timing of sleep had a profound effect on its structure. REM sleep seemed to follow a particularly strict circadian pattern. And sleep-onset REM periods (SOREMPs), previously seen almost exclusively in infants and people with narcolepsy, could be triggered in healthy adults by changing the time they went to sleep. Further research would be required to better understand these findings—but one practical lesson was that the use of SOREMPs to diagnose narcolepsy needed to be reexamined. At the very least, the timing of the test should be considered.[25]

The ninety-minute-day study set the stage for later investigations that helped shape sleep science's approach to issues ranging from medication side effects to school scheduling to traffic safety. For Carskadon, however, the study also had a more personal significance. As she plotted the chart of the sleep stage-versus-sleepiness correlation, she experienced an epiphany: "This beautiful 'U' shape emerged," she recalled. "It was just a perfectly beautiful set of data. I thought, *Oh my! This is so cool!* It lit the lightbulb: *This is what I want to do.*"[26]

———

By the time Carskadon had her revelation, growing numbers of ordinary people were also becoming fascinated by sleep science—a trend reflected at the highest levels of pop culture. On December 10, 1974, the guest lineup on *The Tonight Show* included comedian Flip Wilson, character actor James Coco, *Charlie's Angels* star Jaclyn Smith, and the director of the Stanford University Sleep Disorders Clinic.[27] The host, Johnny Carson, teed off the last segment with a double entendre. His guest, he announced, was the author of a book called *Some Must Watch While Some Must Sleep.* Carson raised an eyebrow: "We'll find out what *that's* all about!" The audience roared. "'I like to watch, myself.' No, it kills me to introduce a doc with that line. He's a distinguished gentleman. This is no yo-yo! Welcome, Dr. William Dement."[28]

"Nice to be here," the distinguished gentleman replied, smiling gamely.[29] In reality, the book he'd come to promote was an account of his experiences in sleep science over the two decades since he'd first knocked on Nathaniel Kleitman's door. Aimed at a popular audience, it was meant not only to introduce readers to the latest findings but also to infect them with its author's enthusiasm for his strange and wondrous field of study. "Admittedly, the average person would not, at first blush, pick watching people sleep as the most apparent theme for a spine-tingling scientific adventure thriller," Dement wrote. "However, there is a subtle sense of awe and mystery surrounding the 'short death' we call sleep, to which people who reflect for a moment almost always respond."[30]

Despite Carson's low-brow intro, he'd clearly done his reading. He asked Dement what drove him to do his work. (Answer: "There are many, many people in the United States who have insomnia, who have sleepiness during the day, who can't function because they're not properly alert. And the answers to these problems, I think, will come from research.") He asked if there were really such a thing as a "day person" or a "night person." (Answer: "Very much so.") He asked why flying through different time zones scrambled people's biological clocks. (Answer: "Circadian rhythms.") And he asked the most fundamental question of all: Why do people need to sleep?[31]

Dement's response was a kind of Zen koan: "What you need sleep for is to avoid being sleepy." This got a laugh. "No," he protested, "to be sleepy is very unpleasant." He went on to describe some of the horrors that brought patients to his clinic: blackouts, fugue states, memory loss, learning problems, inability to focus.[32]

Dement had revealed a new direction in sleep science, inspired by growing evidence that out-of-sync sleep rhythms could severely disrupt people's waking lives. The field would increasingly focus on pathological sleepiness: what caused it, how to prevent it, and how to treat it. This theme bound together research into the neurochemistry and chronobiology of sleep, as well as the mechanics of sleep apnea, the genetics of narcolepsy, the treatment of insomnia, and the way we order our days and nights.

The goal was to gain deeper insights into slumber's ills—and to keep them from blighting the other side of the cycle of existence.

14

Seeing Patterns

By the summer of 1975, Mary Carskadon was hungry to make a mark on the field she had wandered into accidentally and fallen in love with belatedly.[1] An awareness of passing time may have sharpened her appetite. She was approaching thirty; her mother had died at the start of the year.[2] (Dement accompanied her to Pennsylvania for the funeral.)[3] To pursue an independent career, Carskadon knew, she would need a PhD. One day—perhaps over a rubber of bridge, a diversion she and Dement shared whenever they could manage it—she confided that she was looking into graduate schools. She asked for his advice on where to apply and for a letter of recommendation.[4]

Her mentor, too, was grappling with the unexpected turns of the path he had stumbled onto in his twenties. At forty-seven, Dement was enjoying the kind of success that few scientists achieve. The field of medicine he championed was beginning to take wing. He'd become a public figure, living in a style far removed from the shabbiness of his Depression-era childhood. After the family's year at Cedro, they had moved into a five-bedroom, Spanish-style mansion on the Stanford campus, with a swimming pool and a giant oak tree in the backyard. Yet he didn't get to spend much time there: the vortex of work was inescapable.[5]

When Dement wasn't in the lab or clinic or office or classroom, he was often traveling to conferences or on lecture tours. He had a small army of staffers and postdocs to wrangle, and the minefields of university and organizational politics to navigate. His three kids now ranged in age from twelve to seventeen; he was thrilled to see them blossoming, but he saw them largely on the fly. The NIH had recently denied him a crucial grant, and he was making up for it by writing five more proposals (with Carskadon's help). Meanwhile, he was enduring endless meetings with insurance executives to negotiate coverage for sleep tests and treatments, and hosting nerve-racking site inspection visits for potential funders. In a few months, he would be hospitalized with stress gastritis.[6]

It would have been understandable, then, if Dement had simply given Carskadon his thoughts on suitable PhD programs and wished her the best of luck. Instead, he urged her to consider Stanford's new Neuro- and Biobehavioral Sciences Program, and offered to act as her dissertation supervisor.[7]

Dement's motives may not have been wholly unselfish. Although he doubtless believed that the program would be a good fit for his protégé, he may also have grown too fond of her to easily let her go. He probably sensed, too, that keeping this talented and driven researcher in his stable would benefit his own larger mission. As for Carskadon, anxious about facing the rigors of a neuroscience doctorate six years after graduating college with a liberal arts degree, the prospect of staying on familiar turf felt comforting. And the opportunity to dig deeper into the scientific soil she'd begun tilling—with Dement's continued support—was too compelling to pass up. She applied and was accepted for the following September.[8]

While waiting for the term to begin, she decided to tackle a problem that had cropped up during the ninety-minute-day study: how to measure daytime sleepiness in a precise way. Carskadon and her team had used the most advanced tool available—the Stanford Sleepiness Scale, developed by Dement's postdoc Eric Hoddes in 1972. The SSS required subjects to rate their alertness at a given moment on a scale from 1 (wide-awake) to 7 (drifting into sleep). But the test was too subjective, she'd soon recognized, to be fully reliable. Different respondents could have varying levels of sensitivity to their own fatigue. Some might downplay their sleepiness to demonstrate toughness or exaggerate it to gain sympathy. Others (such as small children) might lack the verbal ability to give meaningful responses.[9] This posed a stumbling block for any study in which it was important to know exactly how drowsy subjects were—a question that would prove central to Carskadon's own groundbreaking work, as well as to the field as a whole.

In the spring of 1976, Carskadon and a small group of collaborators set out to develop the first *objective* measure of sleepiness. Its design grew out of a correlation that she and Dement had noticed during the ninety-minute-day studies: the degree of sleepiness subjects reported usually predicted the rapidity with which they fell asleep when given a chance—a metric known as "sleep latency." Because sleep latency varied over the course of a twenty-four-hour period, Carskadon reasoned, taking this measure just once or twice would be insufficient. The most accurate way to judge subjects' sleepiness would be to measure their sleep latency at regular intervals throughout the day.[10]

Carskadon named this new tool the Multiple Sleep Latency Test, or MSLT. She designed it around a somewhat less hectic schedule than the ninety-minute-day format, with four to six opportunities to nap over a twelve- to fourteen-hour period. Every two hours, volunteers were put to bed in a dark, quiet room, hooked up to a polygraph machine, and asked to lie still with their eyes shut and try to fall asleep. If sleep did not appear within twenty minutes, the session was terminated. If the subject did doze off, the session was ended after a brief snooze[11] (usually ninety seconds).[12] Each run was scored from 0 to 20, with the numbers indicating minutes before sleep onset; the top score meant the session had passed without a wink.[13]

When the researchers tried the MSLT on volunteers who'd been deprived of different amounts of sleep, they found the next day's average latency scores declined in direct proportion. After extensive trials, they classified a score of 5 or less as signifying extreme sleepiness, in which reactions were slowed and a person was likely to drift off in class or while driving. A score of 5–10 signified borderline impairment, 10–15 meant manageable sleepiness, and 15–20 represented full alertness.[14]

Over the next few years, Carskadon and Dement published a series of papers—coauthored with psychologist Merrill Mitler and an intellectually precocious premed named Gary Richardson, among others—demonstrating the MSLT's sensitivity and specificity. They found that it was superior to previous screening methods for narcolepsy (sleep-onset REM periods during two consecutive tests was a sure sign) and that it could detect carryover effects from sleep medications as well.[15] By the mid-1980s, it was becoming a standard tool in both clinical diagnosis and drug trials.[16] The MSLT also gave researchers a way to explore how the propensity to sleep, known as "sleep tendency," fluctuated during the day in different population groups and under different conditions.

Carskadon would use this new tool for a pivotal expedition into slumber's terra incognita: the decade-long suite of studies known as the Stanford Summer Sleep Camp. Along the way, she created the first detailed map of the sleep patterns of children and adolescents—and uncovered hints of a sleepiness crisis that few adults had suspected.

————————

The idea for Sleep Camp was born around the same time as the MSLT. Dement had just received a large grant from a private foundation to study sleep in children, which had never been systematically investigated using the resources of modern sleep science.[17] At a staff meeting, Stanford's chief of child psychiatry, Thomas Anders, proposed that the lab launch a summer camp where researchers could study the relationship between sleep and attention deficit hyperactivity disorder (ADHD).* Dement suggested that such a camp—if participants returned over several summers—would also be ideal for investigating the onset of narcolepsy, which typically emerges during adolescence.[18]

————————

* At the time, the disorder was known as "hyperkinetic reaction of childhood." See "The History of ADHD: A Timeline," Healthline, https://www.healthline.com/health/adhd/history#1952.

Carskadon saw the potential for a broader set of studies, examining basic questions about sleep and human development: How much sleep do children really need? How do those requirements change throughout their youth? How do variations in bedtime and wake time, such as kids experience during their elementary and high school years, affect their daytime alertness and mood? And how does sleep—and the need for it—differ in the elderly?[19]

She realized at once that she had hit upon an excellent PhD project. Dement agreed, and Anders (known for his own pioneering research on infant sleep) offered to become one of her dissertation mentors. Biostatistician Helena Chmura Kraemer soon signed on as well. A key decision that Carskadon made early on was to group children based on their stage of puberty, determined using a clinical tool called the Tanner scale rather than chronological age. This enabled her to correlate sleep need with other aspects of her subjects' physiological development.[20]

That April, Carskadon conducted a preliminary study—a survey of students at local elementary and middle schools, designed to provide baseline data on how kids slept in the real world. As other researchers had reported, she found that bedtimes grew later with age and that between the ages of ten and thirteen, children began to get significantly less sleep on school nights than on nonschool nights. They slept late on weekends, perhaps to compensate for deficits during the week. The expert consensus was that adolescents needed less sleep than younger children and that their tendency to burn the midnight oil reflected school and social pressures rather than biological ones. But Carskadon wondered whether this was really the case—and whether "chronic partial sleep loss" might be harmful to teens' cognitive, emotional, and physical well-being.[21]

She would explore this question, among many others, at Sleep Camp. For the project site, Dement commandeered Lambda Nu House—a Spanish Colonial–style dormitory overlooking a former irrigation pond known as Lake Lagunita. Aside from its view of the water, the house's most distinctive feature was a Grateful Dead skull logo over the fireplace. (Today, the structure is called Jerry House, for bandleader Garcia; the lake is dry for much of the year.) It was Carskadon who corralled four polysomnography machines and set them up in the dorm's library, with cables running down the hall to a console placed above each bed. In later years, she would train undergraduate counselors to apply electrodes, run the machines, and care for the campers. That first summer, however, she did most of the technical work herself. She designed the trials, analyzed the data, wrote up the reports, and oversaw every other aspect of the camp's operations.[22]

The kids were recruited mainly from the families of Stanford faculty and local physicians. They started arriving in June 1976—batches of four or five ten- to twelve-year-olds, who stayed for at least three days. All subjects were put to bed at 10:00 each night and were awakened at 8:00 the next morning. At 9:30 a.m., they underwent the first of six daily MSLTs; they took the SSS every half hour and batteries of other tests at four- to six-hour intervals. In between, Carskadon and her crew led recreational activities. The kids went bowling at the student union at 4:00 p.m., with the electrode leads clipped into topknots on their heads. There were volleyball matches after lunch and dinner, Techs (the counselors) versus 'Trodes (the campers). Other breaks were filled with Ping-Pong, pool, cards, board games, or TV. Carskadon did her own sleeping from midnight to 7:00, in a room on the second floor, but she was on call for emergencies whenever they arose.[23]

When the camp's first session ended that July, it was far too soon to draw any conclusions from the data—except that, in the one boy who'd been diagnosed with ADHD, tests confirmed Anders's hunch that daytime sleepiness exacerbated hyperactivity.[24] Within a few more summers, however, patterns began emerging that would upend established notions about sleep in young humans.

Meanwhile, Carskadon continued with her PhD program. Grad school proved to be as challenging as she had feared. But instead of floundering, she found herself exhilarated by the vistas unfolding before her.[25]

"Some of it was quite magical," she later recalled. In neuroanatomy, for example, the guest lecturers included some of the top scientists in the field. The students got to make their own electrodes, using a glass-puller to create the filament. They learned to perform surgery on animals, drilling holes through the scalp and suturing incisions. (Once, when Dement cut his foot on broken glass while she was visiting, Carskadon volunteered to sew up the wound. "I could see he was in excruciating pain," she said, "but he let me finish, and his foot didn't fall off.") She even loved statistics class, where the instructor explained how famous clinical trials were designed and analyzed—including the Salk polio vaccine trial, in which Carskadon had participated as a child. What excited her most, however, was learning the esoteric vocabulary of neuroscience. "The cerebellar peduncle! The locus coeruleus! And then when we did dissections and had an actual brain to look at, there it all was."[26]

For three years, Carskadon juggled her academic assignments and Sleep Camp duties with paid labor. (Lacking fellowship funding, she had little choice.) Every fall quarter, she scored and analyzed the records from the previous summer's campers. In the winter, she served as chief TA for Dement's Sleep and Dreams course; her duties included booking guest speakers, organizing audiovisual materials, and writing lecture scripts. In the spring, she taught her own course on sleep research technology, training undergrads—mostly veterans of Sleep and Dreams—to perform polysomnography and other tests. (Many of these students went on to work at Sleep Camp.) During spring break, she ran studies at Dement's new Center for Human Sleep Research, a suite of labs in a glass-and-concrete megalith on Welch Road. On the side, she wrote grants and papers, planned for the next summer, and tried to recruit more counselors and campers.[27]

And each summer, she moved back into Lambda Nu House. She measured the returning campers' sleep patterns and daytime alertness as they grew from prepubescents to teenagers, testing their responses to sleep deprivation and sleep extension as well as normal sleeping schedules. She also conducted some of the earliest studies of daytime sleepiness in the elderly, testing whether it resulted from insufficient sleeping hours, poor sleep quality, or other factors.[28] Slowly, she began to get a picture of each cohort's proclivities and needs.

———————

As Carskadon recorded her early results from Sleep Camp, a bicoastal pair of scientists were embarking on their own pattern-recognition project— one that looked at abnormal rhythms of sleepiness and wakefulness in the context of biological clocks. At Montefiore Medical Center in New York City, Elliot Weitzman had been striving to unite the realms of sleep research and chronobiology since publishing his first studies on REM and the rhythms of cortisol secretion in the mid-1960s. When a Stanford-based prodigy named Charles Czeisler joined his team, that quest began to gain new momentum.[29]

A paunchy Newark native, with shaggy hair and heavy features, Weitzman was known for his piercing intelligence and his delight in applying it to the toughest problems he could find. "He once told me that his grandfather took him on his knee and said, 'Elliot, you have to be the smartest man in the world,'" recalled neurologist Jon Sassin, who worked with him at Montefiore. "He really tried to live up to that." In 1974, Weitzman took his sabbatical year at Stanford, where he also received treatment for a recently diagnosed lymphoma. He spent part of his time in Dement's lab, collaborating on narcolepsy research with Merrill Mitler and debating the mechanics of obstructive sleep apnea with Christian Guilleminault. (Weitzman argued that OSA was likely caused by abnormal muscle contractions rather than sagging tissue in the airway. Using a fiber-optic camera to shoot videos inside patients' throats, among other methods, Guilleminault eventually proved him wrong.) Weitzman also frequented the circadian rhythm lab of chronobiologist Colin Pittendrigh, originator of the concept of multiple circadian clocks,* who'd joined the faculty five years earlier.[30]

———————

* See Chapter 8.

Czeisler, then twenty-two, had just entered Stanford's MD-PhD program after earning a bachelor's in biochemistry and molecular biology from Harvard. The son of a Chicago physician, he was tall and weedy, with wire-rim glasses and a scruffy beard—and like Weitzman, he was usually regarded as the smartest person in any room he walked into. "You just get goose bumps when you try to describe his intellect," said Mitler. "He operated at a different level." Czeisler had written his undergraduate thesis on the circadian timing of cortisol release. Now he was continuing that research, working with both Pittendrigh and Dement.[31]

Weitzman became a mentor to Czeisler during his time at Stanford, helping to shift the latter's interest from the rhythms of hormone secretion to those of sleep. And when the older man returned to Montefiore in 1975, the two began a cross-country collaboration.[32] Later that year, Weitzman established a sleep clinic based on Stanford's—the second such facility in the United States.[33] The connection between the two centers quickly led to the discovery of a new sleep disorder and to the development of the first chronobiological treatment method for such a condition.

In early 1976, a young man arrived at the Stanford clinic complaining of severe insomnia. No matter how tired he felt, he couldn't fall asleep until 5:00 or 6:00 a.m. He was unable to get out of bed before noon and typically slept until 3:00 or 4:00 p.m. A graduate student in physics, he hadn't met with his professor—who operated on a more conventional schedule—for a year. The clinic had seen such cases before, but the only recourse had been to assign a family member to pry the sufferer out of bed each morning. Suspecting that the student's problem stemmed from a malfunctioning biological clock, Dement called in Czeisler, who had become the sleep lab's resident expert on such matters.[34]

Czeisler hypothesized that the patient's insomnia reflected an inability to adjust his internal clock when it fell out of step with the outside world. Because human circadian rhythms are programmed to follow a cycle slightly longer than twenty-four hours, our biological clocks must be set back a bit each day to stay in sync with the planet's rotation. Usually, the *zeitgebers* of daylight and nightfall entrain our rhythms to a twenty-four-hour day. With this young man, however, something was preventing that from happening. The trouble might have started with a stint of late-night studying, Czeisler guessed. For most people, going to bed earlier for a few nights would reestablish a normal sleep-wake cycle. But what if, due to a genetic or biochemical defect, the metaphorical "hands" on someone's inner clock couldn't move backward far enough to restore synchrony? In that case, the best way to reset his clock would be to move the hands forward until they'd circled the dial.[35]

In consultation with Dement and Weitzman, Czeisler decided to place the student on a twenty-seven-hour cycle and have him go to sleep progressively later each day. He conducted the experiment at Lambda Nu House during Sleep Camp, in a part of the dorm that Carskadon and her team weren't using. On the first day, Czeisler and his team put the student to bed at 6:00 a.m. and woke him eight hours later, at 2:00 p.m. The man then stayed awake for nineteen hours and went to bed at 9:00 a.m. He awoke at 5:00 p.m., went to bed at noon the next day, got up at 8:00 p.m., and so on—until he reached the point where he went to bed at 9:00 p.m. The following day, the team decided to freeze his bedtime at 10:00 p.m. The gambit worked: from then on, he slept soundly on a schedule that suited both his physiological and his social needs.[36]

For the first time, a sleep disorder had been brought to heel by tinkering with a patient's biological clockwork.

Czeisler and his colleagues—also including Stanford psychologist Richard Coleman and premed Gary Richardson—named the student's ailment "delayed sleep phase syndrome" and called the new treatment "chronotherapy." Over the next few years, Weitzman and his team at Montefiore tested the method successfully on dozens of other patients, many of whom had found no relief from sleeping pills or other remedies. The Stanford and Montefiore groups reported their results jointly in a series of journal articles, with Czeisler and Weitzman trading off as lead author.[37]

In 1977, Weitzman built his own time-isolation lab, modeled on Jürgen Aschoff's bunker.* Occupying the fifth floor of an old hospital wing, the unit held three one-bedroom suites and a control room. Czeisler flew out to New York frequently to help run studies there. He and Weitzman placed ads in the local newspapers to recruit potential subjects—people willing to spend one to six months in a windowless apartment, cut off from contact with anyone but the facility's staff, in exchange for free room and board and a few hundred dollars a week. Many applicants were artists, writers, graduate students, or other individuals with long-term projects to work on and a tolerance for solitude. Candidates underwent psychological exams to weed out those likely to break down under the experimental conditions.[38]

Weitzman and Czeisler hoped that by observing the interactions of multiple circadian rhythms in the absence of time cues, they could better understand the factors that influence the structure of sleep. Like Aschoff, they allowed subjects to go to bed and get up whenever they chose. But they tracked the volunteers' physiological cycles in more detail than any previous sleep study. Along with rectal probes to monitor their temperatures, participants wore catheters in their arms so that lab technicians could take small blood samples every twenty minutes around the clock. They donned electrodes for polysomnography before going to sleep. They provided multiple urine samples daily and submitted to frequent tests of their alertness and cognition. Caffeine, alcohol, and other substances that could disrupt the rhythms of sleep and wakefulness were banned.[39]

No clocks or watches were permitted inside the apartments. To ensure that subjects remained ignorant of the passing hours, days, and weeks, they were forbidden to make phone calls, watch TV, listen to the radio, or read current periodicals. A computer assigned staff members to work in random shifts so that volunteers couldn't keep track of their comings and goings. Male technicians and doctors shaved before interacting with subjects, to prevent their stubble from offering circadian clues. Staffers

---

* See Chapter 11.

always greeted volunteers with "hello," never "good morning" or "good evening."[40]

The subjects were cast adrift without compasses on the sea of time. In the control room, the researchers watched to see how far the currents would carry these sailors from their chronobiological home waters.

———————

An initial report on these voyages arrived in 1978, when Czeisler's freshly completed doctoral dissertation began circulating among a small coterie of sleep scientists. It was a revelatory document. The young researcher appeared to have found a master key to several puzzling phenomena: the tendency of some subjects in time-isolation experiments to develop ultra-long sleep periods at seemingly random times, the propensity of healthy subjects in Carskadon's ninety-minute-day study to enter REM sleep more quickly at certain times of day, and the proclivity of REM periods in normal sleepers to grow longer toward the end of a night's slumber.[41]

At the Montefiore isolation unit, Czeisler reported in his unpublished manuscript, five of the first eleven volunteers became "internally desynchronized"—that is, their sleep-wake rhythms fell out of phase with their cycles of body temperature and hormone secretion. As in Aschoff's bunker experiments and similar studies, these subjects began staying awake or asleep for marathon stretches, falling into cycles that averaged about forty hours. They often toggled between long sleep sessions and short ones—twenty-one hours one time, four the next—with no awareness that they were doing so.[42]

Czeisler began searching for explanations. As he pored over his charts of subjects' physiological cycles, he noticed that one set of curves remained constant: the rhythms of body temperature, cortisol secretion, and alertness. However irregular the volunteers' sleep-wake patterns became, these three measures rose and dipped in unison, over a period slightly longer than twenty-four hours. And when he graphed one subject's sleep patterns and temperature cycles in relation to each other, using a chart known as a *raster plot*, he found a hidden pattern: long sleeps typically began at or just after high points in the temperature cycle, and short sleeps at or just after low points. He made raster plots for his other subjects and found that the same rule applied. He analyzed data previously published by Aschoff's group and by teams in France and England. The rule held fast: sleep duration depended on what point in the temperature cycle a subject fell asleep.[43]

Besides explaining why subjects in time-isolation studies slept for wildly varying lengths of time, this finding shed light on a real-world mystery: why shift workers or exam crammers who go to bed in the early morning often find it hard to sleep, no matter how sleepy they feel. The problem, Czeisler's analysis suggested, is that they're trying to drift off just as their temperature curve begins to climb, acting as a physiological alarm clock.[44]

Next, he turned to REM sleep. In people entrained to a twenty-four-hour day, the first REM period typically occurs about ninety minutes after sleep onset; the episodes grow longer in the early morning, toward waking. In his free-running volunteers, REM arrived sooner than in entrained subjects, and the episodes were longer near the beginning of sleep than near its end. What could explain this reversal? Czeisler revealed the cause with yet another raster plot: REM occurs most frequently, and at greatest length, just after the lowest point of the body temperature cycle. In entrained subjects, this normally occurs in the hours around dawn. Free-running subjects, however, tend to fall asleep during their temperature trough—a pattern reflected in their REM cycles.[45]

Czeisler's preliminary research led him to an overarching theory—that alertness, sleep duration, and sleep's internal structure are governed by the same biological clock that drives body temperature. If his findings held up, they carried an ominous message for a we-never-close society: people couldn't just go to bed at any time they (or their employers) chose and expect to sleep well.[46]

Among the few researchers aware of its existence, Czeisler's dissertation created a considerable stir. Some found the opus so provocative that they tried to replicate its results.[47] One of them was Carskadon, who reanalyzed her ninety-minute-day study in light of temperature cycles. Sure enough, she found, changing bedtimes altered sleep structure in predictable ways. REM occurred most often in her subjects when they went to sleep during the rising phase of their temperature curve. And a whopping 74 percent of REM periods clustered around the low point of the cycle.[48]

Carskadon, however, was less interested in the mechanisms of sleep than in how much sleep developing humans need. By the spring of 1979, she'd collected enough data on that question to complete her own dissertation, which held the seeds of the quest that would drive her forever after. Besides the Sleep Camp studies—the longest section, at sixty-six pages—the document covered several other investigations, including her survey of schoolchildren's sleeping habits, a small study of sleep and sleepiness in elderly subjects, and studies of total and partial sleep deprivation in young adults. Two of her findings would later be recognized as seminal.[49]

One came from the study of partial sleep deprivation. Previous research, which relied on subjective assessments of alertness, had suggested that people gradually adjusted to sleep restriction over time. But when Carskadon limited volunteers to five hours a night for seven nights, MSLTs revealed that they became progressively sleepier as the trial wore on. This was the first solid evidence that people who go several nights with insufficient slumber develop a rising "sleep debt," which must be paid off for their sleepiness levels to return to normal—a potentially serious problem for students, industrial workers, drivers, pilots, and anyone else for whom alertness was crucial.[50]

The second bombshell finding in Carskadon's dissertation emerged from Sleep Camp. At that point, she'd evaluated twenty-seven campers over three summers, ranging in age from ten to fifteen years. Some of the experiments had involved sleep deprivation or restriction—the first ever done with children. (As might be expected, they were measurably sleepier the next day.) In another trial, she'd used the MSLT to look for early signs of narcolepsy among a group of eight campers who had a parent with the disorder. Two of the kids developed daytime sleep-onset REM periods before more obvious symptoms appeared, showing that the test could serve as an effective warning system for disease onset among such children.[51]

But the study that would ultimately change science's view of children's sleep involved healthy campers on a normal schedule. It explored the question of why teenagers often seemed so sleepy, nodding off in class and spending half the day in bed on weekends. "Most parents are familiar with this stereotype and consider the phenomenon just another developmental phase, although others worry about health problems or the possibility that the child is taking drugs," she wrote. "What is the basis of this change in behavior?"[52]

Like most experts, Carskadon had assumed that teens need less sleep than younger kids and that their daytime drowsiness stemmed from staying up late for homework and social activities.[53] The campers' MSLTs, however, told a different story. Every kid went to bed and got up on a schedule that allowed for 10 hours of sleep; on average, polysomnography showed, they got 9.5. The youngest campers exhibited no sign of daytime drowsiness on that regimen. Yet once they reached mid-puberty, they began to demonstrate increased sleepiness—particularly in the afternoon. Prepubescent children fell asleep during only 8 percent of sleep latency tests throughout the day; those in advanced puberty dozed off during a stunning 52 percent of tests. The experts seemed to have it backward: To remain alert, older kids needed *more* sleep than younger ones.[54]

This finding had potentially explosive implications for the way Americans (and people in many other countries) raised their children. Typically, high schools and middle schools started the day earlier than elementary schools. Older kids also had more after-school commitments, including jobs and extracurricular activities, adding to their time pressures. Could they make up for lost shut-eye by sleeping in on weekends, or did their schedules saddle them with a sleep debt that could never be repaid? If they did carry such a debt, how seriously did it affect their schoolwork and their overall well-being? Further research would be needed to answer such riddles, Carskadon knew—a task that would become her all-consuming mission.[55]

That June, she received her degree—with distinction—and marched beside Czeisler in Stanford's commencement parade. Bill and Pat threw a party for them and Dement's other newly minted PhD, neurobiologist Lloyd Glenn, beneath the great oak in the backyard. Carskadon's father and two brothers traveled from Pennsylvania for the occasion, and her friends from the lab and clinic came to raise a toast. She couldn't help but hope that her mother was somewhere watching.[56]

———————

For all the forward momentum in sleep science and sleep medicine at the end of the '70s, a familiar specter returned to haunt the field: evidence that the latest form of sleep medication, touted as eliminating the dangers of earlier drugs, carried an array of serious risks.

In April 1979, the United States Institute of Medicine, part of the National Academy of Sciences, published a massive report on sleeping-pill use. The study had been prompted by research showing that barbiturates—first introduced in the early twentieth century—were involved in almost five thousand deaths a year, mostly due to accidental overdose and suicide. Many experts argued that these sedatives should be supplanted by benzodiazepines, which had been introduced in the '60s as a safer alternative. The IOM report, however, found the advantages of those drugs to be highly overrated.[57]

The authors noted that use of barbiturates had plummeted since 1970 when Congress classified them as controlled substances. Flurazepam, a benzodiazepine, was now the most popular sleep medication. Yet drug suicides had declined only slightly, in large part because benzos could be as deadly as barbiturates when combined with alcohol. Benzodiazepines did have the advantage of not being physically addictive, but the psychological dependence they created in many patients could be equally difficult to overcome. (Doctors trained in sleep medicine tried to avoid prescribing them long term, but others too often gave unlimited refills.) And in one respect, benzos were *more* dangerous than barbiturates: They remained in a user's system far longer. The drugs could impair a user's alertness for days, leading to car crashes and other accidents—especially in the elderly.[58]

The report asserted that most of the 25.6 million sleeping pills taken annually by 4 to 6 million Americans were unnecessary, citing studies showing that the medications were only marginally effective in helping patients fall asleep faster or sleep longer. It called for physicians to prescribe "only a very limited number of sleeping pills for use for a few nights at a time, to aid in specific situations," such as travel or transient stress.[59] And it urged the federal government and the medical profession to support "new, more sophisticated multidisciplinary research...to test the efficacy of various pharmacological, psychotherapeutic, behavioral, and psychosocial treatment approaches to insomnia."[60]

In fact, this kind of research was already gathering steam—as demonstrated strikingly that September, with the release of the first *Diagnostic Classification of Sleep and Arousal Disorders*. Published jointly by the APSS, the recently established Association of Sleep Disorders Centers (ASDC), and the European Society for Sleep Research, it filled an extra-thick issue of the journal *Sleep*. A committee of experts led by Howard Roffwarg had spent three years compiling the 154-page nosology, which described sixty-eight separate disorders. Many of these conditions had been identified over the past decade as the proliferation of sleep clinics attracted patients whose ailments might otherwise have gone undocumented. Others had been described in scientific journals years earlier but were unknown to the vast majority of physicians.[61]

The list was divided into four categories:

1. disorders of initiating and maintaining sleep (insomnias)
2. disorders of excessive somnolence
3. disorders of the sleep-wake cycle

4. dysfunctions associated with sleep, sleep stages, or partial arousals (parasomnias)[62]

Unlike manuals for, say, oncology or cardiology, the *Diagnostic Classification of Sleep and Arousal Disorders* was based mostly on symptoms, not signs; as in psychiatry (where most of the era's sleep clinicians got their start), ailments were identified by how they made patients feel and behave rather than on evidence such as cancerous cells or an arterial blockage. Also as in psychiatry, only a few of these conditions could be adequately controlled by medication or surgery. For most sleep disorders, alternatives ranging from talk therapy to changes in personal habits or societal arrangements were potentially more effective.[63]

In response to the IOM report, policy-makers began to pay attention to sleep for the first time. In December 1979, US surgeon general Julius Richmond launched Project Sleep, designed to promote the reforms the institute had called for.[64] The initiative would be "a major educational and research effort aimed at increasing the level of knowledge by physicians, their patients, and the public about the nature of insomnia and sleep disorders and their treatment," Richmond declared.[65] Project Sleep generated enormous excitement in the sleep science community—and enormous disappointment when funding was eliminated by the incoming administration of President Ronald Reagan just over a year later.[66] Nonetheless, the field was on the verge of another set of transformations.

One harbinger was a two-issue package of articles in *Sleep*, published in the spring and summer of 1980. Titled "REM Sleep: A Workshop on Its Temporal Distribution," the series heralded sleep science's belated but passionate embrace of chronobiology, two decades after Cold Spring Harbor. It included six papers with Czeisler as lead or coauthor (with Weitzman, neurophysiologist Martin Moore-Ede, and others), presenting the REM-related findings from his isolation-lab studies. It also featured Carskadon's chronobiological reanalysis of her ninety-minute-day study, and several other articles on related topics.[67] Czeisler and his East Coast colleagues published a landmark paper in *Science* that year as well, reporting his discovery of the relationship between sleep length and body temperature rhythms.[68]

Another sign of things to come was a second article by Carskadon (with Stanford coauthors Dement, Thomas Anders, adolescent medicine pioneer Iris Litt, pediatrics fellow Paula Duke, and undergraduate lab assistant Kim Harvey) in the summer 1980 issue of *Sleep*. Reporting her Sleep Camp findings on daytime drowsiness in adolescents, it was the first scientific paper to hint at a sleep crisis among teenagers.[69]

These publications pointed toward new approaches to sleep medicine, not only in the therapeutic realm, but also in the arenas of public health and workplace policy. The field would soon be in a position to affect millions of lives and to spawn a multibillion-dollar industry. The study that nudged it into place, however, came from an entirely unexpected direction.

15

Beyond Pills and Scalpels

When Colin Sullivan launched Australia's first sleep lab in 1979, most doctors still thought of obstructive sleep apnea as a rare disorder that mostly affected obese men, and few grasped its potentially devastating impact on patients' health. Sullivan's colleagues at the Royal Prince Alfred Hospital in Sydney responded with "mirth and even contempt," he recalled, to the focus of his research: an ailment whose principal symptom was snoring.[1]

To Sullivan, however, apnea was no laughing matter. At thirty-four, he'd just returned from a postdoctoral fellowship at the University of Toronto under Eliot Phillipson, one of the world's leading researchers on the mechanisms of sleep-related breathing disorders.[2] Phillipson's favored experimental animals were dogs—particularly short-faced breeds such as pugs and English bulldogs, among the few creatures besides people to suffer from OSA.[3] Like humans with the syndrome, these animals experienced disabling drowsiness, cardiovascular complications, and sometimes sudden death.[4]

In Phillipson's lab, Sullivan had learned to measure dogs' respiratory airflow using sensors inserted through a tracheostomy, a hole in the throat fitted with a breathing tube. Back in Australia, the young pulmonologist devised an improved method: molding a fiberglass mask to fit over the animal's snout and sealing it in place with silicone adhesive. The new setup enabled him to monitor the whole respiratory tract, including the nose, with sensors placed in the mask itself.[5]

Sullivan's next brainstorm came in 1980, during an international conference on sleep and breathing that he organized in Sydney. As he watched Christian Guilleminault's endoscopic videos of the upper airway collapsing during apnea episodes, it occurred to him that air pumped through a mask could be used to keep a patient's throat open, acting as a sort of splint. If the idea worked, it could provide an alternative to a tracheostomy—still the only treatment then available for OSA.[6]

Sullivan tried this technique on a few apneic dogs, observing their slumber while they were hooked up to the device. Polygraph readings showed that it eliminated their sleep disturbances. But would such a gadget work on humans? That June, he set out to find the answer.[7]

Sullivan's first subject was a forty-three-year-old construction worker whose daytime sleepiness was so severe that he would doze off while perched on scaffolding; he'd recently been forced to quit his job. Like many people diagnosed with OSA, the man refused a tracheostomy, but he leaped at the chance to test a new approach.[8]

Sullivan—a tinkerer since childhood—threw together a prototype in a single afternoon. For the blower, he used the motor from a vacuum cleaner, adding a dial that allowed him to adjust air pressure precisely. He improvised the connections and mask assembly from parts available in the hospital workshop. Air flowed from the machine into a corrugated polyethylene hose, which connected to an orange plastic chamber that would be attached to the patient's nose. From there, two short tubes entered the nostrils. Exhalations would flow out through another hose, escaping at a controlled rate through a valve at the end.[9]

One night soon afterward, Sullivan checked the man into a hospital room, connected him to a polygraph, and glued the chamber beneath his nose with silicone gel to make an airtight seal. The patient fell asleep immediately and began experiencing apneas within minutes. When Sullivan turned up the pressure dial, the apneas halted. When he decreased the pressure, the apneas returned. After a few more cycles, it was clear that the machine was doing its job. Sullivan then set the pressure at the lowest effective level and sat back to see how long the results would last.

The man slept deeply for seven hours, with a massive REM rebound that testified to his previous deprivation. The next morning, he declared that he felt well-rested for the first time in years.[10]

Over the following months, Sullivan tested his system on four more patients—including a company executive who often fell asleep during important meetings, and a thirteen-year-old boy who'd been deemed "mentally retarded" because he couldn't stay awake at school. The machine performed as it had for the construction worker. After one night of treatment, Sullivan wrote, each volunteer "awoke spontaneously, was alert, and remained awake unprompted through the rest of the day." After three consecutive nights, the improvements held fast.[11]

Sullivan reported his findings in the *Lancet* in April 1981, in a paper titled "Reversal of Obstructive Sleep Apnoea by Continuous Positive Airway Pressure Applied Through the Nares."[12] His technique, abbreviated as CPAP, initially aroused more skepticism than excitement. The method was hard to get right using jury-rigged devices. In a trial at New York Hospital–Cornell Medical Center, one patient couldn't fall asleep when fitted with the apparatus; another needed constant nursing attention and quit after a few nights; the third had 134 apneas per hour, even on the highest pressure setting; and the fourth slept fitfully for three hours, then sat up with a scream and ripped off his mask. Over a year passed before researchers at NYU confirmed Sullivan's results.[13]

At first, Sullivan himself had limited ambitions for CPAP. "We viewed it as a short-term rescue therapy," he said. "We thought we'd use it to get people better—give them a week's sleep—and that would give us time to work out a surgical way of treating the problem." That changed shortly before the *Lancet* paper was published when a patient who'd been part of a trial of the device at the hospital asked if he could try it at home. Sullivan cobbled together another machine, and it proved as effective in a domestic setting as it had been in a medical one.[14]

Sullivan secured a patent on his concept and set about developing a commercially viable product. The earliest version of the airflow generator used a sixty-watt AC motor to drive a blower designed for a spa bath;

the contraption weighed fifteen pounds and was so noisy that it had to be kept outside the bedroom during use. The first-generation face masks were custom-made of fiberglass from a plaster cast of each patient's nose.[15] By 1985, about one hundred Australians were gluing on those masks each night.[16] (Physicians who were early adopters kept a cupboard full of nose molds for their apnea patients.)[17] That year, the Pennsylvania-based Respironics company—then a struggling start-up—began selling its own CPAP devices in the United States, setting the stage for a global boom.[18]

In 1988, Sullivan licensed a large US medical equipment manufacturer, Baxter, to mass-produce an updated version of his original model, with a strap-on mask that didn't require adhesive. When the deal collapsed the following year, Sullivan and his Australian partners launched a firm called ResCare. By the early '90s, sales had reached the tens of thousands and were doubling annually. ResCare soon changed its name to ResMed,[19] moved to San Diego,[20] and overtook Respironics to become the largest CPAP manufacturer in the world.[21]

The CPAP explosion reflected the size of the patient population: the first epidemiological study of OSA, in 1993, found that the disorder affected 24 percent of adult men and 9 percent of women—an estimated thirty million people in America alone.[22] And equipment makers weren't the only entities that profited from that market's pent-up demand. The medical historian Kenton Kroker has written that CPAP "quickly turned the sleep laboratory into a lucrative opportunity for enterprising physicians." As the technology caught on and insurance issues were ironed out, labs found they could charge thousands of dollars for the several nights of polysomnography often required to establish a patient's optimal air pressure.[23]

CPAP altered the balance of power in sleep medicine, drawing legions of cardiopulmonary specialists into the ranks and undoing the dominance of the psychiatrists and neurologists who'd established the discipline. Although that change created tensions, it would ultimately broaden the field's support in the medical community and add to its political clout. The technology also helped fuel an explosion of sleep clinics—from an estimated thirty-four in 1980 to more than one thousand a decade later. CPAP reshaped the priorities of sleep research as well. In the late 1970s, the proportion of journal papers published on insomnia, narcolepsy, dreams, sleep neurophysiology, and sleep apnea were roughly equal. By 1986, apnea accounted for fully half of such articles.[24]

Yet these developments formed only one stream in a larger cascade of change that swept sleep science in the 1980s. As research revealed an ever-broader array of factors that enabled or hindered healthy sleep, it also suggested remedies that fell outside the realm of pills and scalpels. Those insights increasingly challenged the status quo within the field and in society's approach to slumber. Long-established ways of treating sleep disorders—and thinking about them—were losing their monopoly.

Early in the decade—before CPAP had proven itself or become widely available—the question of how best to treat obstructive sleep apnea was very much in play. Although tracheostomies could be miraculously effective, they placed restrictions on a patient's waking life: no showers; no swimming.[25] They could also cause blockages of the trachea, chronic inflammation of the incision site, lung infections, and what experts called "psychosocial" side effects. (Studies reported postoperative marital troubles, depression, or job loss in some individuals.) In light of these risks, the procedure was recommended only for OSA cases with serious cardiopulmonary complications. But because sleep apnea was progressive, that meant leaving people untreated until the damage to their health was potentially irreversible.[26]

In 1981, a team led by otolaryngologist Shiro Fujita at Henry Ford Hospital in Detroit unveiled a new approach to apnea surgery: uvulopalatopharyngoplasty.[27] First introduced in the '60s to control snoring, UPPP enlarged a patient's airway by removing the uvula (the grape-shaped structure at the opening to the throat), the tonsils, and portions of the soft palate.[28] Because people with sleep apnea tended to snore loudly—and because research suggested that individuals with narrower upper airways had a higher risk of OSA—Fujita and his colleagues hypothesized that the surgery could also alleviate the disorder. And unlike a tracheostomy, a successful UPPP would enable patients to resume their normal lives after the incision healed.[29]

Initial results were encouraging: Of Fujita's first twelve patients, nine experienced relief of daytime sleepiness, and eight showed objective improvement in sleep quality on the polygraph. Within five years, an estimated five thousand people had undergone the grueling procedure. UPPP's efficacy, however, proved highly uneven. Overall, about 40 percent of patients enjoyed a significant reduction in apnea symptoms; the rest experienced little or no improvement. Complications could include scarring of the airway, difficulty swallowing, and nasal regurgitation of liquids and solid foods. Surgeons found it difficult to predict which patients would benefit and which would not.[30]

Eventually, other surgical options became available. The two most effective procedures were developed in the mid-1980s by Stanford surgeons Robert Riley and Nelson Powell, working with Guilleminault. In the first operation, genioglossus advancement with hyoid myotomy (GAHM), the tongue muscle and the hyoid bone to which it's anchored were moved forward to widen the upper airway. In the second, maxillo-mandibular osteotomy (MMO), the mid-face, palate, and lower jaw were all moved forward; a more complex surgery, it was usually reserved for patients who didn't respond to other approaches. Success rates were about 70 percent for GAHM and 95 percent for MMO.[31]

Although CPAP would eventually eclipse these approaches, they remained valuable alternatives for patients who didn't benefit from that treatment—often because of anatomical abnormalities in the airway—or couldn't tolerate sleeping with a mask strapped to their face.[32] That turned out to be a large proportion of apnea sufferers: Studies have shown that as many as two-thirds of people who own CPAP machines stop using them, and up to 83 percent undergo treatment for less than the recommended minimum of four hours a night.[33] Still, the risks and discomforts of any type of surgery made it an unattractive choice for many people with OSA.

Another set of options grew popular in the '80s: devices designed to control the disorder by less intrusive means than surgery *or* CPAP. Oral appliances to prevent snoring had been available since the nineteenth century, but the first one aimed at OSA was developed by Chicago psychiatrist Charles Samelson. To control his own snoring problem, Samelson had crafted a wax sleeve that held his tongue forward during sleep; he patented a plastic version in 1979. Two years later, he re-patented the device, expanding its purpose to include "avoidance of obstructive sleep apnea." Samelson brought his invention to psychologist Rosalind Cartwright, who'd recently launched a sleep disorders center at Rush University Medical Center—and who tried the gadget on several of her apnea patients.[34] In 1982, the pair published a small study in *JAMA*, which reported significant reductions in apnea symptoms and daytime sleepiness in fourteen users.[35]

Samelson's device gave rise to the field of oral appliance therapy (OAT), which also came to include gizmos that advanced the lower jaw rather than the tongue. None of these instruments were as effective as CPAP in controlling sleep apnea, and none were particularly pleasant to wear. Still, studies showed that most people who tried both approaches preferred OAT. If patients were likelier to use oral appliances regularly, some experts argued, higher rates of compliance might compensate for any gaps in performance.[36]

An even less invasive OSA treatment appeared shortly after OAT: sleep position training, or SPT. As far back as the seventeenth century, physicians had noted an association between back sleeping and snoring; during the American Revolution, soldiers wore rucksacks during slumber to keep from rolling over and making noises that could alert the enemy. In the early 1980s, researchers found evidence that back sleeping also worsened apnea symptoms in many patients and suggested that they be taught to avoid this position.[37]

Cartwright was a pioneer in this area as well. In 1984, she published a study in *Sleep* showing that the rate of apnea episodes doubled when patients with OSA slept on their back.[38] The following year, she introduced the concept of SPT in the same journal.[39] Cartwright and her team at Rush had developed a gravity-sensitive monitor and alarm system that fit in a small box strapped to a sleeper's chest; a buzzer sounded whenever the person stayed in a supine position for more than fifteen seconds. The team trained ten male patients with the device for a single night, then sent them home to practice without it. During the training night, the percentage of side sleep rose from 48.6 to 97.9 percent. The group's apnea episodes fell by half overall, with seven of the ten men achieving near-normal sleep. Three months later, a follow-up test showed that four of the patients still slept on their sides throughout the night, though the rest had relapsed. The device had promise, Cartwright and her team concluded, but more frequent training might be needed to make the lessons stick.[40]

From these beginnings, a plethora of SPT systems arose, ranging from the low-tech (a tennis ball sewn into the back of a T-shirt) to the middling (sensor-equipped wearables that vibrated when they detected a forbidden position).[41] As with oral apnea appliances, these devices were less effective than CPAP, but they could reduce symptoms enough in some patients to potentially ward off the worst consequences of the disorder.[42]

It's worth noting that Cartwright had never previously shown an interest in respiratory issues; she was one of the few remaining sleep researchers whose principal concern was the relationship between dreaming and mental health. (She is remembered today for books such as *Crisis Dreaming* and *The Twenty-Four Hour Mind*, and for her studies of how dreams can serve as a tool for solving emotional problems.)[43] Her detour into OSA shows how central the syndrome had become to sleep science in the decade since Guilleminault identified it.

As the transformation of apnea therapy gathered momentum, a quiet revolution was also underway in the treatment of insomnia.

For centuries, humanity's main method of battling sleeplessness had been to ingest soporific substances—from calming teas to laudanum to benzodiazepines. The milder ones were of limited efficacy, however, and the stronger ones tended to carry significant downsides. The first modern attempt at a nonpharmacological remedy for insomnia (and for stress in general) was Edmund Jacobson's "progressive relaxation" technique, introduced in the 1930s, but its narrow focus on muscle tension limited its usefulness. In the 1960s, researchers began investigating other factors that could contribute to the disorder. Allan Rechtschaffen and his students at the University of Chicago reported that self-described poor sleepers showed signs of depression and anxiety, as well as increased physiological arousal. Anthony Kales and his group at Penn State found that chronic insomnia sufferers shared certain neurotic "character styles" as measured by standardized personality tests. Most experts came to agree that resolving the sufferer's underlying psychological conflicts (including anxiety over sleeplessness itself) was essential to effective treatment. If a patient needed a few nights' worth of pills to break the grip of a keyed-up nervous system, so be it. A long-term solution, however, would typically require an extended course of psychotherapy.[44]

In the 1970s, some sleep scientists rebelled against this view. Northwestern University psychologist Richard Bootzin argued that insomnia often arose from Pavlovian conditioning rather than deep-seated neuroses. By doing things like watching TV, eating, or studying in bed, many people weakened the power of that piece of furniture to act as a stimulus for sleep. If they then spent night after night tossing, turning, and worrying, the bed became associated with *not* sleeping—and a vicious circle was set in motion. To reverse it, Bootzin developed a technique called *stimulus control therapy*, the first nondrug treatment designed expressly for insomnia. SCT laid down a set of simple rules:

1. Use the bed only for sleep and sex. Do not get into bed until you are ready for sleep.
2. If you are not asleep within about twenty minutes, get out of bed.
3. Return to bed when sleepy (or ready to go to sleep).
4. Repeat steps 2 and 3 if necessary.
5. Get up and out of bed at the same time every day.
6. Do not nap.[45]

Bootzin's simple method proved surprisingly effective, outperforming progressive relaxation and placebos in several studies. But it still didn't help everyone.[46] In 1977, psychologist Peter Hauri added another weapon to the arsenal of nonpharmacological insomnia therapies: an approach he dubbed "sleep hygiene."[47] The more cumbersome term *hygiene of sleep* had been used since the late nineteenth century to describe commonsense practices believed to promote healthy slumber.[48] Some of Hauri's tips, indeed, echoed those offered in the 1920s by the popular sleep expert Donald Laird.* Hauri, however, was able to draw on an additional fifty years of research. His initial list (which would be modified over time) consisted of eleven precepts:

1. Sleep as much as needed to feel refreshed and healthy during the following day, but not more. Curtailing the time in bed seems to solidify sleep; excessively long times in bed seem related to fragmented and shallow sleep.

2. A regular arousal time in the morning strengthens circadian cycling and, finally, leads to regular times of sleep onset.

3. A steady daily amount of exercise probably deepens sleep; occasional exercise does not necessarily improve sleep the following night.

4. Occasional loud noises (e.g., aircraft flyovers) disturb sleep even in people who are not awakened by noises and cannot remember them in the morning. Sound-attenuated bedrooms may help those who must sleep close to noise.

5. Although excessively warm rooms disturb sleep, there is no evidence that an excessively cold room solidifies sleep.

6. Hunger may disturb sleep; a light snack may help sleep.

7. An occasional sleeping pill may be of some use, but their chronic use is ineffective in most insomniacs.

---

* See Chapter 3.

8. Caffeine in the evening disturbs sleep, even in those who feel it does not.

9. Alcohol helps tense people fall asleep more easily, but the ensuing sleep is then fragmented.

10. People who feel angry and frustrated because they cannot sleep should not try harder and harder to fall asleep but should turn on the light and do something different.

11. The chronic use of tobacco disturbs sleep.[49]

Although studies failed to confirm that sleep hygiene on its own could control insomnia in those already suffering from the disorder, clinicians embraced it widely as an evidence-based preventive regimen. Equally important, it laid the groundwork for a broader-based assault—one that drew from multiple strategies rather than attacking sleeplessness from a single angle.[50]

The tipping point came in 1979, with the publication of the aforementioned *Diagnostic Classification of Sleep and Arousal Disorders*. One of the newly identified disorders described in the nosology was psychophysiological insomnia—the first type of insomnia that was not defined as "associated with" some other condition. Psychophysiological insomnia didn't emerge from underlying neuroses or physical ailments; it resulted from normal psychological or bodily processes, such as people's tendency to translate worry into muscle tension or to feel amped up when they're overstimulated. It was a problematic response to a given set of circumstances—which meant it could be treated by changing either the circumstances, the response, or both.[51]

Over the next dozen years, researchers developed a series of techniques based on that insight. In the early '80s, Hauri found that biofeedback devices—which use visual or auditory signals to train patients to regulate their brain waves or EMG impulses—could reduce the hyperarousal that often contributes to insomnia. This method proved highly popular, perhaps because of its calming effect, though studies of its impact on sleep produced mixed results.[52]

Soon afterward, psychologist Arthur Spielman unveiled the 3P Model for assessing insomnia, which held that the disorder stemmed from the interaction of three elements: predisposing factors (such as depression or anxiety), precipitating factors (stressors that trigger an initial bout of sleeplessness), and perpetuating factors (behaviors, habits, or thought patterns that prolong the problem). The model emphasized the importance of the third P as a basis for treatment, focusing on the factors that perpetuated a patient's insomnia rather than those that initially caused it.[53]

Spielman, a cofounder of Elliot Weitzman's sleep center at Montefiore, introduced another important innovation in 1987—sleep restriction therapy. SRT, he later wrote, "was based on the documented and near universal experience of sleep loss producing a compensatory enhancement of subsequent sleep." The technique was straightforward, if counterintuitive: when other methods failed, a person with insomnia would be required to spend less time in bed.[54]

In their initial study,[55] the results were favorable. Although some participants found the routine too rigorous and dropped out of the trial, most of those who stuck it out experienced significant improvements: They fell asleep faster and stayed asleep for a greater proportion of their time in bed. These changes persisted, in large part, after a thirty-six-week follow-up. And the gains weren't limited to patients with psychophysiological insomnia. Those whose sleeplessness was associated with psychiatric disorders and some other ills benefited as well. After further studies replicated Spielman's findings, sleep restriction therapy came into widespread use for treating stubborn insomnia.[56]

Eventually, SRT, SCT, and sleep hygiene would be incorporated into a powerful new approach to the disorder—cognitive-behavioral therapy for insomnia. Introduced in 1993 by Canadian psychologist Charles Morin, CBT-I bundled those earlier therapies with cognitive-behavioral therapy, a form of psychotherapy that focuses on identifying and correcting harmful patterns of thought and feeling, rather than analyzing, the roots of those patterns. Patients typically undergo six to eight sessions, with follow-ups as needed. Studies have shown this multimodal form of therapy to be as effective as sleeping pills for chronic insomnia, with far more durable results.[57]

Besides providing a side effect–free alternative to those medications, CBT-I resolved a question that experts had debated for more than a century: Is insomnia a physical or psychological ailment? The answer is both—and that the two categories are less distinct than they seem. The disorder often encompasses the mind, the body, and behavior, with an array of elements interacting in a feedback loop. The best way to treat it may be to address the whole circle.

————————

At the start of the 1980s, the growing alliance between sleep science and chronobiology helped researchers open other nonpharmacological approaches to sleep problems—not only in the clinical arena but in the workplace as well. Charles Czeisler was at the forefront of these explorations, with a handful of his colleagues on both coasts. In the fall of 1980, as he started his final year of medical school at Stanford, he got a phone message from the production manager of the Great Salt Lake Minerals and Chemicals Corporation in Ogden, Utah. "I have a hundred and thirty men who can't sleep," the note read. "Can you help?"[58]

The company, Czeisler learned, harvested potash from solar evaporation ponds around the clock, using gigantic front-end loaders to dump the mineral salt into an endless parade of trucks. The workers had recently launched a union drive, and the manager, whose name was Preston Richey, had been assigned to find out why. Their biggest source of complaints turned out to be shift work. "Family problems, disrupted sleep, fatigue, indigestion, and downright despair were recounted graphically," Richey later recalled. He tried to convince the workers that the existing schedule was as good as any in the industry. But the more he gave this answer, the less comfortable he became with it.[59]

Richey searched in vain for information on the health impacts of shift work or how to ease them. Then he ran across a news item about Czeisler's 1980 *Science* paper on sleep and circadian rhythms, whose last sentence mentioned that its findings might help explain "the sleep-wake patterns in shift workers." On that slender clue, he reached out to the scientist.[60]

Czeisler flew to Utah with two of his colleagues—Martin Moore-Ede, a Harvard neurophysiologist who often collaborated with Elliot Weitzman's team at Montefiore, and Stanford psychologist Richard Coleman.[61] The researchers soon recognized one likely problem: Great Salt Lake's shift schedules ran "counterclockwise," a common practice at the time. Employees worked the night shift (midnight to 8:00 a.m.) for a week, then the evening shift (4:00 p.m. to midnight) for a week, then the day shift (8:00 a.m. to 4:00 p.m.) for a week.[62]

The trouble with this sequence, the scientists told the management team, was that it forced workers to bed down and wake up earlier and earlier.[63] They were in a state of permanent jet lag—as if they were flying eastward week after week, from Utah to Paris to Tokyo.[64] Not only were many workers falling asleep on the job, but their other biological rhythms were thrown off-kilter as well.[65] "It really disrupts your metabolism," one foreman said. "Have you ever had your bowels do a double somersault in midstream, stop and say to you, 'Just which way do you want us to move, Jack?' I know this sounds far-fetched, but it isn't."[66]

The company's shift rotation schedule violated the rules of the circadian timing system, Czeisler and his team explained. Time isolation studies had shown that most people had a free-running sleep-wake cycle of about twenty-five hours.* They could easily entrain to a twenty-four-hour day, which required them to set back their inner clock only slightly. But larger backward movements—known as "phase advances"—were much more difficult to accommodate. People found it easier to go to bed and get up two or three hours *later* on successive cycles, which was why jet lag typically wasn't as severe in travelers flying from Tokyo to Paris to Utah.[67]

---

* Czeisler later revised this figure downward to 24.18 hours, based on experiments using a technique called *forced desynchrony*, which involves putting subjects on a sleep-wake cycle that diverges sharply from circadian rhythms—as in Dement and Carskadon's ninety-minute-day studies. See Charles A. Czeisler et al., "Stability, Precision, and Near-24-Hour Period of the Human Circadian Pacemaker," *Science* 284 (June 25, 1999): 2177–2181.

The direction of the rotation wasn't the only red flag, however; another was its speed. When the researchers conducted a survey, 81 percent of the respondents said it took two or more days for their sleep schedule to adjust after each phase advance; 26 percent said they were never able to adjust. So Czeisler and his colleagues designed a trial in which the rotators were divided into two groups. Both were placed on a clockwise rotation, with phase delays instead of phase advances—but one continued to change shifts every week while the other did so every twenty-one days. Before the new schedules went into effect, all workers and managers attended a presentation on circadian rhythms, with tips on how to adjust their sleep time to their shift assignment. They also received a booklet to help them navigate the changes.[68]

After three months, both groups took another survey. Most of the workers preferred the clockwise rotation, but improvements in schedule satisfaction and health scores were far greater in the twenty-one-day rotators than in the weekly group. The researchers then placed all shift workers on the twenty-one-day phase-delay rotation and tracked a wide range of metrics at the company as the study continued.[69]

Unlike Nathaniel Kleitman's experimental shift schedules of the 1940s (which he designed before the links between sleep and circadian rhythms were even faintly understood),* the new plan worked. Equipment operators made fewer errors. Morale improved. Medical complaints dropped, as did employee turnover. Productivity rose by 22 percent, and profits increased by $800,000 in the first year alone.[70]

When Czeisler and his two colleagues reported their findings in *Science* in 1982, the paper unleashed a flood of media attention. Hundreds of companies contacted the researchers, pleading for similar help. All three men launched their own circadian consulting firms, and Moore-Ede soon left academia behind. Czeisler joined the faculty at Harvard Medical School, but he continued to advise clients ranging from NASA to professional sports teams to the Rolling Stones.[71]

---

* See Chapter 4.

Sadly, the scientist who'd midwifed this branch of sleep medicine reaped little of the resulting glory. Soon after the *Science* paper appeared, Weitzman moved from Montefiore to New York Hospital–Cornell Medical Center in White Plains, New York, where he founded the Institute of Chronobiology. The lymphoma for which he'd been treated eight years earlier recurred almost immediately. He died in June 1983, just fifty-four years old.[72]

16

Crossroads

One of the paradoxes of sleep science (and, perhaps, most other sciences) is that it often violates the precepts of Occam's razor—the principle that between two competing theories, the simpler explanation is to be preferred.[1] In fact, the trend has generally been toward greater complexity, as in the shift from theories that sleep is triggered by reduced cerebral blood flow to those involving electrochemical feedback loops among multiple nerve centers in the brain.

In March 1982, a professor of pharmacology at the University of Zurich named Alexander Borbély added a new kink to the theory that human sleep-wake cycles are governed by the same biological clock that controls the circadian rhythms of body temperature and alertness. That notion had always left a few important things unexplained. For example: Why do people typically feel ready for sleep around 11:00 p.m., even though their temperature-and-alertness curve is only a little past its peak? Why do they wake up seven or eight hours later, even though that curve has just begun to rise from its low point? And why do they become increasingly sleepy after several nights of insufficient slumber, even though nothing in their circadian cycles would account for that phenomenon?[2]

Borbély offered an answer to all these questions: Circadian rhythms weren't the whole story. Sleep in mammals was governed by *two* physiological mechanisms, which could reinforce each other or come into conflict. He laid out his "two process model" in the journal *Human Neurobiology*, citing evidence ranging from nineteenth-century sleep deprivation studies to the latest chronobiological experiments—including some in his own lab.[3]

Borbély described the first factor as a "sleep-dependent process," which he called Process S. Throughout each day, he theorized, a sleep-promoting chemical accumulated in an animal's brain. This substance (hypothesized since the 1890s, but not yet identified)* created a pressure for slumber that increased each moment the creature stayed awake. During slow-wave sleep, the brain worked to clear away the chemical, normally completing the task by morning. The substance then began building up again, as inexorably as sand trickling through an hourglass. On a graph, Process S formed choppy waves, with each day marked by a sharp crest, a sudden downslope, and a V-shaped trough.[4]

The second factor in Borbély's model was the circadian cycle, or Process C. This process operated independently of sleep, but it made slumber easier or more difficult at certain times—easier as the temperature-and-alertness curve descended, harder as the curve rose higher. On a graph, Process C formed a gently rolling undulation, which peaked and dipped a few hours behind Process S.[5]

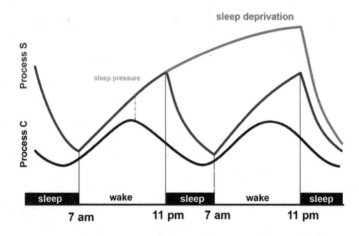

Borbély's two-process model. The top line shows how sleep pressure builds when a person pulls an all-nighter. © *Taylor & Francis, 2021.*

---

* Stanford researchers Joel H. Benington and H. Craig Heller later found evidence pointing to adenosine, a by-product of energy consumption in the brain, as the likely substance. See Joel H. Benington and H. Craig Heller, "Restoration of Brain Energy Metabolism as the Function of Sleep," *Progress in Neurobiology* 45 (1995): 347–360; Joel H. Benington, Susheel K. Kodali, and H. Craig Heller, "Stimulation of A1 Adenosine Receptors Mimics the Electroencephalographic Effects of Sleep Deprivation," *Brain Research* 692 (1995): 79–85.

Under ideal circumstances, the two processes worked together to promote healthy sleep rhythms. A favorable confluence arrived each night around 11:00 when the Process S curve soared high enough to make people feel drowsy and the Process C curve dipped low enough not to keep them awake. Over the next few hours, body temperature reached its minimum, then slowly began to climb. By 6:00 or 7:00 a.m., the brain had finished its neurochemical housecleaning, and the circadian curve had risen just enough to stimulate arousal. The sleeper awoke refreshed and ready to meet the day.[6]

The two-process model explained not only normal human sleeping patterns but also what happens when the patterns go awry. After a shift change, for example, workers might have to wake up before the Process S chemicals were gone and then go to bed at an hour when Process C was on the rise. They'd feel drowsy all day but too wired to drift off the following night. If people got too little shut-eye on a regular basis, those S molecules could build up in their brains indefinitely, creating ever-deeper levels of exhaustion. The fluctuations of the circadian cycle would make sleep-deprived individuals feel more alert at some hours, but would worsen impairment at others.[7]

Borbély's brainchild soon became the dominant model for understanding the basic dynamics of sleep rhythms as it remains to this day.[8] It also helped fuel a growing sense of anxiety among sleep scientists over the state of slumber in modern society.

***

For Mary Carskadon, the two-process model shed a disturbing new light on the sleeping patterns of American teenagers.[9]

After receiving her PhD, Carskadon had stayed on at Stanford as a research associate,[10] aiming to track her Sleep Campers through their late teens before hitting the job market. Since then, she'd published two important papers based on her dissertation research. In 1981, she reported that in young adults allowed only five hours of sleep per night, daytime sleepiness grew steadily worse over the course of a week. In 1982, expanding on the same data, she suggested that sleep habits in this age group often lead to chronic sleep debt.[11] Although the concept of such a debt first arose in the late 1950s, Carskadon was the first to prove its existence—and to quantify it using a tool as precise as the Multiple Sleep Latency Test.[12]

Her study's findings, however, had been far from dire. After Night 7, when the volunteers were allowed to sleep for up to ten hours, their MSLT scores returned to normal. If they went back to getting eight hours a night the next week, she wrote, there was little likelihood of lasting consequences.[13]

Borbély's two-process manifesto made Carskadon suspect that for kids in mid- to late puberty—who needed more slumber to function normally—the impact of school-year sleeping patterns could be more insidious. If they built up a sleep debt of ten or twelve hours over the school week, they couldn't make up for it on Saturday and Sunday; the somnogenic chemicals in their brains might never be completely cleaned out. And sleeping late on weekends would throw Process S and Process C ever further out of sync. Like shift workers on a counterclockwise rotation, Carskadon surmised, these children would experience a phase advance each Monday. No one knew what the health effects might be—but as with the shift workers, they probably weren't good.[14]

At that point, however, Carskadon wasn't in a position to investigate those impacts. Sleep Camp didn't operate during the school year, and she lacked the funding to launch a major study of her own.[15] (At Stanford, only tenure-line faculty members were permitted to apply for such support.)[16] Besides, she had too much other work to do. To begin with, she was helping Dement with drug trials and sleep apnea studies, both of which were highly demanding.[17]

Carskadon was also focusing increasingly on sleep in the elderly, thanks to a grant her mentor had secured from the newly formed National Institute on Aging.[18] In her dissertation research, she'd studied the relationship between daytime sleepiness, overall sleep quality, and nocturnal variables such as respiration, leg movements, and electrocardiograms in people over sixty, but her sample size had been too small to draw statistical conclusions.[19] Now, she began to delve further into these questions. One of her first discoveries was that the drowsiness commonly seen in older people is driven more strongly by sleep fragmentation—brief arousals, apneas, and other interruptions—than by the length of a night's sleep.[20]

She was soon joined in this area of inquiry by psychologist Donald Bliwise, a former student of Allan Rechtschaffen's who was doing his postdoc with Dement. A genial New Jerseyite with a luxuriant mustache, Bliwise had been drawn into sleep science by a psychiatrist uncle who'd worked with Rechtschaffen in the early '60s. At the University of Chicago, he had become keenly interested in how slumber changes over the human life span; in his PhD research, he'd found that slow-wave sleep (already known to dwindle in seniors) declined even between the ages of eighteen and twenty-three. He'd come to Stanford to learn more. "Bill said, 'Mary has collected some data on old people. Why don't you pick it up from there?'" Bliwise recalled. "That's the way he was. He wouldn't say, 'Go do A and then B and then C, but divide C into C1 and C2.' He would just say, 'Go.'"[21]

With Carskadon's help, Bliwise began gathering a cohort for a long-term study. And at Dement's request, he included Nathaniel Kleitman among his subjects. The father of modern sleep science was eighty-six when he arrived for his first all-night session in late 1981.[22] His wife, Paulena, had died four years earlier, but he was still actively involved in the field. (He would publish his last paper, on his theory of the basic rest-activity cycle, a few months later.)[23] He flew north from Los Angeles on his own and appeared in Bliwise's lab dressed in a suit and tie—tiny, wizened, briskly cordial, and intent on the business at hand.[24]

Kleitman stayed with the Dements during these and other visits. When he wasn't undergoing a sleep test, he often passed the evening playing bridge with Bill and Pat, as he'd first done a quarter century earlier. Carskadon sometimes served as his partner—a role she initially found terrifying, though her hosts' easy banter with the great man put her at ease. When Dement's widowed mother was visiting from Walla Walla, she sat in, too.[25]

Away from the card table, Dement himself was playing a strangely mixed hand. A decade after launching the field of sleep medicine, he remained in many ways at its forefront. Yet when it came to insomnia, he was doubling down on a traditional approach: pharmaceutical research.

In part, this reflected a genuine clinical need. Although other therapies were emerging in the early 1980s, medication was still an accepted short-term treatment for the disorder. To maximize both safety and efficacy, greater knowledge of how different drugs affected different patient groups was essential—particularly for older adults, who accounted for a disproportionate share of users. So was the search for that elusive grail: a pill that would provide something close to natural sleep, without carryover effects of drowsiness or confusion.[26]

One promising candidate was triazolam, a benzodiazepine whose half-life (the time it remained in a patient's system) was far shorter than those of its predecessors. Starting in 1982, Dement's team ran an extensive series of studies comparing triazolam with flurazepam—the most popular of the older benzos—using the MSLT to measure both drugs' impact on daytime sleepiness. Triazolam won consistently in men and women of all ages and degrees of insomnia, leaving them more alert during the day than either flurazepam or a placebo.[27] Those findings helped the new pill become wildly popular after the US Food and Drug Administration approved it that November.[28] Marketed as Halcion, triazolam would make a fortune for its manufacturer, the Upjohn Company, before a string of murders allegedly committed under the drug's influence cast a pall on its reputation.[29]

Perhaps a more compelling motivation for Dement's sleeping-pill work, however, was the need to keep his own enterprise afloat. "There were numerous studies," Don Bliwise recalled. "Berlex had a drug called brotizolam. Wyeth had an old benzodiazepine called oxazepam. Schering had quazepam. Abbott had estazolam. Bill did clinical trials for all these companies because his lab needed the money." Leftover cash from such commissions helped pay for equipment, supplies, staff salaries, travel—all the expenses of an ambitious research program that were difficult to cover with grants alone.[30]

That income was especially crucial during the first years of the Reagan era, when spending on biomedical science in general took a sharp hit.[31] Funds from other quarters were also hard to come by. Stanford contributed little to Dement's coffers; like most American universities, it required researchers to rely primarily on outside support. The sleep clinic had become self-sustaining, but it didn't yet generate enough surplus to help sustain basic research. Nor were private philanthropists a reliable wellspring. Dement spent lavish portions of his energy and charm in wooing them, but the results were sporadic at best.[32]

Another source of financial strain was the hefty overhead fee that Stanford demanded from every grant funder. Government regulations allowed all universities to recover the indirect costs of research (such as construction, building maintenance, and administrative support) through such a surcharge. Stanford's commission rose from 46 to 58 percent between 1975 and 1977, then leaped to 69 percent in 1983—one of the highest rates in the nation. To receive a $100,000 grant from a donor, Dement and his colleagues had to request $169,000.* That made it tougher than ever to land desperately needed funds.[33]

To get around these hurdles, Dement often routed private donations to the Sleep Disorders Foundation, a nonprofit he'd set up in the early '70s. But that practice raised frictions with the university administration,

---

* In 1991, federal investigators alleged that Stanford had overcharged the government by up to $200 million over the preceding decade, using the funds to cover expenses such as a yacht and antique furniture purchased by university president Donald Kennedy—who resigned in the wake of the scandal. No wrongdoing was found, however, and the matter was dropped after Stanford paid a $1.2 million settlement. Bill McAllister, "Stanford Chief Quits Over Billings Flap," *Washington Post*, July 30, 1991, https://www.washingtonpost.com/archive/politics/1991/07/30/stanford-chief-quits-over-billings-flap/defc068b-9712-4224-acd3-ea8248706a6e/; Barry Bozeman and Derrick M. Anderson, "Public Policy and the Origins of Bureaucratic Red Tape: Implications of the Stanford Yacht Scandal," *Administration and Society* 48, no. 6 (June 21, 2014), https://journals.sagepub.com/doi/10.1177/0095399714541265.

which objected to being denied its cut. Meanwhile, Dement's most cherished projects—particularly a narcoleptic dog colony and the research on the disease's brain chemistry and genetics that arose from it—teetered perpetually at the brink of insolvency. The drug-company money kept them from sliding over the edge.[34]

In September 1982, Dement hired an attorney named Stuart Rawlings as secretary of the foundation and administrator at large for the sleep clinic. As a Stanford undergrad, Rawlings had been one of a series of students who helped the Dements with childcare and errands in exchange for room and board; later, he'd worked briefly in the Quonset hut lab at the Menlo Park VA. The scion of a wealthy San Francisco family, he'd also spent a few years trekking through Peru and Brazil, toiling as a voting rights organizer in Mississippi and an aid worker in Vietnam, and serving as a consumer advocate for Ralph Nader. Rawlings had remained close to the Dements throughout his adventures and had recently been married in their backyard. His new job, whose modest salary was paid through the foundation, involved as much moral support as bureaucratic wrangling. "Bill was at his wit's end when I arrived," Rawlings recalled. "He said, 'Stuart, I'm in way over my head. I need your help getting through the day.' "[35]

When Dement was in high spirits, he would slide down the hall in his stocking feet. When he was down, he would invite Rawlings into his office and ask for a hug. And when he needed to blow off steam, he would call in the administrative staff and scream, "I can't take it anymore!" Then he'd rip open a large bag of popcorn, toss the contents around the room, and stalk out. Everyone knew him well enough to laugh.[36]

These rituals, however, could not dispel the pressures that assailed Dement through the first half of the '80s. Besides his funding troubles, he faced the need for more space to accommodate the clinic's growing clientele, and the administration's reluctance to provide it.[37] There were tensions with Christian Guilleminault, who seemed unwilling to accept that, having left his homeland to help launch the sleep-medicine revolution, he would forever be relegated to second-in-command.[38] There were outbreaks of illness in the dog colony, which prompted Dement to remove the animals from the custody of the university's Division of Laboratory Animal Medicine and place them in private kennels—triggering more conflict with Stanford.[39]

Meanwhile, the rise of the animal rights movement was sowing fear in the hearts of scientists who ran experiments on dogs, cats, rodents, birds, or nonhuman primates.[40] In recent years, a variety of regulations had emerged to protect lab animals from suffering. The 1966 Animal Welfare Act required the US Department of Agriculture to inspect animal housing conditions at research facilities; a 1970 amendment set standards for humane experimental procedures.[41] Such measures, however, did not always prevent abuses—and a growing number of activists insisted that virtually *all* animal research was cruel and unnecessary. The movement's militant wing, spearheaded by the UK-based Animal Liberation Front, was using guerrilla tactics to bring such research to a halt.[42]

The wave began in England, where the ALF vandalized or burned down several labs, before reaching America in 1979.[43] In 1981, an activist with People for the Ethical Treatment of Animals (PETA) went undercover at the Institute of Behavioral Research in Maryland, exposing alleged abuse of monkeys. Arrested for animal cruelty, psychologist Edward Taub was convicted on six counts, though the verdicts were later overturned on appeal.[44]* At the same time, Dement—whose relatively gentle cat experiments in the '60s had made him a target—received frequent death threats. He worried constantly that someone would abduct his narcoleptic dogs and erase years' worth of work.[45]

One morning, he sat down at his desk and typed out a confession. "I must be honest with myself. The load appears too heavy. My wife cried

---

* Taub's research—which involved severing the nerves in a monkey's arm or leg, then forcing the animal to move it—led him to develop constraint-induced movement therapy, a rehabilitation method that enables stroke patients to regain the use of long-paralyzed limbs. His case epitomizes the ethical quandaries raised by animal experimentation. See "Constraint Induced Movement Therapy," Physiopedia, https://www.physio-pedia.com /Constraint_Induced_Movement_Therapy.

and cried last night. She never sees me. Too many problems. Not enough rewards.... I must face that I am absolutely at my limit."[46]

It's not clear whether he showed the note to anyone before placing it in a file folder. At some point, however, he added a headline in felt-tip marker: "EARLY 1984."[47]

———————

Some of Dement's troubles eased a bit over the following year. A new source of income arrived that October when he launched a training program for polysomnographers. Headed by one of the country's leading practitioners—Sharon Keenan, who'd previously served as the sleep clinic's chief technologist—the program drew students from around the world.[48] In the spring of 1985, after intense negotiations with the university, the clinic moved into roomier digs at the Hoover Pavilion, an art deco ziggurat built in the 1930s.[49] In August, after hiring a new director of the Division of Laboratory Animal Medicine, Stanford opened a state-of-the-art, $11 million housing complex for nonhuman experimental subjects.[50] Satisfied with both its biological and physical security, Dement relocated his dogs to the underground facility.[51]

Still, his sense of beleaguerment lingered—and he had come to see that his struggles were not unique. Sleep medicine was at a crossroads, he judged, growing explosively yet still too marginalized to achieve its full potential.[52] To remedy the situation, the Association of Sleep Disorders Centers (a national clinical society led by Dement) had recently formed a committee for insurance and government relations. Its director was Merrill Mitler, who'd left Stanford in 1978 and now ran his own sleep center at the Scripps Research Institute in La Jolla, California. Mitler hired a lobbying firm, the Health and Medicine Counsel of Washington, to help the ASDC make the right connections.[53]

And so it was that in May 1985, two scientists addressed Congress on the topic of "Funding for Sleep Disorders Research." The first was Mitler, who spoke to the Senate Subcommittee on Labor, Health and Human Services, and Education. His speech emphasized facts and figures: the one hundred million Americans with sleep disorders; the $200 million spent annually on sleep medications;* the 23 percent cut in funds for sleep research in the proposed budget for the coming year. He asked for an increase reflecting "the financial and social impact that sleep pathology has on our society."[54]

Later that month, Dement appeared before the equivalent subcommittee of the House of Representatives. His tone, in contrast to Mitler's, was sweeping and impassioned. "[As] our tiny band of dedicated sleep specialists draws back the curtain of darkness to reveal the scores of illnesses that are experienced *only* when we sleep," he proclaimed, "the medical establishment is unwilling to allocate precious resources to research on these newly discovered problems." He called on Congress to authorize funding that would "elevate the sleeping brain and its many illnesses to a status equal to the waking brain." And he ended with a ringing appeal: "Sleep research and sleep disorders medicine has been a foster child for all its existence. However, sleep disorders medicine and the study of the sleeping brain is now a fully mature adult, and absolutely must have its independence. Independence requires a means of support, a means of education and discipline, identity, and a specific effort to resolve its great problems."[55]

Despite their differences, Dement's approach had a similar effect to Mitler's—or lack of one. The legislators listened politely, then did nothing.

As Dement made his case to Congress, Carskadon, too, was craving independence. After observing her campers for nine seasons, she'd established that kids' need for sleep remains elevated through late adolescence. Now, she wanted to probe that phenomenon—and its effects on teens' lives— more deeply than was possible in a few weeks each summer.[56] At the Stanford School of Medicine, however, only MDs were eligible for tenure, and

---

* The figure for sleeping pills represents $537 million today, a fraction of that market's current value.

thus for independent research funding. She loved working with Dement, but she was eager to test her wings. "At some point, one questions one's identity," she later recalled. "Am I Mary Carskadon, or am I Dement and Carskadon? Can I succeed without the ampersand?"[57]

Yet job openings for sleep researchers were scarce, and she soon discovered that her cutting-edge PhD in neuro- and biobehavioral sciences was not considered a compelling qualification. The ads for faculty positions called for a degree in psychology or neurobiology. What she possessed was something in between. Her stellar publication record, she feared, wouldn't get her to the starting gate.[58]

Deliverance came that spring when she received a call from one of her dissertation advisers. Thomas Anders had recently been appointed chief of child and adolescent psychiatry at Brown University, in Providence, Rhode Island. He'd also become academic director of the affiliated Emma Pendleton Bradley Hospital, the nation's oldest psychiatric institution for children, where he hoped to expand research operations. He asked Carskadon if she would like to start a sleep lab of her own.[59]

In September 1985, after the tenth and final session of Sleep Camp, the Dements threw her a going-away party in their backyard. Bill presented her with a gold pocket watch as a memento of her fifteen years under his tutelage. "How will I ever replace Mary?" he cried before giving her a farewell hug.[60]

Soon afterward, Carskadon climbed into her VW Beetle, with Pat Dement as her traveling companion, and drove east.[61]

17

Catastrophe

Emma Pendleton Bradley was seven years old in 1886 when an attack of encephalitis left her with intellectual disability, epilepsy, cerebral palsy, and frightening behavioral disturbances. Her wealthy parents, financier George Lathrop Bradley and his wife, Helen, embarked on a worldwide search for therapies that could help her but found that knowledge about such ailments in children was vanishingly scarce. The best the couple could do was to provide Emma with round-the-clock nursing care, which she received until her death in 1907.[1] Her father had passed away a year earlier, leaving instructions in his will for the establishment of "a place for the care, treatment, relief, and support of poor and needy [children] afflicted with nervous and other chronic diseases." The facility, George stipulated, would also foster research in pediatric neurology and mental health. "Out of the misfortunes of our only child," he wrote, "has grown the purpose and the hope that from the affliction of this one life may come comfort and blessing to many suffering in like manner."[2]

The Emma Pendleton Bradley Home opened its doors in 1931. Besides serving as the model for generations of pediatric neuropsychiatric hospitals, it produced some groundbreaking research in its inaugural decade.[3] Psychologist Herbert H. Jasper built one of the earliest EEG machines in its laboratory and published the first North American paper on the new technology.[4] His successor, Donald B. Lindsley, conducted the first systematic survey of brain wave development from the fetal stage to puberty.[5] Medical director Charles Bradley (George's great-nephew) made the epochal discovery that amphetamines could be used to treat ADHD.[6]

Bradley Hospital (as it had been renamed) became affiliated with Brown University in 1973.[7] By the time Mary Carskadon signed on, however, the hospital had lost much of its academic luster. Thomas Anders was determined to restore it, and he'd hired her as part of that effort. As a vote of confidence, Anders proposed Carskadon as an associate professor of psychiatry at Brown, skipping the assistant professor stage. At the hospital, her title was director of chronobiology, reflecting his firm grasp of recent developments in sleep science. Her assignment, he'd explained, would be to map the connections between sleep and mental health in children and adolescents.[8]

Carskadon was far from certain that she could fulfill this objective. To begin with, she had little training in psychiatry and no experience running an independent research program; though she'd written countless grants for senior investigators, she'd never composed one for herself. Moreover, her ten summers' worth of data from Sleep Camp didn't really address the topic Anders had in mind—and no models existed for steering her investigations in that direction. Studying chronobiology in children was particularly tricky: to avoid risks to their physical and mental development, as well as to their education and socialization, they couldn't be kept in a time-isolation lab for weeks at a time. To understand how sleep problems affected kids in the real world, she would have to find new ways to measure both their slumber and their waking lives.[9]

The scene that greeted Carskadon's arrival in September 1985 seemed to embody the challenges she faced. Hurricane Gloria had torn through Providence the day before, and downed tree limbs lay everywhere. The power was out in the rented condo that would be her new home, as it was through much of the state; with emergency crews filling the local hotels, she landed at a dingy motor lodge ten miles out of town. Days passed before she could access her lab at Bradley Hospital—a space in the basement consisting of a bedroom, a control room, an office for herself, and a smaller one for a part-time research assistant. The van transporting her files, research library, and equipment from Stanford (including three polygraph machines that Dement had gifted her) was delayed as well.[10]

Once she got physically settled, Carskadon came to a startling realization: She was virtually alone. At Stanford, she'd been surrounded by colleagues studying sleep from every angle. Now, aside from Anders, there was no one in the immediate vicinity who shared her passion. Sleep science was still a relatively small club, despite its accelerating growth, and she was its first member at this posting.[11]

It occurred to Carskadon that one way to ease her isolation was to light a spark in students. Although her position didn't come with teaching duties, she persuaded the psychology department to let her launch a course modeled on Dement's Sleep and Dreams. She also began looking for fellow travelers farther afield. Carskadon learned that Rhode Island Hospital, another Brown-affiliated facility, had just hired a pulmonologist named Richard Millman to establish a sleep disorders clinic. She invited him to lunch, and the two reached out to other researchers and clinicians across the region. The group they founded, the Northeastern Sleep Society, held its first meeting in April 1986. Over the years, the society's membership would grow from a few dozen to several hundred.[12]

Carskadon was flying solo in another sense as well. At thirty-eight, she was romantically unattached, without any intention of altering that status. "Research is my life," she later told an interviewer. "No partners, no kids."[13] As a teenager, she'd considered herself too unattractive to attempt dating. In her twenties, she'd decided that if the choice was between domestic and career fulfillment, she would choose the latter.[14] The contrast with her male progenitors was striking: Kleitman, whose wife proofread his manuscripts, and whose daughters served as experimental subjects; Aserinsky, whose son's eye movements first revealed the patterns of REM; Dement, whose spouse's social skills eased his passage to prominence and whose children added to his motivation for success. For men, family life stoked the fires of professional accomplishment. For women, the physics tended to work differently as the numbers attested: In 1985, they made up just 17 percent of the PhD-holding science labor force.[15] Carskadon aspired to build an alternate web of nurturance and affection—one that would not require her to sacrifice the lab for the kitchen or nursery.[16]

First, however, she would have to figure out how to begin the next chapter of her work. As with Kleitman when he started out in Chicago or Dement when he arrived at Stanford, it would take her some time to find an approach that would move the plot forward.

———

In the autumn of 1985, Dement was grappling with a different kind of disaster from the one that had paralyzed Providence. That October, the Senate passed the Gramm-Rudman-Hollings Act, which called for draconian budget cuts over the next six years, with the aim of reducing the federal deficit from $180 billion to zero by 1991.[17] Fearing that federal funding for sleep research would soon disappear, he set out to find a more reliable source of support. He placed his initial bet on private industry—specifically, on the manufacturer of Halcion, America's most popular sleeping pill.[18] Partnerships between university researchers and biotech companies were becoming common in the '80s, but this would be the first one involving sleep.[19]

Dement wasn't content merely to do more product testing for the Upjohn Company. Rather, he envisioned a center for basic as well as clinical sleep research, which would be funded by Upjohn in the interest of developing new medications and growing its market. He first broached the idea in November, on a visit to company headquarters in Kalamazoo.[20] His anxiety was apparent in the follow-up note he sent to Robert Purpura, head of the company's psychopharmacology unit. "I can see a few investigators surviving, but I am certainly not optimistic about federally funded multidisciplinary basic sleep research," he wrote. "[It] seems to me that if the Upjohn Company cannot be persuaded to invest in the future…other companies might be considered."[21]

Dement would eventually get his center.* But it was another catastrophe that set sleep science on a course toward financial sustainability—and real power in the public sphere.

On January 28, 1986, at 11:38 a.m., the space shuttle *Challenger* rose from its launch pad at Cape Canaveral, Florida, atop a pillar of flame. Although this was the *Challenger*'s tenth voyage, the presence on the crew of Christa McAuliffe, the first schoolteacher to venture into space, made the liftoff a gargantuan media event; across the country, some forty million people watched on television from their homes, offices, or classrooms.[22] In just over a minute, the 2,250-ton craft was nine miles above Earth, a speck against the clear blue sky. Then, without warning, it blew apart. As the vast audience looked on in horror, a fireball bloomed across their screens, shooting streamers of white smoke.[23] The seven astronauts survived for almost three more minutes until the intact crew cabin slammed into the Atlantic.[24] Debris was scattered over hundreds of square miles.[25]

That June, a presidential investigative commission identified the disaster's cause: A gasket known as an *O ring*, which sealed one of the rocket boosters, had failed due to the cold weather at launch time. Yet human error was involved as well. Engineers at Morton Thiokol, the company that built the boosters, had long warned that the O rings could malfunction below fifty degrees. The night before the launch, meteorologists had predicted subfreezing temperatures—rare in Florida—and a debate erupted during a phone conference between the engineers and NASA officials over whether to scrub it. According to the commission report, the

---

* The Stanford-Upjohn Center for Sleep/Wake and Circadian Processes launched in 1988. In 1994, battered by falling sales and lawsuits over Halcion's psychiatric side effects—which some users claimed had spurred them to kill loved ones—the company declined to renew the contract. Author interview with H. Craig Heller, who co-led the center; Anne K. Finkbeiner, "Getting Through the Sleep Gate," *Sciences*, September/October 1998, 14–18; "Upjohn Wins Another Legal Challenge over Halcion," *Los Angeles Times*, June 3, 1994, https://www.latimes.com/archives/la-xpm-1994-06-03-fi-48-story.html.

decision "should have been based on engineering judgments. However, other factors may have impeded or prevented effective communication and exchange of information." A crucial one: two key NASA managers had been awake for twenty-three hours straight and had slept for less than three hours the night before that.[26]

The *Challenger* explosion traumatized a generation of schoolchildren and changed the culture of safety and accountability in the US space program.[27] Less widely noticed was how it altered the course of sleep science. The catastrophe served as the crystallizing event for a new conception of sleep as a public health issue—one that had to be addressed not merely through the development of better therapies but through transformations in social policy, cultural attitudes, and the way we collectively arrange our lives.

This paradigm had begun to take shape months earlier, hinted at in Dement's 1985 speech to the House Subcommittee on Labor, Health and Human Services, and Education. But the shuttle disaster pushed the threat of mass destruction to center stage as an argument for paying more attention to sleep. For the sake of economic and technological development, society had fostered practices detrimental to slumber; now, advocates asserted, those practices threatened the very progress they were meant to promote.

At the 1986 APSS meeting in Columbus, Ohio, Merrill Mitler formed a committee to explore the role of sleep and circadian rhythms in catastrophes caused by both medical crises and human error. The members included Dement, Carskadon, Charles Czeisler, and two other scientists whose interests dovetailed with the topic at hand:[28] University of Pennsylvania psychologist David F. Dinges, an authority on sleep restriction and napping;[29] and neuropsychologist R. Curtis Graeber, a fatigue management expert at the NASA Ames Research Center who'd contributed to the *Challenger* investigation.[30]

As the catastrophes committee got to work, other researchers were beginning to show how a disregard for humans' endogenous rhythms could lead to disasters large and small. An important set of findings involved patterns of sleepiness and alertness. It had long been recognized that humans were at their groggiest around the low point of their circadian body-temperature cycle—a period that typically occurred between 3:00 and 4:00 a.m.—with a lesser period of sleepiness in the midafternoon. But new data was emerging on the structure of those fluctuations and on their potentially devastating consequences in the real world.

In May 1986, Israeli psychologist Peretz Lavie published a study suggesting that the sleep-wake curve was bumpier than had been previously recognized. Lavie traced sleep tendency in unprecedented detail, using an approach even more rigorous than the ninety-minute day pioneered by Dement and Carskadon. He placed small groups of young adults on a twenty-minute sleep-wake cycle: seven minutes in bed in a darkened room, followed by thirteen minutes out of bed in a lighted room, over a twenty-four-hour period. One group was instructed to try to sleep during their time in bed; another was told to try to stay awake. A third group was placed on the twenty-minute schedule for an additional twelve hours, under the same conditions as Groups 1 or 2. When Lavie charted the groups' sleep-wake patterns, he found that they were remarkably similar. During each twenty-four-hour period, participants experienced two "sleep gates" when they were likeliest to doze off—roughly corresponding with the familiar nighttime and afternoon drowsiness periods. Each gate appeared with remarkable abruptness, and its timing was consistent for each participant when the tests were repeated at a later date. In addition, Lavie detected a previously unknown "forbidden zone," during which sleep almost never occurred. This began, on average, between 8:00 and 10:00 p.m.—two to four hours before the nocturnal sleep gate.[31]

These findings added key pieces to the puzzle of why shift work, jet lag, and other socially driven disruptions in sleep schedules were often so deleterious. It wasn't just that these changes conflicted with endogenous rhythms of sleeping and waking; they also collided with a biologically programmed period each day when it was particularly hard to fall asleep, and two periods when it was especially difficult to stay awake.

Lavie's findings echoed a discovery Carskadon had made during the last summer of Sleep Camp, when she studied young adults on "constant routine"—a regimen developed by chronobiologists to study circadian rhythms with as few external variables as possible. (Her study had not yet been published when Lavie's came out.) Participants were kept awake for forty hours, propped up in bed in a dimly lit room and fed portions of a nutrient drink every sixty minutes. During the last twenty hours of the experiment, polygraph readings showed that the subjects experienced occasional "microsleeps," fleeting episodes of slumber of which they were usually unaware. Carskadon found that these occurred most often during the afternoon and nighttime sleep-gate periods, and almost never within the evening forbidden zone. She detected a second forbidden zone as well, somewhat shorter and shallower, in the midmorning.[32]

Carskadon's study proved that afternoon sleepiness did not result from the effort of digesting lunch as many experts had assumed. It also showed that sleep-deprived people were likelier to drift off at two points in their daily cycle (around 4:00 to 5:00 a.m. and 2:00 to 3:00 p.m.), even under conditions meant to enforce wakefulness—and that they were unlikely to do so at two other points (10:00 to 11:00 a.m. and 9:00 to 10:00 p.m.), no matter their degree of exhaustion.[33]

In Czeisler's lab at Harvard Medical School, mathematical biologist Richard E. Kronauer and his grad student Steven Strogatz wondered how these "ultradian" rhythms—shorter than a day but longer than an hour—related to circadian rhythms. They analyzed over three hundred records from time-isolation experiments (at their own facility and elsewhere) during which participants' sleep-wake rhythms became desynchronized from their body-temperature cycles. When Strogatz graphed the frequency of sleep onsets over a twenty-four-hour period, he found two valleys, each two or three hours wide, centered approximately five hours after and eight hours before body temperature reached its daily minimum—a perfect match for the forbidden zones that Lavie and Carskadon had described. Strogatz also found that desynchronized subjects had two peaks of sleep onset: the already-known "zombie zone" around the temperature trough, and a second one nine or ten hours later. In people on a normal schedule, this would correspond to a siesta time of 2:00 to 3:00 p.m. The ultradian rhythms of sleepiness, it seemed, were driven by the circadian rhythms of body temperature.[34]

Soon after reaching these conclusions, Strogatz attended a lecture on single-vehicle truck accidents. There, he learned that researchers had found such accidents occurred most often between midnight and 6:00 a.m.; the second most likely time was between 1:00 and 4:00 p.m. These times, he recognized, lined up neatly with the chronobiological zombie zones. The fewest crashes occurred at 10:00 a.m. and 9:00 p.m., within the forbidden zones for sleep.[35]

In a 1987 paper in the *American Journal of Physiology*, Strogatz, Kronauer, and Czeisler drew together all these findings. Schedules that forced people to go to bed during their forbidden zones likely contributed to insomnia and sleep deprivation, they suggested; schedules that required them to do certain kinds of work during the sleep-gate periods increased their risk of errors or accidents. In many instances, such dangers could be avoided by adjusting the timing of activities to account for these ultradian cycles. Another possibility, however, was to develop clinical approaches that could help people match their internal rhythms to those of their environment.[36]

One such therapy, recently introduced by Czeisler, involved using precise doses of light to reset an individual's biological clock. Scientists had known for a quarter century that artificial light could speed up or slow down the circadian rhythms of many animals, but experiments on humans had failed to demonstrate a similar result. Then, in 1980, National Institute of Mental Health psychiatrist Alfred J. Lewy had shown that if the light was intense enough (at least 2,500 lux, about five times the level of ordinary indoor illumination), it would suppress secretion of melatonin, a hormone produced almost exclusively in nighttime darkness. Lewy's study proved that bright light affected the central circadian pacemaker in humans—the suprachiasmatic nucleus, which regulated not only the rhythms of melatonin production but also those of cortisol, body temperature, and sleep.[37]

In August 1986, Czeisler and his colleagues reported that they'd used light to rejigger the circadian cycles of a sixty-six-year-old woman with an unusually short body-temperature rhythm of 23.7 hours. In the time-isolation lab, she sat for four hours each evening facing a bank of sixteen four-foot fluorescent bulbs producing 7,000 to 12,000 lux—equivalent to outdoor sunlight around twilight. After two sessions, the woman's circadian rhythms had been delayed by a full six hours.[38] Further tests showed that light therapy could move the hands of a patient's inner clock in either direction: forward when used before bed, or backward when used just after awakening. By contrast, chronotherapy (the clock-resetting technique Czeisler developed in the '70s) could be used only to move the hands forward. Before long, Czeisler and his team would show that light therapy could also aid workers in adjusting to shift rotations on the job.[39]

Around the time of the *Challenger* disaster, other researchers were investigating orally administered melatonin as a way of resetting circadian rhythms thrown off-kilter by the pace of modern life.[40] Dement and his team were testing Halcion and other benzodiazepines for the same purpose.[41] All these studies reflected a growing awareness among sleep scientists that misalignments between patients' inner clocks and their alarm clocks were a principal cause of daytime drowsiness—a condition, surveys suggested, that affected up to 5 percent of the US population.[42]

For the field's leaders, the space shuttle explosion transformed the drowsiness epidemic from a worrisome trend into an urgent crisis. That shift was well underway by December 1986 when Dement chaired a session titled "The Neuropsychopharmacology of Daytime Alertness" at the annual meeting of the American College of Neuropsychopharmacology in Washington, DC. In their presentations, he and his colleagues—including Mitler and Thomas Roth, director of the sleep disorders center at Detroit's Henry Ford Hospital—discussed recent findings on sleep gates, forbidden zones, and microsleeps. They noted that daytime sleepiness had been implicated in large numbers of auto fatalities, aviation accidents, and industrial injuries. Nonetheless, Dement remarked, drowsiness was "unbelievably tolerated" in a nation that imposed strict measures against drunk driving.[43]

In a *Boston Globe* feature pegged to the conference, headlined WAKING UP TO A MODERN PLAGUE: DROWSINESS, an array of sleep researchers expanded on this theme. "We almost never see a fully alert person, except for children," Dement said. Along with inadequately treated disorders such as sleep apnea and narcolepsy, he explained, the relentless demands of industrialized society contributed to a "national sleep debt"—a shortfall exacerbated by overuse of substances meant to mitigate its effects, including alcohol, caffeine, and sedating prescription drugs.* The article quoted Carskadon on her studies of chronic sleepiness in adolescents and the elderly; David Dinges on the masses of Americans who habitually got less than seven hours of slumber; and Stanford researcher Richard Coleman on a sleep-deprived cockpit crew that had overshot Los Angeles International Airport by one hundred miles, almost running out of fuel over the Pacific before realizing their mistake.[44]

The *Globe* story also cited the draft report of the APSS catastrophes committee, which linked not only the *Challenger* tragedy but several nuclear power plant accidents—including that spring's calamitous Chernobyl meltdown—to errors by operators working between midnight and daybreak. The drowsiness plague even afflicted presidents, the reporter wrote. Ronald Reagan, known for his tendency to nod off during cabinet meetings, was "not alone among world leaders who have been observed struggling against sleepiness stemming from hectic tours across several time zones."[45]

In retrospect, this might have been an ideal moment for Dement to escalate his congressional lobbying. But other priorities clamored for his attention. He continued to negotiate with Upjohn over the proposed research center.[46] He was pressing the dean of Stanford Medical School to liberate the Division of Sleep Research from the confines of the Department of Psychiatry.[47] He was launching a jazz program at the university,

---

* Dement's deepening ties to the pharmaceutical industry didn't stop him from warning against sleeping-pill dependency. But he also insisted it was less widespread than commonly believed, citing studies showing that of the one-third of Americans who experienced insomnia, only 5 percent sought medical treatment of any kind. William C. Dement and Merrill M. Mitler, "It's Time to Wake Up to the Importance of Sleep Disorders," *JAMA* 269, no. 12 (1993): 1548–1550.

with saxophonist Stan Getz as its first artist in residence.[48] He was advising the Black Pre-Medical Association and hosting recruiting parties for the football team in his backyard.[49] He was traveling to conferences and invited lectures.[50] He was teaching Sleep and Dreams and mentoring grad students. He was writing journal articles, grant proposals, project reports, and correspondence—fundraising letters, thank-you notes, get-well cards, memos praising postdocs for a well-prepared presentation or chiding staffers for letting him lose track of time when he had a plane to catch.[51]

And then came the disaster that made all these matters seem unimportant.

———————

In February 1987, Dement's middle child, Elizabeth, was twenty-six years old and living in Salt Lake City, where she was beginning a career as an artist and designer. One day, she was walking along a busy street when her small dog slipped the leash and darted into traffic. Liz ran after it. Struck by a car, she landed on her head, sustaining massive brain injuries.[52]

Dement, Pat, and their children Cathy and Nick flew to Utah, where they gathered at Liz's bedside. Doctors initially gave her less than a 50 percent chance of survival, and she spent three weeks in a coma. When she awoke, she was paralyzed and unable to communicate, her muscles clenched in painful spasticity. After she had spent three months of acute care at the University of Utah Medical Center, her parents moved her to a rehabilitation facility in San Jose, California. She had begun to recover some motion in her left arm and leg; soon afterward, she entered a period of continuous involuntary movement and vocalization. Only in late August was she deemed stable enough to go home. By then, she could indicate yes or no by lifting a finger.[53]

The family set up a hospital bed in the living room, with an IV pole for nutrition and medication. Liz's first laugh since the accident came on her first day home, when Mary Carskadon, who'd flown in from Providence, took a spin in the patient's wheelchair; her second laugh came when Carskadon jumped fully clothed into the backyard pool. Over the following months, there were incremental improvements. By fall, Liz had learned to stand with support; she could eat solid food, enabling her stomach tube to be removed. In November, she began speaking again—single words, then full sentences, though what she said was often difficult to understand. At Thanksgiving, she was able to sit by the table for most of the meal. "She has developed an absolutely wonderful sense of humor," Dement wrote to Liz's doctor. "There is either some acceptance of her condition or she is not fully aware of her impairment. She can be entertained rather easily and appreciates it very much."[54]

For Dement, however, equanimity was harder to come by. That fall, he sent a note to his friends Woody and Debra Myers, former residents of Cedro House, who'd married and become physicians. (Woody was now health commissioner for the state of Indiana; Debra was a pulmonologist specializing in sleep medicine.)[55] "As you know," Dement wrote, "the past months have been incredibly difficult. In addition to the daily stress, panic, anxiety, etc., I found myself becoming very depressed at a certain point and losing all concerns and energy. Although I have managed to maintain a presence and to fulfill my family obligations, I stopped doing all sorts of other things, which included writing letters."[56] In a draft memo to Stanford colleagues, apparently unsent, he made the point more bluntly: "1987 has been the worst year of my life."[57]

At the ASDC meeting in September, held in San Francisco, the organization changed its name to the American Sleep Disorders Association (ASDA).[58] The event also marked another transition: Dement handed over the presidency to Tom Roth after twelve years in office. In recognition of his service, the association presented him with its highest honor, the Nathaniel Kleitman Award—a milestone he neglected to report to Kleitman until a few weeks later. "I was overcome with modesty the other day when you called to ask about Elizabeth," Dement explained. "She continues to make very slow progress and has very far to go to return to normal function. Nonetheless, our spirits are soaring because in the last couple of days (knock on wood) there has been a marked decrease in the daily pain which she experiences."[59]

Although he returned to the office full-time in October, his days continued to revolve around Liz's care. Each morning, he toiled in his study from 6:00 to 7:30. From 7:30 to 9:30, he and Pat did rehabilitation therapy with their daughter. From 9:30 to 10:00, he ate breakfast and showered. At 10:00 a.m., he headed across campus to the Sleep Disorders Center. At 4:00 p.m., he returned home for another therapy session. From 6:00 to 7:00, he did more work in his study. Then came dinner, followed by two hours of free time. He was in bed by 10:00, and the cycle started again eight hours later.[60]

"I don't know how it seems to you," he wrote in a staff memo, "but my schedule is now more regular than at about any time in my life."[61]

It would not stay that way for long. Dement was about to enter the most sociopolitically significant period of his career—as was Carskadon. The science they championed was on the brink of a new transformation.

18

Awakening

In March 1988, two years after the *Challenger* explosion, the APSS catastrophes committee finally published its blockbuster report—long delayed by Dement's family crisis—in the journal *Sleep*. Titled "Catastrophes, Sleep, and Public Policy," it was the first study to connect the dots between the risk of disaster, the epidemic of daytime sleepiness, and the societal forces that made that condition ubiquitous. And it called for change.[1]

The committee explained that two separate but interacting factors were often superimposed on the normal, two-peak pattern of sleepiness: sleep deprivation associated with shift work, and sleep disruption resulting from sleep disorders. The effects of such sleep loss were cumulative, creating a growing sleep debt. Most people coped with such a buildup through physical activity and stimulants such as caffeine. As the debt mounted, however, so did the danger that an individual would slip into a "microsleep"—potentially while conducting an activity in which a moment's inattention could be calamitous.[2]

Accidents were not the report's only topic. The committee also examined years of data on medical catastrophes and found a similar correspondence with daily fluctuations in alertness. Deaths from all causes peaked between 4:00 and 6:00 a.m., with a smaller summit between 2:00 and 4:00 p.m.. Likewise, heart attacks crested twice a day, though somewhat later—between 6:00 and 10:00 a.m., and less steeply between 6:00 and

8:00 p.m. The authors called for more research to identify the factors behind these patterns and the population groups at greatest risk.[3]

But the paper's most sensational findings concerned catastrophes—and close calls—arising from human error. The authors cited the findings of the *Challenger* commission, as well as numerous studies of nuclear mishaps, single-vehicle car crashes, truck accidents involving hazardous materials, meter reading errors at gas works, and emergency braking incidents on railroads. In all these categories, disasters and near disasters peaked between 1:00 and 8:00 a.m., when circadian rhythms hobbled alertness. There was a smaller uptick between 2:00 and 6:00 p.m., when afternoon drowsiness struck the sleep-deprived.[4]

"The committee recognizes that inadequate sleep, even as little as 1 or 2 hours less than usual sleep, can greatly exaggerate the tendency for error during the time zones of vulnerability," the report concluded. The team recommended that industries and services address both the physiological needs of their personnel and the safety requirements of society at large. They urged management to limit working hours to ensure that employees got enough sleep between shifts and to implement educational programs aimed at fostering "the physiologically sound use of scheduled rest time." In addition, they wrote, further research was needed to develop "effective countermeasures to minimize the often catastrophic consequences of sleep-related brain processes. Such countermeasures will be essential if society is to continue its relentless push to around-the-clock operations in all aspects of life both on the planet and as we broach the frontiers of space."[5]

With the space shuttle and Chernobyl disasters still fresh in public memory, the "Catastrophes" report made a sizable splash. It was covered by newspapers across the country, with headlines like ALARMING LINK TIES DISASTER TO DROWSINESS and MAKE THE WORLD SAFER: TAKE A NAP.[6] "Society has got to be aware of the body's natural processes," committee chairman Merrill Mitler told the *Los Angeles Times*. "In the old days, falling asleep could mean a bent fender. Now, with the world so complex and interdependent, the risks are just too great."[7]

Here, finally, was an argument that might sway Congress: simple, punchy, backed by clear evidence, and powered by the threat that the consequences of inaction could be costly. Dement and his allies had the ammunition for the next phase of their assault.

That summer, he and lobbyist Dale Dirks visited the office of Massachusetts senator Ted Kennedy to discuss strategies. Kennedy agreed to sponsor legislation creating a National Commission on Sleep Disorders Research, which would investigate the impact of sleep problems on society and make recommendations to the federal government.[8] Led by Dement, the commission held public hearings in eight cities in 1990 and 1991, where a parade of ordinary Americans described their struggles with disordered slumber.[9]* Witnesses with narcolepsy recounted how sleep attacks and cataplexy had circumscribed their lives. Those with severe apnea testified to falling asleep on the job and at the wheel.[10] A father spoke of an evening just after Christmas, when his wife found their baby daughter dead in her crib, a victim of SIDS. "She was my perfect little angel," he lamented. "In a matter of a few moments, her life of promise was gone, and our road of grief was just beginning."[11]

In September 1992, Dement submitted the commission's report to Congress, calling for the establishment of a National Center on Sleep Disorders Research at the NIH.[12] And the following May, Congress passed a bill doing just that.[13] In June 1993, when Dement showed a video of the signing ceremony at the APSS meeting in Los Angeles, the 1,700 attendees stood and cheered.[14]

It took two or three years for the center to become fully operational. Between 1996 and 2003, however, NIH spending on sleep research nearly tripled, from $76 million to $203 million.† The number, scale, and

---

* The commission's ten members included Carskadon; James P. Kiley, chief of the airways branch of the National Heart, Lung, and Blood Institute; pulmonologist Norman Edelman, dean of the Robert Wood Johnson Medical School at Rutgers University; and former Dement student (and Cedro House resident) Debra Myers, who'd become head of the sleep center at Methodist Hospital of Indiana.

† The center's budget has since risen to almost $500 million.

ambition of sleep studies expanded accordingly. Dement's vision of sleep science as a thriving, independent discipline was coming true at last.[15]

Meanwhile, Mary Carskadon was also finding what she needed to move forward. In the summer of 1988, she received her first independent research grant—a $6,600 Trustee's Award (about $16,000 today) from the Grass Foundation. What made the award doubly meaningful for Carskadon was its connection to the history of sleep science: She'd been nominated by EEG pioneer and former Bradley Hospital psychologist Donald Lindsley, a trustee of the foundation since the 1950s. She also received $89,790 from the National Institution of Mental Health as a subcontractor for a study of sleeping and waking behavior in adolescents, with Dement as the principal investigator.[16]

By then, Carskadon had found a cheap but effective way to study the real-world impact of sleep disruption on kids: conducting surveys. A few months earlier, she'd begun publishing a series of papers based on preliminary data from questionnaires distributed to more than 4,500 students at New England public and private schools.[17]

Carskadon was aided in her investigations by the research assistant she'd brought along from Stanford, Joan Mancuso; by a growing cadre of undergrad volunteers from her sleep course at Brown; and by her first postdoctoral fellow, Mark Rosekind. The son of a San Francisco motorcycle cop who was killed in the line of duty, Rosekind had caught the sleep-science bug as a Stanford sophomore in Dement's Sleep and Dreams class. He'd later become a TA for the course and met his future wife when they both assisted another researcher on a study of sleep quality on waterbeds at Lambda Nu House during Sleep Camp. After college, Rosekind had briefly managed Dement's human research lab before heading to Yale for a PhD in clinical psychology and psychophysiology. Toward the end of his studies there, he'd gotten a call from Carskadon: "What are you doing next?"[18]

Rosekind arrived in Providence with his wife, Debra Babcock—a pediatrician who'd left behind a good job in Connecticut—and their toddler son. Carskadon helped the couple get settled in a nearby suburb and worked her clinical connections to find a practice for Deb to join. "It was a new start for everybody, and everybody got adopted," recalled Rosekind, whose subsequent career would include stints as director of NASA's Fatigue Countermeasures Program, member of the National Transportation Safety Board, and administrator of the National Highway Traffic Safety Administration. "Mary was looking after us, making sure we were going to be okay." She mentored, he added, as she had been mentored. Like Dement, "she was tough to work for, because she had such high standards. But it was all part of a family with her."[19]

The summer of 1988 marked another milestone for Carskadon—she was starting to see patterns in those survey responses. She presented two disquieting findings in June, at the APSS meeting in San Diego: 30 percent of high school students reported falling asleep in class at least once a week, and 50 percent had dozed off at some point during the last year.[20]

The hints of a crisis that she'd first sensed at Sleep Camp were popping up in the outside world.

In early 1990, Carskadon published her first two major papers on the potentially calamitous impact of typical school-year sleep schedules on young people's lives. The first, blandly titled "Patterns of Sleep and Sleepiness in Adolescents," was published in the journal *Pediatrician*. Drawing on data from her Sleep Camp studies and her recent school surveys, the paper examined how a range of factors affected teenagers' sleeping patterns, and how those patterns affected their behavior, performance, and mood.[21]

The big news had to do with school start times and after-school activities. The custom in most public school districts, Carskadon observed, called for earlier starting times and therefore earlier rising times as children got older—even though, as she had previously shown, teens needed *more* sleep than younger kids. Her new data indicated that students adjusted to these shifts by shortening the interval between getting up and leaving the house, risking tardiness to squeeze in a few more minutes of slumber. For many, however, that was far from enough to make up for the deficit.[22]

Teenagers with after-school jobs, Carskadon reported, had it especially hard. Those who worked twenty or more hours per week, dubbed the "high-work" group, stayed up later and slept less than their "low-work" counterparts. They reported more symptoms of daytime sleepiness, including a greater tendency to get to school late because of oversleeping and to fall asleep in class. Among low-work boys, for example, 20 percent dozed off in the afternoon at school at least once a week; 24 percent of high-work boys did so. Alarmingly, high-work teens also reported greater use of substances such as caffeine, tobacco, and alcohol. The proportion of students who smoked cigarettes every day was 33.9 percent for high-work girls compared with 24.4 percent for low-work girls; the proportion of students who drank alcohol every day was 25.5 percent for high-work boys versus 16.7 percent for low-work boys.[23]

"By far the most striking conclusion from the available information about adolescent sleep," Carskadon wrote, "is that many adolescents do not get enough sleep. School schedules require waking up earlier and earlier, while academic work loads, social obligations, and work patterns obligate staying up later and later. Compressed between these daily bookends is an increasingly narrow window for sleep."[24]

Carskadon also revealed evidence of a link between insufficient sleep and emotional problems. She and her team had asked thirteen boys in grades nine through twelve to reduce their sleep by two hours a night over five consecutive nights. The teens completed a standardized mood checklist each evening and filled out a questionnaire on depressive symptoms once a week. Both measures showed significant negative changes. "Although only a small preliminary study, this finding suggests that a portion of the moodiness of adolescents may be a consequence of insufficient sleep," she noted. Poor sleep patterns might interfere with teens' ability to cope with daily stresses and impair their relationships with peers and adults.[25]

She expanded on these findings—and their frightening implications—in her second groundbreaking paper of 1990, titled "Adolescent Sleepiness: Increased Risk in a High-Risk Population." Published in the journal *Alcohol, Drugs and Driving*, the article focused on motor vehicle use, and added illicit drugs to the list of substances that might affect the performance of sleep-starved teens.[26]

This paper was based on Carskadon's survey of 2,300 Rhode Island high school students. About 60 percent of these students held part-time jobs, and many participated in extracurricular activities, such as sports, clubs, musical groups, or service organizations. Carskadon divided the students into four cohorts. Group 1 consisted of kids who worked and participated in extracurriculars for less than twenty hours a week (low extracurricular, low work). Group 2 represented students who participated in extracurriculars for at least twenty hours but worked for less than twenty hours (high extracurricular, low work). Group 3 contained students who participated in extracurriculars for less than twenty hours but worked for at least twenty hours (low extracurricular, high work). Group 4 comprised students who participated in extracurriculars *and* worked for at least twenty hours a week (high extracurricular, high work).[27]

Carskadon found that sleep time on school nights was greatest in Group 1 (averaging 7.7 hours for boys and 7.5 for girls) and lowest in Group 4 (7 hours for boys and 6.6 for girls). Conversely, daytime sleepiness, substance use, and hazardous driving were lowest in Group 1 and highest in the cohorts with the greatest time commitment to extracurriculars and/or work. Among girls in Group 4, 8.3 percent used cocaine at least once a week—fully twenty times the rate for Group 1. Weekly marijuana use climbed from 7.6 percent of boys in Group 1 to 11.6 percent in Group 4. About 35 percent of girls and 25 percent of boys in Group 4 reported sometimes struggling to stay awake while driving a car versus 10 percent of girls and 15 percent of boys in Group 1. And a terrifying 10 percent of boys in Group 4, compared to 2 percent in Group 1, said they'd actually fallen asleep while driving.[28]

The last section of the article presented a composite portrait of the members of each cohort, adding a human dimension to the statistics. Perhaps the most poignant was the profile of a teen in Group 4, a college-bound student who devoted many hours each week to a job and extra-curricular pursuits. This student's susceptibility to crashes "may manifest on the way to school in the morning, having been dragged out of bed after only six hours of sleep; or it may occur on the way from school to an after-school job ... or a catastrophe may strike late at night on the way home from work or a party," Carskadon wrote. "This bright, energetic young person may succumb to a life style that has set him or her up for disaster."[29]

Carskadon concluded with a summation and a warning. Teens were naturally inclined toward risk-taking, she noted, but excessive sleepiness reduced the margin for safe experimentation—adding to the peril posed by a can of beer or a puff of weed. "As long as the common perception holds that adolescents 'need' less sleep, as long as teenagers remain ignorant of the principles of proper sleep/wake function, as long as there is a failure to understand and acknowledge the risks of excessive sleepiness, so long will a large number of youngsters be needlessly vulnerable to tragic accidents."[30]

These studies raised a question that scientists alone could not answer: How to engineer a better fit between teens' sleep needs and society's demands?

———————

Soon afterward, Carskadon made another crucial advance in her quest to understand how teens' sleep schedules affected their cognitive and mental health. To solve that puzzle fully, she needed to address a riddle that had dogged parents through much of human history: Why do teenagers insist on staying up so late, and why is it so hard to get them out of bed in the morning? If teens' so-called delayed phase preference were driven by social forces, as many experts believed, they should be able to adapt to the rhythms of a typical school day by simply turning in earlier. If biological factors dictated their phase preference, as Carskadon had long suspected, it would be wiser to adjust the school day to teens' inner clocks.[31]

To test the latter hypothesis, Carskadon designed her most elaborate survey yet. This one targeted *pre*teens, an age group in which many children metamorphose into adolescents. In the January 1991 teacher's issue of *Super Science Blue*, a science magazine for fourth through sixth graders, she published a note asking readers to contact her if they were interested in having their students participate in a research project examining physical maturation and body clocks. More than ninety arranged for their pupils to take part, at seventy-eight schools across the country. Students completed a questionnaire about their own sleep-wake patterns, then took home a permission form and question sheet for their parents to fill out. Data was collected from 955 boys and 1,060 girls. Of these, 238 boys and 313 girls were sixth graders, eleven or twelve years old—kids likely to be entering puberty.[32]

The children's questionnaires were designed to tease out respondents' "morningness" or "eveningness"—terms used by sleep scientists to describe a person's phase preference. For example: "Imagine: School is canceled! You can get up whenever you want to. When would you get out of bed?" The multiple-choice answers ranged from 5:00 a.m. to noon. Other questions concerned participants' energy level at different times of day, what time they performed best on exams or in gym class, and what time they started feeling ready for bed. Participants were also asked about their actual bedtime and wake time, the reason they typically went to bed (for example, "My parents set my bedtime," or "My TV shows are over"), and their physical development.[33]

When Carskadon and her team tallied the responses, they pointed strongly to biology—especially for girls, who tend to reach puberty earlier than boys. Sixth-graders scored higher on eveningness and lower on morningness than younger children, even though they occupied the same social milieu as their younger classmates. This pattern was more pronounced in kids whose physical development was more advanced. These children attended the same schools as prepubescent children, and their parents still largely dictated their bedtimes, yet their slide toward teenage phase preferences was unmistakable.[34]

The study was published in April 1993, in the journal *Sleep*. "If biological factors in early adolescence initiate a phase preference delay," Carskadon wrote, "then certain assumptions about teenagers' sleep patterns may need to be reexamined." The most important of those assumptions, she added, concerned school schedules. The widespread practice "for the opening bell to ring earlier at high schools than junior high schools, and earlier in junior high schools than primary schools, may run precisely counter to children's biological needs."[35]

Carskadon's paper made headlines nationwide and was picked up overseas as well.[36] (BODY CLOCK BLAMED FOR LIE-IN TEENAGERS, trumpeted London's *Daily Telegraph*.)[37] Her research was beginning to find a new audience: worried parents.

Over the next couple of years, Carskadon developed more direct ways to gauge the impact of physical development on adolescents' sleep needs. She began to use a new, chronobiological technique: measuring melatonin levels in saliva, which rise each evening and fall in the morning in a rhythm driven by the brain's circadian clock. Because bright light can suppress melatonin secretion, Carskadon kept participants in a dimly lit room for several twenty-four-hour cycles, taking spit samples at regular intervals. In a study of eleven- to fourteen-year-olds, she found that melatonin rhythms grew later as they entered puberty—confirming the evidence of a biological shift that she'd found in her earlier investigations.[38]

She also launched a study using the MSLT to see what happened when teenagers moved from ninth grade at a middle school whose start time was 8:25 a.m. to tenth grade at a high school whose start time was 7:20. Even the ninth-graders showed signs of a chronic sleep debt, falling asleep within an average of 11.4 minutes during the four daily tests. For the tenth-graders, however, average sleep latency was just 8.5 minutes—bordering on the pathological. Scarier still, nearly half of the tenth-graders had at least one sleep-onset REM period during the MSLT, a symptom that would ordinarily serve as a possible warning sign of narcolepsy. By comparison, 12.5 percent of the ninth-graders, and fewer than 7 percent of younger kids in Carskadon's studies, had SOREMPs.[39] "The students may be at school, but their brains are at home on their pillows," she later wrote.[40]

Carskadon reported on her early results at the 1995 meeting of the APSS, held in Nashville. By then, she had begun sharing her concerns with experts outside the sleep science community as well. In recent months, she'd testified to the National Institute for Occupational Safety and Health (NIOSH) on "Adolescent Sleep Patterns: Associations with Work Schedules and Job Safety." She'd also spoken at a symposium on school start times at Technion University in Haifa, Israel, at the annual meeting of the Society for Research in Child Development in Indianapolis, and at the International Symposium on Night and Shiftwork in Ledyard, Connecticut.[41]

But her most consequential lecture may have been one she gave two years prior, in 1993, at the Minnesota Regional Sleep Disorders Center in Minneapolis. In the audience was Maurice Dysken, whose teenage daughter attended a school that started at 7:20 a.m., requiring her to board a bus before dawn. "You could see the stars at that hour," he recalled. "I thought, 'This is nuts.'" Besides being a parent, Dysken happened to be president of the Minnesota Psychiatric Society. After the talk, he spoke with the center's director, neurologist Mark Mahowald, about how they could alert the community to the toll taken by early school start times. They decided to raise the issue with the Minnesota Medical Association, which quickly passed a resolution calling for classes to begin no earlier than 8:30 a.m. That fall, the MMA sent a mailing to all the state's school superintendents, informing them of its recommendation and explaining the rationale behind it.[42]

In 1994, when the association sent out a follow-up questionnaire, they found that no one had acted on the advice. That mailing, however, caught the eye of Ken Dragseth, superintendent of the Edina School District near Minneapolis. He had recently been approached by Edina High School's principal and health teacher, who'd urged that start times be pushed back after discussing the idea with other faculty and students. For Dragseth, the medical association's recommendation confirmed his own gut feeling that a change was needed.[43]

Getting other stakeholders to go along, however, was another story. School start times before 8:30 or 9:00 a.m. were a relatively recent development in the United States, sparked by postwar suburban sprawl and accelerated by the energy crisis of the 1970s, when communities found they could save fuel costs by using the same buses for elementary, middle, and high schools on tiered shifts. Yet they'd become ubiquitous by the '90s—and they were difficult to alter once they were in place. Commuters built their morning routines around getting the kids out of the house early. Sports teams played in the early afternoon, when classes ended; if one school changed its schedule, whole leagues would have to adjust theirs. Beyond all that, daytime sleepiness had come to be seen as normal for teens—the inevitable price of balancing schoolwork and extracurriculars with an active social life. Parents often reacted defensively when told that the status quo might be harming their children.[44]

Dragseth led an outreach campaign to parents, coaches, and others who would be affected by any schedule changes. After some heated debates, most opponents came around. The district evaluated twenty-two different scenarios, then settled on a plan that would keep transportation costs from rising and have the least impact on other schools' schedules. And in the fall of 1996, three years after Carskadon's talk, Edina High School moved its start time from 7:25 to 8:30—the first school in the country to act on the data that she'd begun amassing two decades earlier.[45]

The immediate results were promising. Dragseth heard from teachers whose students were more alert and from custodians who found fewer Coke cans and empty coffee cups in the hallways. "I love it," a varsity football and basketball player told the local newspaper. His grades had risen, and he no longer felt like a "zombie" during first period.[46]

Hoping to spread the experiment farther, Dragseth contacted Kyla Wahlstrom, associate director of the Center for Applied Research and Educational Improvement at the University of Minnesota. In consultation with Carskadon, Wahlstrom launched a longitudinal study of the new policy's impact in Edina, as well as a survey of existing conditions in seventeen neighboring districts. Meanwhile, the Minneapolis school district—also working with Wahlstrom and Carskadon—began investigating the possibility of a change for its fifty-one thousand students.[47]

By May 1997, preliminary results from the Edina High study showed that fewer kids were seeing counselors or nurses for depression or physical complaints. The principal reported that the halls and cafeteria were calmer. Parents said their kids were easier to live with. Encouraged by those findings, the Minneapolis school board voted to delay the starting bell from 7:15 to 9:40 a.m. at middle schools and to 8:40 at high schools.[48]

A movement had been born. It would slowly spread across the country—and around the world. Carskadon would remain at its scientific center, joined by a growing cadre of researchers whose work would have been impossible without her own.[49]

Carskadon (seen here in 2007) mentors a lab assistant. *Courtesy of Mary A. Carskadon.*

In June 1995, at the same APSS meeting in Nashville where Carskadon gave her talk on the teenage sleep crisis, the organization celebrated
Nathaniel Kleitman's one-hundredth birthday with a symposium at the
Opryland Hotel. Bill Dement acted as master of ceremonies; the speakers
were Eugene Aserinsky, Michel Jouvet, and the guest of honor himself. A
crowd of two thousand filled the banquet hall.[50]

Too frail to stand at the podium, Kleitman delivered his lecture from
the speaker's table. His face, framed by thick glasses, looked almost
mummified, but his mind remained astoundingly clear. He delivered
a twenty-minute disquisition that began with his apprenticeship with
Anton Carlson and raced through what he saw as the highlights of his
long career: his early sleep-deprivation experiments, his lab's discovery of
REM sleep, and his formulation of the theory of the basic rest-activity
cycle—the ninety-minute rhythm of alertness that he believed to be continuous through sleeping and waking. Remarkably, there was no mention
of the famous cave expedition, the monumental textbook, or the decades
as sleep science's only full-time practitioner, toiling to shape the field into
a coherent form.[51]

What was stunning about the speech (besides the fact that a centenar-
ian was able to deliver it) was its radical modesty. At bottom, Kleitman
had always seen himself as a simple researcher, sitting at his lab bench and
trying to interpret the data. If he sensed something larger in his mission—
its grandeur, its audacity—he kept it to himself. His audience perceived it,
however. Their applause was long and warm.[52]

Next, Dement invited Aserinsky onstage. The last time the pair had
met, at the 1965 APSS conference in Bethesda, Maryland, his former
collaborator had been testy and brusque. This time, he'd been somewhat
friendlier, perhaps because Dement had brought along his copy of Aserinsky's PhD thesis and asked for its author's autograph.[53] Still, that gesture
couldn't erase the painful history between the two men. Dement had

taken Aserinsky's discovery and leveraged it into a magnificent career. Aserinsky, meanwhile, had continued to struggle. After returning to sleep science from self-imposed exile, he had trouble getting grants and collaborating with other scientists. Although he became a well-loved professor and chair of physiology at Marshall University in West Virginia, retiring in 1987, he'd never gotten over his resentment at the way things had worked out.[54] "He spent the rest of his life feeling like somehow he had found the pot of gold and dropped it," his son, Armond, later recalled.[55]

With his black coat and tie, black mustache, and fringe of black hair, Aserinsky seemed as tightly coiled as a matador entering the ring. As he told his own version of the discovery of REM sleep, his voice had the flat calm of someone willing himself not to scream. His words, however, were magnanimous, self-deprecating, and often funny. He recounted his tedious early studies of blinking in infants, and his desperate battle to master the ancient polygraph machine he'd salvaged from Abbott Hall's basement. He gently teased Dement about the night he dozed off in the sleep lab while his subject was REMing in the next room. And he concluded with a tribute to his old adviser. "Just as Dr. Kleitman had selected me as a graduate student, I had chosen him," Aserinsky said. "I made a good choice. I hope he thinks he made a good choice, too."[56]

The room erupted in a standing ovation. Aserinsky beamed. The prodigal son of sleep science had finally been welcomed home.[57]

Jouvet had been given a tough act to follow, but the seventy-year-old Frenchman had never lacked for confidence. Dapper in a blue blazer with brass buttons, he traced his own path to the discovery that animals are paralyzed during REM sleep—the insight that spurred Dement to switch his focus from REM's psychological impact to its physiological mechanics, and set him on the path to inventing sleep medicine.[58]

After Jouvet finished speaking, hotel workers wheeled out a giant artificial cake with light bulb candles, and Dement led everyone in two choruses of "Happy Birthday to You." At the reception that followed, there was a real cake. Admirers swarmed Kleitman, asking him to sign copies of

Sleep and Wakefulness and pose for photographs. They did much the same for Aserinsky.[59] "It was one of the highlights of his life," his daughter Jill Buckley later said.[60]

At his one hundredth birthday celebration, Kleitman shared the limelight with (from left) Jouvet, Dement, and Aserinsky. *Courtesy of the Hanna Holborn Gray Special Collections Research Center, University of Chicago Library.*

Aserinsky's life ended three years later, in July 1998, when he drove his car into a tree near his home in Escondido, California. He was seventy-seven, an age when people are prone to daytime drowsiness; it's possible that he fell asleep at the wheel.[61]

Kleitman followed him in August 1999, at 104, succumbing to pneumonia at a rehabilitation center in Beverly Hills.[62] He had never quite accepted the importance of the accidental discovery Aserinsky made under his tutelage. "Kleitman died," Dement said, "still believing there was only one state of sleep."[63]

Yet the forces that both men unwittingly set in motion were gaining momentum as never before. In the quarter century between that celebration in Nashville and the present day, sleep science's breakthroughs (whether conceptual, clinical, institutional, or societal) have come every two or three years—and often more frequently than that. The field's burgeoning success can be measured in many ways, including the numbers of researchers and clinicians involved. The annual APSS meeting now draws twice as many attendees as it did back then.[64] There are approximately 7,500 board-certified sleep specialists in the United States alone, three times the figure for 1995. Membership in the American Academy of Sleep Medicine, as the former ASDA is now known, has more than quadrupled to 11,000.[65]

Still, the challenges facing sleep science remain vast. The most basic questions about slumber still lack definite answers. As Jouvet put it in his book *The Paradox of Sleep*, "[W]hy has evolution built a brain that…is periodically subject to a mechanism that delivers fantastic images, paralyzes our muscles, suppresses our homeostatic systems, and gives us an erection?"[66] Therapeutic puzzles persist as well, including the cures for devastating sleep disorders such as narcolepsy and fatal familial insomnia. Common insomnia can be almost as daunting: Even the most advanced medications don't work well for all patients, nor do alternatives ranging from herbal tinctures to melatonin capsules to behavioral therapies. For sleep apnea, CPAP is far from a perfect solution; many patients find it so unpleasant and inconvenient that they quit after a few nights. Its alternatives have drawbacks, too.[67] And there still aren't enough sleep doctors to go around; in the United States, the ratio is one for every forty-three thousand people.[68]

Perhaps the greatest challenge is society's ongoing, and ever-escalating, assault on sleep. Despite decades of studies showing that adults need seven to nine hours for optimal health, large swaths of the world's population get less than the recommended minimum. Among low-income Americans, especially in communities of color, the problem is worse. Our growing attachment to digital devices makes it harder to disconnect from waking consciousness, and the blue light from their screens throws our circadian clocks into confusion.[69] Around 20 percent of us are shift workers, but only a minority of workplaces have embraced rotations designed to reduce the health impacts of such schedules.[70] And despite continued progress toward later start times—including a 2019 law making California the first state to join the trend—over 75 percent of US high schools still drag kids to their desks before 8:30 a.m.[71]

The work is far from over. And so Kleitman's heirs toil on.

Epilogue

In August 2020, as the COVID-19 pandemic brought a halt to mass gatherings, the annual meeting of the Associated Professional Sleep Societies was held virtually, with participants logging in via Zoom from around the planet.* The centerpiece of the conference was a tribute to Bill Dement, who'd died of heart disease a month earlier, at ninety-one. The program was called Blues Night, in honor of the jam sessions that had long been his favorite event at the meetings—and in which he'd often played his stand-up bass.[1]

Between sets by the Chicago Blues All-Stars, an array of friends and colleagues shared their memories of Dement, who'd officially retired nearly two decades earlier but had continued to teach his Sleep and Dreams course into his mid-eighties. Mary Carskadon spoke of how he'd changed her life. So did Clete Kushida—who'd taken the class as an undergraduate, was guided by Dement as a novice researcher and clinician, and went on to succeed him as director of Stanford's sleep research and sleep medicine divisions. Other eulogizers included Mark Rosekind (who'd also taken Sleep and Dreams and wound up running the National Highway Traffic Safety Administration), sleep clinician Rafael Pelayo (who'd befriended the octogenarian scientist as a young professor and eventually took over teaching the course), psychologist Tom Roth and pulmonologist Meir Kryger (who'd created the first sleep medicine textbook with Dement), and psychologist James Walsh (who'd lobbied with him to establish the National Center on Sleep Disorders Research).[2]

---

* This organization, founded in 1986, is a joint venture of the Sleep Research Society (as the former Association for the Psychophysiological Study of Sleep is now known) and the American Academy of Sleep Medicine. It inherited its acronym from the old APSS.

Dement's daughter Cathy told how, on the night he began working on his first REM-rebound study in their apartment on Riverside Drive, her mother went into labor. "At that moment, fortuitously, Al Rechtschaffen knocked on the door," she recalled. When Rechtschaffen asked if there was anything he could do, Dement instructed him to record the subject's sleep while he rushed Pat to the hospital—where Cathy entered the world.[3]

Now, the family of sleep science had lost the last of its fathers.[4]

———

A few months later, Carskadon invited me for a glimpse of how sleep science's future was emerging from its past. She was still mourning her mentor, but she also had something to celebrate: She'd just been awarded a $10 million grant by the NIH to develop the world's first research center focused on the connections between sleep, circadian rhythms, and mental health in children and adolescents.[5] I flew out to Providence to visit the center's headquarters—the modest lab that had been Carskadon's professional home for three decades.[6]

The facility occupied a white clapboard house with black shutters, an overhanging second story, and a pair of small cottages attached at one end. Carskadon, seventy-two, met me in a conference room on the ground floor, where (in deference to the lingering coronavirus) we bumped elbows and sat at opposite ends of a long table before unmasking. Her pale gray eyes, framed by a round face and short white bangs, had an air of intense alertness; she brought to mind a snowy owl with a supernaturally high IQ. Those eyes showed a hint of tiredness, too. When I congratulated her on the grant, she said her first priority was to hire another senior faculty member to help mentor the early-career researchers who were besieging her with project proposals. "I mean, sleep is pretty hot right now," she said with a laugh. "My bandwidth is getting used up."[7]

The rationale behind the new Center for Sleep and Circadian Rhythms in Child and Adolescent Mental Health was simple: While it had long been known that sleep disruptions can exacerbate conditions ranging from depression to schizophrenia, and vice versa, little research had been done on the links between sleep and psychiatric disorders in kids. Understanding such connections was crucial to improving prevention, diagnosis, and treatment of such disorders in pediatric populations.[8]

The NIH award came from the agency's Centers of Biomedical Research Excellence (COBRE) program, which supports innovative, long-term studies aimed at strengthening an institution's research capacities. For Carskadon, that meant nurturing a new generation of investigators at Bradley Hospital, who would go on to share their knowledge and skills with the wider world. The COBRE program provides awards for three five-year phases. In phase one, the Bradley center aimed to sponsor two or three major studies annually, plus an equal number of small pilot projects. It would also host training in the assessment of pediatric mental health, sleep and circadian theory, and data collection and analysis, as well as career mentorship for project leaders. "Our main goal isn't to do research per se," Carskadon explained. "It's to fund young investigators to do research and to help them learn to spread their wings."[9]

So far, the center had taken on two studies. The first, led by epidemiologist Diana Grigsby-Toussaint, would use a novel approach to examine how access to green space, such as parks and forests, influences children's sleep and mental health. Groups of elementary school students from a range of socioeconomic and ethnic backgrounds would wear smartwatch-like devices known as *actigraphs*, which would gauge their sleep duration and timing, activity levels, and light exposure. The proximity of grass and trees would be detected using a GPS system programmed to identify areas of vegetation. By linking data from these two sources and from standard psychological tests, Grigsby-Toussaint hoped to learn how spending time in verdant environments affects kids' slumber, mood, and stress levels.[10]

The second study was headed by cognitive neuroscientist Jared Saletin—a former postdoc of Carskadon's who was now associate director of her lab. Saletin planned to investigate the circadian mechanisms influencing daytime sleepiness and cognitive problems in children with ADHD. His study would use cutting-edge technology (as well as some older techniques pioneered by Carskadon) to find associations between ADHD symptoms, sleep patterns, and brain chemistry. First, the kids' sleep and activity levels would be tracked using an Actigpatch—a new type of actigraph, developed by Carskadon's former doctoral student Eliza Van Reen, which takes the form of an adhesive patch that can be worn anywhere on the body. Then they would spend two nights in the sleep lab, undergoing polysomnography, MSLTs, dim-light melatonin sampling, and cognitive tests. Finally, Saletin would send them off to Brown University's neuroimaging center for magnetic resonance spectroscopy brain imaging, a technique that analyzes the chemical composition of scanned tissue.[11]

After providing a rundown of these state-of-the-art investigations, Carskadon gave me a tour of the century-old structure that would serve as their command center. Along the way, she pointed out artifacts from every stage of her journey through sleep research.

The stairway to the second floor was decorated with framed posters from APSS meetings dating back several decades—illustrated with a guitar for Nashville, a dolphin for San Diego, a phoenix for Phoenix. At the top, we ran into Saletin, a young man with a shaved head and a dark beard, wearing skinny jeans and flip-flops. "He's the future of sleep science," Carskadon said with a wry smile.

"That's a generous statement," Saletin demurred. But the objects he was carrying—an iPad and an iPod Touch, programmed to enable kids to do cognitive and reaction-time testing in their own homes—suggested that Carskadon was not just being kind.[12]

We ducked into an office under the eaves, where two research assistants were working at computer terminals, signing up subjects for trials. Then we headed downstairs to the basement, where the door to Carskadon's lab was marked by a glazed ceramic sign reading *Laboratoire du Sommeil*. It was a souvenir, she explained, from a road trip to southern France that she took with one of her postdocs, Helen Bearpark, after a conference in Italy in 1996. Bearpark, an Australian who'd done a groundbreaking study on the epidemiology of sleep apnea, was hit by a truck and killed later that year—another sleep researcher who didn't live to see the fruits of her work. "We went to this tile-making studio, and I was dithering, and Helen said, 'Mary, I most regret the things I didn't do, not the things I did,'" Carskadon recalled. "So I decided to have them make a sign for us." She paused, remembering. "Helen was awesome."[13]

Behind the door was the control room—long and beige, lined with blond-wood workstations. When Carskadon first moved into the building, in 1990, this room held two polygraph machines, one for each bedroom. Now the lab had four bedrooms and nothing that could be described as a machine. Instead, there were flat-screen monitors, hooked up to a specialized computer that processed and stored subjects' EEG readings and other physiological data. The walls held racks of multicolored electrode leads and digital thermometers. There was a dedicated refrigerator for saliva samples. But the feature that Carskadon showed off most excitedly ("My favorite place in the lab," she said)[14] was a closet full of games to divert kids during their stays for sleep tests—from checkers to Brain Quest, Candyland to Risk. She was also proud of her developmentally targeted collection of DVDs, which ranged from *Chicken Little* to *Spider-Man* to *Bridget Jones's Diary*.

Carskadon in the control room, 2011. *Courtesy of Mary A. Carskadon.*

The sleeping rooms themselves (decorated with child-friendly artwork from that trip to France) looked much as they would have thirty years ago: a twin bed, an armchair, and a nightstand holding an intercom and a jack box for electrodes. The two more recent additions were ceiling-mounted infrared cameras, so that subjects could be observed even in the dark, and a computer-controlled lighting system that shone at preset levels of brightness or dimness.[15]

Back on the ground floor, Carskadon stopped to greet two ebullient and clearly star-struck young women—participants in the Dement Fellowship program, which she established in 1997 to train undergraduate summer interns in the rudiments of sleep research. (Saletin is one of several former fellows who've gone on to join the field.) Then she led me into a small workroom, where she reached into a bin and pulled out another new device: a research-grade mobile EEG recorder, which enabled detailed sleep testing outside the lab. It consisted of a black headband with a processing unit at its center point and an electrode strip that snapped into the band. Carskadon fastened the gadget around her skull using a Velcro closure. "Jared's project will use these to look for sleepiness in the EEG," she said. "This thing is so simple, but it's also very expensive. So the downside is, if it gets wrecked, it's costly to replace."[16]

Our next stop was Carskadon's office, a sunlit space strewn with evidence of her current work—file folders, reports, stacks of loose paper, sticky notes—and studded with jewels of memory every few feet. Alongside the awards and certificates on the walls was a brightly colored painting based on MSLT charts from Sleep Camp. The bookcase held multiple editions of *The Elements of Style* (plus *The Elephants of Style*, a parody), which Dement had given her over the years. There were photos of Carskadon with her mentor (in one, she wore a fake mustache to match Dement's and Mark Rosekind's real versions), and with a gaggle of other pioneering women sleep scientists (including psychiatrist Ruth Benca, neuroscientist Chiara Cirelli, and zoologist Irene Tobler, whose bylines I'd seen on several seminal studies).[17]

The latter picture reminded Carskadon of a Zoom webinar on women in sleep science that she'd recently done with Benca, psychologist Sonia Ancoli-Israel, and several younger colleagues. "We started talking about, 'And then there was the time Jouvet did this or said that, and then Al Rechtschaffen did this or that,'" she recalled. "And at one point I turned to the other little squares on the screen and said, 'How many of you know what we're talking about when we say these names?' And they were just blank."[18]

In the classes she taught, Carskadon added, she made a point of invoking those and other progenitors. "That's a present I give my students."[19]

It's also a present she gave me, over the many hours of interviews she granted for this book.

As a parting gift that afternoon, Carskadon escorted me to her lab's annex just down the street—an identical white clapboard house that was used for classrooms and storage. In the unfinished basement were half a dozen polygraph machines, some of which she'd brought over from Stanford thirty-six years earlier. Clad in stainless steel, and bristling with knobs and switches, they suggested a cross between a recording studio mixing board and a gas barbecue grill. The oldest was a Grass Model 6 from the early '60s, but they were all antiques by now—relics of an age before flash memory, Wi-Fi, and the World Wide Web.[20]

"Bill said, 'Here, Mary, take these off my hands, because I don't want to put them in *my* cellar,'" she said. "Nobody else needed them. But *I* needed them."[21]

The machines bore a closer resemblance to the ancient Offner Dynograph with which Aserinsky discovered REM sleep than to the boxes of microchips that Carskadon relied on today. Yet all those devices—including the headband that she'd shown me—shared a common strand of DNA. A similar thread, I sensed, linked Carskadon to the sleep scientists who came before her and to those she was now mentoring.

But what was it? I pondered that question as we bumped elbows in farewell.

An answer came a few weeks later, from Aserinsky's son, Armond, now seventy-eight, who'd grown up to be a clinical psychologist. "I think they all had a burning desire to *find out*," he said. "And they were willing to go to hell and back to do it."[22]

Diagram

The Anatomy of Sleep

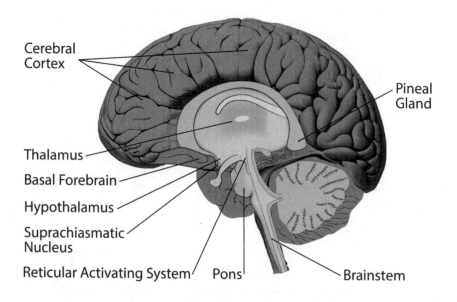

Cerebral Cortex

Pineal Gland

Thalamus

Basal Forebrain

Hypothalamus

Suprachiasmatic Nucleus

Reticular Activating System

Pons

Brainstem

Over the past century, scientists have identified numerous brain structures connected with sleep. These include:

- The **hypothalamus**. This almond-size structure contains several groups of nerve cells that act as control centers for sleep and arousal.
- The **suprachiasmatic nucleus (SCN)**, located within the hypothalamus, is the body's master clock. A cluster of about fifty thousand neurons, the SCN regulates many biological rhythms—including

temperature, hormone secretion, and sleep and wakefulness. Light signals from the eyes reset the clock to match the twenty-four-hour cycle of day and night.

- The **brain stem** communicates with the hypothalamus to control transitions between sleep and waking. Sleep-promoting cells within both structures produce the neurotransmitter GABA, which reduces activity in the brain's arousal centers.

- The brain stem also contains the **pons**, which plays key roles in REM sleep—triggering the brain activity associated with dreaming and inhibiting muscle activity to keep us from acting out our dreams.

- The **thalamus**. During waking, this structure relays sensory information to the cerebral cortex. During sleep, the thalamus is mostly quiet—except during REM, when it relays signals from the pons to the cortex, stimulating dreaming.

- The **pineal gland** (one for each brain hemisphere) receives signals from the SCN that trigger nightly production of the hormone melatonin, which helps bring on sleep.

- The **basal forebrain** is another important relay station for sleep and waking.

- The **reticular activating system** (in the upper brain stem area known as the *midbrain*) is the brain's main arousal center. It sends signals to the cortex that maintain wakefulness until other brain structures order it to suspend operations so that sleep can occur.

Appendix A

A Time Line of Sleep Science

~500 BC—Alcmaeon theorizes that sleep occurs when blood withdraws from the body's surface to internal vessels, filling the brain.[1]

~350 BC—Aristotle proposes that sleep occurs when vapors rise from the stomach to the heart during digestion.[2]

1729—Jean-Jacques d'Ortous de Mairan discovers that heliotropes open their leaves by day and close them by night, even when kept in a dark closet.[3]

1807—London introduces gas streetlamps—a prelude to the spread of cheap and plentiful lighting, and a massive change in human sleep-wake rhythms.[4]

1832—Augustin Pyramus de Candolle reports that plants under constant lighting open and close their leaves at intervals of twenty-two to twenty-three hours rather than following a twenty-four-hour cycle—the first description of an organism's "free-running" rhythm.[5]

1849—Observing cerebral circulation in animals, ophthalmologist Franciscus Cornelius Donders concludes that a reduction of blood flow in the brain triggers sleep.[6]

1862—Ernst Kohlschütter publishes the first study of sleep depth throughout the night.[7]

1864—Bromide salts, previously used to treat epilepsy, are adapted as sleep aids.[8]

1869—Chloral hydrate becomes the first chemically synthesized sleep medication.[9]

1880—Thomas Edison patents the first practical electric light bulb, enabling humans to fully sever sleep-wake cycles from the rhythms of night and day.[10]

1880—Jean-Baptiste-Édouard Gélineau identifies and describes narcolepsy.[11]

1880s—The term *insomnia* is first used to describe chronic sleeplessness.[12]

1890—Ludwig Mauthner theorizes that sleep is controlled by a nerve center in the midbrain.[13]

1891—Eduard Michelson shows that sleep depth fluctuates regularly over a nightlong session—the first observation of what became known as sleep stages.[14]

1894—Maria Mikhailovna Manaseina (a.k.a. Marie de Manacéïne) presents the first study of prolonged sleep deprivation in animals.[15]

1895—Lord Rosebery resigns as prime minister of Great Britain, citing severe insomnia.[16]

1895—Birth of Nathaniel Kleitman.

1896—George Patrick and J. Allen Gilbert publish the first study of sleep deprivation in humans.[17]

1899—Sigmund Freud publishes *The Interpretation of Dreams*.[18]

1903—Barbiturates are introduced as sleep aids.[19]

1909—Kuniomi Ishimori publishes the first study suggesting that chemicals produced by the brain trigger sleep.[20]

1913—Henri Piéron publishes *Le Problème Physiologique du Sommeil*, the most comprehensive book on sleep research to date.

1917—Constantin von Economo identifies and names encephalitis lethargica.[21]

1921—Birth of Eugene Aserinsky.

1923—Ivan Pavlov theorizes that sleep results from "internal inhibition."[22]

1923—Kleitman publishes his first paper, "The Effects of Prolonged Sleeplessness in Man."[23]

1925—Kleitman establishes his laboratory at the University of Chicago.

1926—Von Economo identifies the hypothalamus as the site of lesions in encephalitis lethargica.[24]

1928—Birth of William C. Dement.

1929—Kleitman disputes Pavlov's theory of sleep at the XIIIth International Physiological Congress.[25]

1929—Hans Berger publishes "On the Electroencephalogram of Man," the first description of human EEGs.[26]

1929—Von Economo theorizes that the hypothalamus contains a sleep center and a waking center.[27]

1932—Kleitman publishes the first study showing that humans move frequently during normal sleep.[28]

1933—Kleitman publishes the first study showing that cognitive performance correlates with body temperature.[29]

1933—Kleitman publishes the first controlled study of the effects of alcohol and caffeine on sleep.[30]

1935—Alfred Lee Loomis publishes the first major study of the EEG in sleep.[31]

1935—Frédéric Bremer publishes his "*cerveau isolé*" study, widely (but erroneously) interpreted as evidence that a lack of sensory input to the brain triggers sleep.[32]

1938—Kleitman's Mammoth Cave expedition makes national headlines.

1939—Kleitman publishes *Sleep and Wakefulness*, the first true textbook on sleep science.

1942—US Department of Labor publishes Kleitman's proposal for wartime shift-work schedules.

1947—Birth of Mary Carskadon.

1948—Kleitman spends two weeks studying shift-work schedules aboard the submarine USS *Dogfish*.[33]

1949—Kleitman's experimental shift-work schedule is implemented on the submarine USS *Tusk*, with unsatisfactory results.[34]

1949—Horace Magoun and Giuseppe Moruzzi identify a waking center in the reticular formation of the brain stem.[35]

1951–1952—In Kleitman's lab, Aserinsky makes the first observations of REM sleep.

1952—Dement begins assisting Aserinsky.

1953—Aserinsky and Kleitman publish the first paper describing REM sleep.[36]

1955–1957—Dement and Kleitman publish a series of papers reporting correlations between REM and dreaming.

1957—Dement and Kleitman publish "Cyclic variations in EEG during sleep," describing a regular sequence of REM and non-REM sleep stages throughout the night.[37]

1957—Thalidomide, prescribed as a sleep aid and treatment for morning sickness, is found to cause severe birth defects.[38]

1957—Dement begins his residency at Mount Sinai Hospital in New York City.

1958—Dement describes REM sleep in animals (domestic cats).[39]

1959—Michel Jouvet reports that REM (or "paradoxical sleep") in cats is accompanied by muscle atonia and is controlled by the pons.[40]

1959—Dement identifies the phenomenon of REM rebound.[41]

1959—Kleitman retires, bequeathing the University of Chicago sleep lab to Allan Rechstschaffen.[42]

1960—Gerald Vogel describes sleep-onset REM periods in a patient with narcolepsy.[43]

1960—Dement publishes the first study of REM deprivation.[44]

1960—The Ciba Conference on Sleep is the first international conference on the topic.[45]

1960—Cold Spring Harbor Symposium on Biological Clocks marks the birth of chronobiology.[46]

1960—Benzodiazepine sleep aids are introduced, beginning with chlordiazepoxide (Librium).[47]

1961—Jürgen Aschoff conducts time-isolation experiments to monitor human circadian rhythms.[48]

1961—Dement and Rechtschaffen establish the Association for the Psychophysiological Study of Sleep (APSS).[49]

1963—Dement launches his sleep research program at Stanford University.

1963—Dement and Rechtschaffen report abnormal timing of REM sleep in people with narcolepsy.[50]

1964—Dement begins studying REM deprivation in cats.

1966—Dement and Roffwarg report that infants experience a far greater proportion of REM sleep than adults.[51]

1968—Rechtschaffen and Anthony Kales coedit the first manual of standardized techniques and terminology in sleep research.[52]

1969—Franz Halberg publishes a paper titled "Chronobiology," inspiring the widespread adoption of that term for the study of biological rhythms.[53]

1970—Dement launches the first modern sleep clinic.

1971—Dement begins teaching his Sleep and Dreams course.

1971—Ronald Konopka and Seymour Benzer discover the first "clock" gene in *Drosophila*.[54]

1972—Christian Guilleminault becomes associate director of the Stanford Sleep Clinic.

1972—Researchers identify the circadian "master clock"—the suprachiasmatic nucleus (SCN) of the hypothalamus.[55]

1972—Richard Bootzin introduces stimulus control therapy, the first non-pharmaceutical treatment designed specifically for insomnia.[56]

1973—Carskadon and Dement publish their first "90-minute day" study.[57]

1974—Dement starts breeding the first research colony of narcoleptic dogs.[58]

1974—Elliot Weitzman begins collaborating with Dement and Charles Czeisler on chrono-biologically oriented sleep research.

1975—Weitzman launches the second modern sleep clinic, at Montefiore Medical Center in New York City.

1975—Guilleminault describes obstructive sleep apnea syndrome (OSA).[59]

1975—The Association of Sleep Disorders Centers (ASDC) is formed, with Dement as president.[60]

1976—Carskadon develops the Multiple Sleep Latency Test (MSLT).

1976—Carskadon launches the Stanford Summer Sleep Camp.

1976—Steven Glotzbach and H. Craig Heller publish a study showing that body's thermoregulatory system shuts down during REM sleep.[61]

1977—Peter Hauri introduces "sleep hygiene" for insomnia prevention.[62]

1978—The APSS begins publishing *Sleep*—the first research journal on the topic.[63]

1979—The first *Diagnostic Classification of Sleep and Arousal Disorders* is published.[64]

1980—Carskadon reports that adolescents need more sleep than younger children.[65]

1981—Carskadon and Dement publish "Cumulative Effect of Sleep Restriction on Daytime Sleepiness," quantifying the phenomenon of "sleep debt."[66]

1981—Colin Sullivan introduces continuous positive airway pressure (CPAP) to treat sleep apnea.[67]

1981—Shiro Fujita introduces uvulopalatopharyngoplasty for sleep apnea.[68]

1981—Czeisler describes delayed sleep phase syndrome and introduces chronotherapy.[69]

1982—Alexander Borbély publishes his "two process model" of sleep.[70]

1982—Czeisler et al. report on a chronobiologically based shift-work schedule for industrial workers.[71]

1982—Rosalind Cartwright and Charles Samelson introduce the first oral appliance for sleep apnea.[72]

1982—Kleitman publishes his last scientific paper: "Basic Rest-Activity Cycle—22 Years Later."[73]

1983—Rechtschaffen theorizes that the cause of death in prolonged sleep deprivation is breakdown of the thermoregulatory system.[74]

1985—Carskadon starts her lab at Emma Pendleton Bradley Hospital.

1986—Zopiclone, the first nonbenzodiazepine "Z drug," is approved in Europe.[75]

1986—Investigators report that the *Challenger* disaster was caused, in part, by sleep deprivation.[76]

1986—Peretz Lavie reports finding "gates" and "forbidden" zones for sleep.[77]

1987—Steven Strogatz and his team show that sleep gates and forbidden zones are tied to body temperature cycles.[78]

1987—Arthur Spielman introduces sleep restriction therapy for insomnia.[79]

1988—Merrill Mitler and colleagues publish "Catastrophes, Sleep, and Public Policy," detailing connections between sleep deprivation and disasters.[80]

1988—Congress passes legislation establishing a National Commission on Sleep Disorders Research.[81]

1989—Rechtschaffen reports that prolonged REM deprivation is fatal in rats.[82]

1989—Publication of *Principles and Practice of Sleep Medicine*, the field's first clinical textbook.[83]

1990—Publication of the first *International Classification of Sleep Disorders*.[84]

1990—Carskadon publishes a series of studies showing harmful effects of early school start times on adolescents.

1990—Czeisler and colleagues introduce light therapy to reset disordered circadian rhythms in shift workers.[85]

1990—NASA uses light therapy on crew of space shuttle *Columbia* to ease adaptation to night work.[86]

1990—The National Commission on Sleep Disorders Research begins holding hearings.[87]

1993—The first epidemiological study of obstructive sleep apnea finds that OSA affects 24 percent of men and 9 percent of women in the United States.[88]

1993—Carskadon finds evidence that delayed sleep phase in adolescents is driven by biological factors.[89]

1993—President Clinton signs legislation establishing a National Center on Sleep Disorders Research at the National Institutes of Health.[90]

1993—Charles Morin introduces cognitive behavioral therapy for insomnia (CBT-I).[91]

1995—The American Medical Association recognizes sleep medicine as a self-designated specialty.[92]

1995—Joel H. Benington and H. Craig Heller identify adenosine as the chemical primarily responsible for sleepiness after an extended period of waking.[93]

1995—Kleitman addresses the APSS in honor of his one hundredth birthday.[94]

1996—Edina, Minnesota, pushes back school start times.[95]

1997—Carskadon reports that nightly melatonin secretion begins later in adolescents than in younger children.[96]

1997—Minneapolis pushes back school start times—the first major city to do so.[97]

1998—Death of Aserinsky.[98]

1998—Researchers at the Scripps Research Institute and the University of Texas identify the orexins (a.k.a. hypocretins), a set of neurotransmitters that help regulate sleep.[99]

1999—Emmanuel Mignot at Stanford reports that narcolepsy in dogs results from a mutation in an orexin receptor gene, opening new paths for research on the disease in humans.[100]

1999—Death of Kleitman.[101]

1999—Carskadon becomes the first woman president of the Sleep Research Society (as the former Association for the Psychophysiological Study of Sleep is now called).[102]

1999—Eve Van Cauter reports that healthy young men who undergo sleep restriction for six nights metabolize glucose as poorly as diabetics.[103]

2000—Carol Everson and Linda Toth identify immune dysfunction as the primary cause of death in rats after prolonged sleep deprivation.[104]

2007—The American Board of Internal Medicine takes over exams for the sleep medicine subspecialty from the American Academy of Sleep Medicine, completing the field's incorporation into the US medical establishment.[105]

2009—The NIH's Sleep Heart Health study finds that mortality from all causes is higher in adults with sleep apnea, increasing with the condition's severity.[106]

2011—A poll by the National Sleep Foundation finds that US high school seniors sleep an average of 6.9 hours a night, down from 8.4 in sixth grade; 72 percent bring their cell phones to bed.[107]

2011—Two Maryland mothers—health journalist Terra Ziporyn Snider and blogger Maribel Ibrahim—establish a nonprofit called Start School Later, the first organization of its kind.[108]

2012—Neuroscientist Maiken Nedergaard discovers the "glymphatic" system in the brain, which flushes out toxic metabolites such as beta amyloid during sleep.[109]

2014—The FDA approves upper-airway stimulation (UAS), a less invasive surgical approach to obstructive sleep apnea.[110]

2014—The FDA also approves suvorexant for people with insomnia—the first sleeping pill designed around the orexin system.[111]

2014—Two large studies find improvements in attendance, grades, and test scores, and reductions in depressive symptoms, sick days, and automobile crashes in schools with later start times.[112]

2014—The American Academy of Pediatrics issues a policy statement recommending that high schools and middle schools ring the opening bell no earlier than 8:30 a.m.[113]

2014–2016—The American Psychological Association, the US Centers for Disease Control and Prevention, and the AMA follow suit.[114]

2019—California becomes the first state to pass legislation pushing back school start times.[115]

2020—Death of Dement.[116]

2021—Mary Carskadon launches the first research center devoted to sleep and circadian rhythms in child and adolescent mental health.[117]

Appendix B

Meanwhile, in a Parallel Universe:
A Brief Prehistory of Chronobiology

The roots of chronobiology stretch back to 1729 when French astronomer Jean-Jacques d'Ortous de Mairan observed that heliotropes opened their leaves by day and closed them by night, even when kept in a perpetually dark cupboard. De Mairan's experiment revealed the existence of what eventually became known as *circadian rhythms*, governed by a mysterious inner clock.[1]

The first scientist to replicate de Mairan's work was Swiss botanist Augustin Pyramus de Candolle, who in 1832 published a study of mimosa plants in continuous light. Like the closeted heliotropes, these plants unfurled their fronds each morning and folded them each evening without being exposed to sunrise or sunset. De Candolle found that the length of this cycle was slightly shorter than twenty-four hours and varied among individual plants. He theorized that the "sleep-wake movements are connected with a disposition for periodic motion that is inherent in the plant"—the first time a scientist explicitly attributed such behavior to an innate timekeeper.[2]

Charles Darwin endorsed that concept in 1880, in a book titled *The Power of Movements in Plants*. The British naturalist suggested that such rhythmicity might have some sort of evolutionary benefit and was likely inherited.[3]

In the 1920s, American psychobiologist Curt Richter reported that animals, too—specifically, lab rats—followed a daily cycle activity and rest in the absence of environmental cues. Richter coined the term *biological clock* to describe the mechanism that might drive these patterns.[4]

In the '30s, German biologist Erwin Bünning (a colleague of EEG inventor Hans Berger at the University of Jena)* described a similar cycle in fruit fly pupae and showed that their diurnal rhythms were passed down through the generations. To Bünning, this was clear evidence of genetically programmed clockwork. His work failed to get much attention, however, in part because his vocal opposition to the Nazis disrupted his academic career; after his führer-loving students hounded him from campus, he spent a year doing research in Indonesia and was drafted as a foot soldier when he unwisely returned.[5] Many experts, meanwhile, insisted that seemingly endogenous rhythms in animals resulted from as-yet undetectable environmental factors.[6]

Not long after Bünning published his fruit fly paper, Jürgen Aschoff—a young researcher at the University of Göttingen's Physiological Institute—began investigating his own temperature rhythms at his director's request. Aschoff, too, had to grapple with the moral quandaries facing scientists in Nazi Germany. A political moderate who detested Hitler, he nonetheless initially supported his country's war effort, seeing it as his patriotic duty. His early research, under renowned physiologist Hermann Rein, was aimed at improving survival rates of German military pilots and sailors in cold conditions. At a secret 1942 conference on that topic, both men learned that the SS was running lethal hypothermia experiments on prisoners at the Dachau concentration camp. Rein protested to higher-ups after the meeting, to no avail; if not for his international prominence and usefulness to the Reich as a scientist, his transgression might have landed him in a camp himself. Aschoff was deeply shaken—and terrified that he would be assigned similar projects. He was spared that degree of complicity with the regime's atrocities, perhaps thanks to his patron's protection.[7]

Aschoff's experiments involved plunging his limbs into ice water and taking his temperature with a rectal thermometer, with his wife acting as

---

* See Chapter 5.

"technician." After discovering a previously unreported diurnal cycle of heat loss in his extremities, he read everything he could find on the origins of such rhythms. He became convinced that there was only one way to be sure that a daily rhythm was endogenous: if it continued when an animal was cut off from the cues of day and night, but began to deviate from a twenty-four-hour cycle. This would show that the rhythm was primarily driven by genetic programming (which would likely be inexact when operating on its own) rather than by environmental factors.[8]

After the war, Aschoff made that issue a focus of his research.[9] Wondering whether diurnal rhythms might be imprinted by an animal's early experiences, he raised mice and chickens from birth under constant light. Even though they'd never known a moment of darkness, he found, their physiological and behavioral cycles still resembled those of their parents and grandparents. Yet there were small differences—just enough to show that the experimental animals were relying entirely on internal clocks. The subjects' "free-running" rhythms (the term for diurnal cycles occurring in the absence of external cues) averaged a little over twenty-five hours. Their cycles had become desynchronized from those of the outside world.[10]

For Aschoff, this posed another question: How did synchronization happen in the first place? Along with his own findings, experiments by other scientists offered clues. Botanists, he knew, had demonstrated that periodic pulses of light could reset diurnal rhythms in plants. And in mice raised under constant lighting conditions, the *degree* of illumination had been shown to influence the timing of their daily cycles. The rodents' rhythms were at their quickest under total darkness and slowed increasingly as the ambient light grew more intense. These disparate life-forms were configured not only to detect clock-setting stimuli, it seemed, but to respond in preset ways. Perhaps some common rules applied to both sides of the equation.[11]

In 1954, Aschoff published the first in a series of papers outlining his theory of "entrainment"—a term biologists borrowed from physicists to describe the process by which two oscillating systems assume the same rhythm. To match their rhythms to those of the rotating Earth, he suggested, organisms relied on cues he called *zeitgebers* ("time givers"). Light was probably the most prevalent *zeitgeber*, but temperature and other environmental factors could also fill the role. Such signals enabled an organism to entrain to the planet's twenty-four-hour cycle—or, within genetically dictated limits, to other cycles prevalent in its surroundings. (Those limits might explain why Nathaniel Kleitman's subjects were able to adjust only to rhythms close to twenty-four hours. And a genetic variation might explain why Kleitman couldn't adjust at all.) A *zeitgeber*'s effects depended on an assortment of variables, but the response would always be proportionate to the signal's frequency and amplitude. Exposure to continuous light, for example, would shorten the free-running period for animals normally active by day; in nocturnal animals, the reverse would be true. The brighter the light, the stronger those effects would be.[12]

Soon afterward, Princeton evolutionary biologist Colin S. Pittendrigh—a British émigré who'd gotten hooked on diurnal rhythms while studying mosquitos in Trinidad—proposed a more detailed model for how a biological clock might work. Pittendrigh envisioned a system of "endogenous self-sustained oscillators," operating at the cellular level, which kept an organism's rhythms (from body temperature to hormone secretion to sleep and waking) coordinated with the environment and one another. Each of these mechanisms would function like an automated stopwatch, keeping track of time between biologically relevant events. Each would also behave like a pendulum clock, whose swinging weight is known to scientists as a "harmonic oscillator." Ordinary pendulum clocks, it had long been known, could influence each other's oscillations when sitting together on a shelf. Biological clocks, Pittendrigh theorized, could be entrained by the oscillations of an environmental signal, such as the back-and-forth between dark and light; they could form networks as well. Like Aschoff, he argued that the actions of such a system would be governed by mathematical principles—and could be altered in predictable ways by targeted interventions.[13]

Aschoff and Pittendrigh first met in 1958 when the German scientist (who'd moved on to the Max Planck Institute for Behavioral Physiology) visited the United States. They quickly became friends and collaborators and began planning a symposium at Long Island's Cold Spring Harbor Laboratory on the clockwork that enthralled them both.[14] In the meantime, University of Minnesota biologist Franz Halberg coined the term *circadian* (from the Latin *circa*, "about," and *dies*, "day") to describe endogenous cycles lasting approximately twenty-four hours. Understanding these rhythms, a growing number of researchers recognized, was crucial to understanding the mechanics of life itself.[15]

More than 150 scientists and mathematicians attended the 1960 symposium.[16] Erwin Bünning delivered the opening address, in which he summarized the history of the field.[17] Aschoff discussed the evidence for his theory (with examples drawn from algae, bean plants, finches, and lizards) and gave an overview of future areas for investigation.[18] Pittendrigh hypothesized on the complex interactions among internal clocks that would be necessary to keep an organism's physiological rhythms in sync (citing research on euglenas, cockroaches, kangaroo rats, and flying squirrels). He also warned that recently adopted human behaviors—such as shift work and jet travel across time zones—might threaten the healthy functioning of such a system. "It remains to be seen," he said, "whether any stress or even damage is actually imposed by this type of phase shift which the system has never been called upon by natural selection to accommodate."[19] Nearly 50 other experts spoke on subjects including the metabolic, ecological, and (in birds, bees, fish, and crustaceans) navigational aspects of circadian rhythms.[20]

The Cold Spring Harbor symposium established the agenda that would guide chronobiological research for decades.[21] Chronobiology, in turn, would eventually become a key element of sleep science's intellectual arsenal, crucial to solving some of its oldest conundrums and amplifying its impact on the wider world.

That transformation, however, would take nearly twenty years to get fully underway. More discoveries would be required before each side could appreciate what the other had to offer.

Acknowledgments

Many people helped to make this book possible, and I wish they could all be here to read it. Before I thank the living, I want to honor some of those who went before.

I'll start with my grandparents, Harry and Minnie Miller and William and Rose Smith, three of whom came from places near where Nathaniel Kleitman grew up—and fled for similar reasons. (In Grandma Rose's case, it was her parents who did the fleeing.) Their risks and sacrifices enabled their grandkids to do things like write about sleep science, and their difficult journeys informed my understanding of Kleitman's.

Next come my parents, Myron and Miriam Miller, whose curiosity, empathy, and love for a good story sowed the seeds from which *Mapping the Darkness* belatedly blossomed.

I must also beam posthumous gratitude to William Dement, who shared his own extraordinary stories in a trio of interviews before he grew too frail to continue.

Now for the helpers who are still among us:

Thanks to Dan Okrent, who gave me my first full-time writing gig sometime in the previous century, and to Pam Weintraub, who later assigned me the story that inspired this project. My agent, Ethan Bassoff, saw the promise in what began as a vague concept and found my proposal a perfect home at Hachette Book Group. Sam Raim gave the unfinished first draft an incisive edit before leaving to fight the good fight at the Vera Institute of Justice. Lauren Marino and Niyati Patel scrutinized my baggy second draft and located the muscular narrative hidden within it. I'm deeply grateful to this brilliant team for their insight, skill, and sensitivity.

Thanks as well to Sara Robb for brilliant copyediting, to Terri Sirma for a striking cover design, and to Fred Francis for masterful project management.

I'm profoundly indebted to my expert beta readers, Lynne Lamberg and Donald Bliwise, who combed my emerging chapters for errors and omissions, suggested lines of inquiry to follow and people to interview, sent me articles and studies, explained obscure concepts, and acted as go-betweens with a few reluctant sources. Lynne has been known for four decades as America's foremost sleep science journalist, and Don is a renowned pioneer in the study of sleep in the elderly; both volunteered their assistance without being asked. Their knowledge, care, and perspicacity have strengthened this book immeasurably. Any inaccuracies, of course, are mine alone.

Another key benefactor was Stuart Rawlings, who shared observations gathered during a fifty-year friendship with Dr. Dement. Stuart also shared two crucial works that he'd coauthored: Dement's unpublished memoir, titled *Dreamer*, and the privately published reminiscences of the scientist's mother, *My Life and Times*. These accounts provided details of Dement's ancestry and early life that I could have found nowhere else, and they enabled me to paint a far more vivid portrait than would otherwise have been possible. My heartfelt thanks to Stuart for allowing me to mine these manuscripts and for trusting me to help fulfill his mission in cowriting them: to show the world, as he put it, "the real Bill Dement."

I'm grateful to everyone who sat for interviews, and their names are listed separately in the bibliography. But I'd like to single out Mary Carskadon, who spent an hour or two on the phone every Saturday for three months—with follow-up sessions afterward—answering questions about her work, her life, and the science she has helped to shape. This book has benefited incalculably from her openness and thoughtfulness and her willingness to tolerate my intrusiveness. Thanks also to Cathy Dement Roos, Nick Dement, Armond Aserinsky, Jill Aserinsky Buckley, and Hortense Kleitman Snower, who put up with my endless calls, texts, and emails seeking just a little more information about their parents. And a shout-out to ace transcriptionist Rebekah Berger, who wrestled many of these dialogues into legible form.

A bevy of archivists helped me unearth the raw historical material that fuels this volume. Thanks to Catherine Uecker, Christine Colburn, Barbara Gilbert, Tanzoom Ahmed, and Alexa Tulk at the Hanna Holborn Gray Special Collections Research Center, University of Chicago Library; Josh Schneider, Tim Noakes, and Hanna Ahn at the Department of Special Collections and University Archives, Stanford University Libraries; Drew Bourn at the Stanford Medical History Center; Sarah Keen at the Special Collections and University Archives, Colgate University Libraries; and Lee Hiltzik at the Rockefeller Archive Center.

Other kind souls contributed hard-to-find and illuminating documents, artifacts, or factoids. Nathaniel Knight, a professor of Russian history at Seton Hall University, helped trace the route Kleitman's parents would have traveled from Bessarabia to Cairo in 1895. Historian Vincent Cannato at UMass Boston described the tests immigrants faced at Ellis Island in 1915. Ranger-historian Colleen Olson and curator Terry Langford at Mammoth Cave National Park furnished arcane details about Kleitman's 1938 expedition. EEG experts Tom Collura and Sharon Keenan offered technical insights on the machine that Eugene Aserinsky used to discover REM sleep in 1952. Psychiatrist Arnold Richards provided unpublished interviews he did with Charles Fisher in the 1980s. Richard Rosenberg (a former senior director of the American Academy of Sleep Medicine), Thomas Heffron (a current senior director), and Steve Van Hout (the AASM's executive director) helped track down a video interview of Kleitman by sleep researcher Gerald Vogel from the '90s. *Smithsonian* magazine honcho Terry Monmaney, who edited Chip Brown's superb 2003 profile of Aserinsky, unearthed a scan of a poignant family photo published with that article (see page 91), whose original had since gone missing.

Two sets of accomplices aided indirectly in this book's creation: my fellow members of the Group Biography and Science Biography roundtables, which operate under the aegis of the Biographers International Organization. Led respectively by Marlene Trestman and Gabriella Kelly-Davies, each of these Zoom-mediated support groups serves a half dozen writers who are grappling with the challenges of making literature out of other people's lives. The advice and encouragement of my virtual tablemates has been a boon and a blessing.

I could never have completed this odyssey without the sustenance of friends, colleagues, and family. Thanks (in alphabetical order) to Bill Allen, Vince Beiser, Joan Berzoff, Bob Brody, Josh Cochin, Lew Cohen, Suzanne Cohen, Michèle Derai, Nicole Dyer, Andrea Egert, Robert Egert, Lorraine Glennon, Jerry Goldberg, Meg Grant, Dave Greenwald, Jody Gross, Paul Himmelein, David Hochman, Bill Horne, Charlotte DeCroes Jacobs, Kimberley Johnson, Holly Korbonski, Becky Lang, M. G. Lord, Bob Love, Christopher "Kit" Lukas, Linda Marsa, Linda Muskat-Rim, Joana Ochoa, Alex Orlando, Josh Palmer, Kira Peikoff, Tom Ranieri, Salley Rayl, Gini Sikes, Laura J. Snyder, Lyndon Stambler, Melissa Stanton, Laren Stover, Gemma Tarlach, Barbara Westermann, and Magnus Widman, all of whom helped me reach the finish line in ways of which they may or may not be aware. I'm grateful to my brothers, Jon and Peter Miller, their spouses, Rebecca Nelson and Val Marcus, my parents-in-law, Jim and Linda Ries, and my siblings-in-law, Ted and Tami Ries, for their warmth, enthusiasm, and steadfastness. My beloved offspring, Leo and Samantha, have kept me stocked with joy, wonder, and a sense of purpose— indispensable forms of energy for an author or any other human being.

And finally, I'm grateful to my wife, Julie Ries, who has nurtured me in every way imaginable since we first met thirty-one years ago. A dazzling writer and an equally gifted reader, she gets the first pass on my pages and the passkey to my heart. Her wisdom, faith, patience, and unstinting presence are gifts that I can never fully repay—but this book's dedication stands as an IOU.

Bibliography

Archives

Nathaniel Kleitman Papers and Association for the Psychophysiological Study of Sleep Records, Hanna Holborn Gray Special Collections Research Center, University of Chicago Library

Rockefeller Foundation Records, Rockefeller Archive Center

Special Collections and University Archives, Colgate University Libraries

Stanford Medical History Center

William C. Dement Papers (SC033), Department of Special Collections and Archives, Stanford University Libraries

Books

Acland, Charles R. *Swift Viewing: The Popular Life of Subliminal Influence.* Durham, NC: Duke University Press, 2012.

Albrecht, Urs, ed. *The Circadian Clock,* 12th edition. New York: Springer, 2010.

Anonymous. *The University of Chicago: An Official Guide.* Chicago: University of Chicago Press, 1916.

Aufses, Arthur H., Jr., and Barbara J. Niss. *This House of Noble Deeds: The Mount Sinai Hospital, 1852–2002.* New York: New York University Press, 2002.

Ben-Menahem, Ari. *Historical Encyclopedia of Natural and Mathematical Sciences.* Berlin: Springer-Verlag, 2009.

Borck, Cornelius. *Brainwaves: A Cultural History of Electroencephalography.* Science, Technology and Culture, 1700–1945. London: Routledge / Taylor & Francis Group, 2018.

Burnham, John, ed. *After Freud Left: A Century of Psychoanalysis in America.* Chicago: University of Chicago Press, 2012.

Buzzati-Traverso, Adriano A., ed. *Perspectives in Marine Biology.* Berkeley: University of California Press, 1958.

Cannato, Vincent. *American Passage: The History of Ellis Island.* New York: Harper Perennial, 2010.

Carlson, Anton J. *The Control of Hunger in Health and Disease.* Chicago: University of Chicago Press, 1916.

Carskadon, Mary A., ed. *Adolescent Sleep Patterns: Biological, Social, and Psychological Influences.* Cambridge: Cambridge University Press, 2002.

Carskadon, Mary A. "Determinants of Daytime Sleepiness: Adolescent Development, Extended and Restricted Nocturnal Sleep." Unpublished PhD thesis, Stanford University, 1979.

Carskadon, Mary A., ed. *Encyclopedia of Sleep and Dreaming.* New York: Macmillan, 1993.

Cartwright, Rosalind Dymond. *The Twenty-Four Hour Mind: The Role of Sleep and Dreaming in Our Emotional Lives.* New York: Oxford University Press, 2010.

Chokroverty, Sudhansu, and Michel Billiard, eds. *Sleep Medicine: A Comprehensive Guide to Its Development, Clinical Milestones, and Advances in Treatment.* New York: Springer, 2015.

Cobb, Matthew. *The Idea of the Brain: The Past and Future of Neuroscience.* New York: Basic Books, 2020.

Coleman, Richard M. *Wide Awake at 3:00 A.M.: By Choice or by Chance?* New York: W. H. Freeman, 1986.

Conant, Jennet. *Tuxedo Park: A Wall Street Tycoon and the Secret Palace of Science That Changed the Course of World War II.* New York: Simon & Schuster, 2003.

Cook, Constance Ewing. *Lobbying for Higher Education: How Colleges and Universities Influence Federal Policy.* Nashville: Vanderbilt University Press, 1998.

Crandall, Kelly B. "Invisible Commercials and Hidden Persuaders: James M. Vicary and the Subliminal Advertising Controversy of 1957." Unpublished undergraduate thesis. University of Florida, 2006.

Cummings, Lindsey. "Economic Warfare and the Evolution of the Allied Blockade of the Eastern Mediterranean: August 1914–April 1917." Unpublished master's thesis. Georgetown University, 2015.

Czeisler, Charles Andrew. "Human Circadian Physiology: Internal Organization of Temperature Sleep-Wake and Neuroendocrine Rhythms Monitored in an Environment Free of Time Cues." Unpublished PhD dissertation. Stanford University, 1978.

Daan, Serge. *Die innere Uhr des Meschen: Jürgen Aschoff (1913–1998), Wissenschaftler in einem beweigten Jahrnhundert.* Wiesbaden, Germany: Reichert Verlag, 2017.

Davenport-Hines, Richard. *The Pursuit of Oblivion.* New York: W. W. Norton, 2004.

Davidson, Jonathan. *Downing Street Blues: A History of Depression and Other Mental Afflictions in British Prime Ministers.* Jefferson, NC: McFarland, 2011.

Davis, Peter, ed. *Contested Ground: Public Purpose and Private Interest in the Regulation of Prescription Drugs.* New York: Oxford University Press, 1996.

Davitt, Michael. *Within the Pale: The True Story of Anti-Semitic Persecutions in Russia.* New York: A. S. Barnes, 1903.

Dement, Kathryn Severyns. *My Life and Times.* Auburn, CA: Sierra Dreams Press, 1990.

Dement, William C. *The Sleepwatchers.* Menlo Park, CA: Nychthemeron Press, 1995.

Dement, William C. *Some Must Watch While Some Must Sleep.* San Francisco: San Francisco Book Company, 1976.

Dement, William C., with Stuart Rawlings. *Dreamer: The Memoir of a Sleep Researcher.* Unpublished manuscript, 2018.

Dement, William C., and Christopher Vaughan. *The Promise of Sleep: A Pioneer in Sleep Medicine Explores the Vital Connection Between Health, Happiness, and a Good Night's Sleep.* New York: Delacorte, 1999.

Derickson, Alan. *Dangerously Sleepy: Overworked Americans and the Cult of Manly Wakefulness*. Philadelphia: University of Pennsylvania Press, 2014.

De Vries, N., Madeline Ravesloot, and Peter Van Maanen. *Positional Therapy in Obstructive Sleep Apnea*. Cham, Switzerland: Springer, 2015.

Diagnostic Classification Steering Committee and Michael J. Thorpy. *International Classification of Sleep Disorders: Diagnostic and Coding Manual*. Rochester, NY: American Sleep Disorders Association, 1990.

Dickens, Charles. *Works of Charles Dickens, Volume 5*. New York: Hurd and Houghton, 1872.

Dubnow, S. M. *History of the Jews in Russia and Poland, Volume 3*. Philadelphia: Jewish Publication Society of America, 1920.

Duff, Kat. *The Secret Life of Sleep*. New York: Atria, 2014.

Dunar, Andrew J., and Stephen P. Waring, *The Power to Explore: A History of Marshall Space Flight Center, 1960–1990*. Washington, DC: US Government Printing Office, 1999.

Ekirch, A. Roger. *At Day's Close: Night in Times Past*. New York: W. W. Norton, 2006.

Empson, Jacob. *Human Brainwaves: The Psychological Significance of the Electroencephalogram*. New York: Stockton Press, 1986.

Erwin, Edward. *The Freud Encyclopedia: Theory, Therapy, and Culture*. New York: Routledge, 2002.

Evered, David, and Sarah Clark, eds. *Ciba Foundation Symposium 117—Photoperiodism, Melatonin and the Pineal*. London: Pitman, 1985.

Fagan, Brian, and Nadia Durrani. *What We Did in Bed: A Horizontal History*. New Haven, CT: Yale University Press, 2019.

Fields, R. Douglas. *Electric Brain: How the New Science of Brainwaves Reads Minds, Tells Us How We Learn, and Helps Us Change for the Better*. Dallas: Ben Bella Books, 2020.

Finger, Stanley. *Minds Behind the Brain: A History of the Pioneers and Their Discoveries*. New York: Oxford University Press, 1999.

Fisher, Charles, and Howard Shevrin. *Subliminal Explorations of Perception, Dreams, and Fantasies: The Pioneering Contributions of Charles Fisher*. Madison, CT: International Universities Press, 2003.

Freud, Sigmund, and James Strachey. *The Interpretation of Dreams*. New York: Basic Books, 2010.

Gastaut, H., et al., eds. *The Abnormalities of Sleep in Man*. Bologna, Italy: Aulo Gaggi Editore, 1968.

General Practitioner, A. *The Illustrated Family Doctor*. London: Waterlow & Sons, 1934.

Gottesmann, Claude. *Henri Piéron and Nathaniel Kleitman: Two Major Figures of 20th Century Sleep Research*. New York: Nova Science Publishers, 2013.

Greene, Gayle. *Insomniac*. Berkeley: University of California Press, 2008.

Guilleminault, Christian, and William C. Dement. *Sleep Apnea Syndromes*. New York: A. R. Liss, 1978.

Guilleminault, Christian, and Elio Lugaresi, eds. *Sleep/Wake Disorders: Natural History, Epidemiology, and Long-Term Evolution*. New York: Raven Press, 1983.

Hale, Nathan G. *The Rise and Crisis of Psychoanalysis in the United States: Freud and the Americans, 1917–1985*. New York: Oxford University Press, 1995.

Hauri, Peter. *The Sleep Disorders*. Kalamazoo, MI: Upjohn Company, 1982.

Hayes, Bill. *Sleep Demons: An Insomniac's Memoir*. Chicago: University of Chicago Press, 2018.

Hirschfelder, Arthur Douglas. *Diseases of the Heart and Aorta*. Philadelphia: J. B. Lippincott, 1918.

Hobson, J. Allan. *Dream Life: An Experimental Memoir*. Cambridge, MA: MIT Press, 2011.

Jacobson, Edmund. *You Can Sleep Well: The ABC's of Restful Sleep for the Average Person*. New York: Whittlesey House, 1938.

Jacobson, Edmund. *You Must Relax: A Practical Method of Reducing the Strains of Modern Living*. New York: Whittlesey House, 1934.

Johnson, Paul. *A History of the Jews*. New York: Harper Perennial, 1987.

Johnson, T. Scott, William A. Broughton, and Jerry Halberstadt. *Sleep Apnea: The Phantom of the Night*. Onset, MA: New Technology Publishing, 2003.

Jouvet, Michel. *The Paradox of Sleep: The Story of Dreaming*. Cambridge, MA: MIT Press, 1999.

Kales, Anthony, ed. *Sleep: Physiology & Pathology—a Symposium*. Philadelphia: J. B. Lippincott, 1969.

Kerson, Cynthia, Tom Collura, and Joanne Kamiya. *Joe Kamiya: Thinking Inside the Box*. Corpus Christi: BMED Press, 2020.

Kety, Seymour S., Edward V. Evarts, and Harold L. Williams, eds. *Sleep and Altered States of Consciousness*. Baltimore: Williams and Wilkins, 1967.

Kleitman, Nathaniel. *Sleep and Wakefulness as Alternating Phases in the Cycle of Existence*. Chicago: University of Chicago Press, 1939.

Kleitman, Nathaniel. *Sleep and Wakefulness, Revised and Enlarged Edition*. Chicago: University of Chicago Press, 1963.

Koren, Yitzchak. *The Jews of Kishinev*. New York: JewishGen, 2019.

Kroker, Kenton. *The Sleep of Others and the Transformation of Sleep Research*. Toronto: University of Toronto Press, 2015.

Kryger, Meir H., Thomas Roth, and William C. Dement. *Principles and Practice of Sleep Medicine*, 1st edition. Philadelphia: Saunders, 1989.

Kryger, Meir H., Thomas Roth, and William C. Dement. *Principles and Practice of Sleep Medicine*, 3rd edition. Philadelphia: Saunders, 2000.

Kryger, Meir H., Thomas Roth, and Cathy A. Goldstein. *Principles and Practice of Sleep Medicine*, 7th edition. Philadelphia: Elsevier, 2022.

Kumar, Vinod, ed. *Biological Timekeeping: Clocks, Rhythms and Behavior*. New Delhi: Springer, 2017.

Laird, Donald Anderson, and Charles Geoffrey Muller. *Sleep: Why We Need It and How to Get It*. New York: John Day, 1930.

Lamberg, Lynne. *Bodyrhythms: Chronobiology and Peak Performance*. New York: William Morrow, 1994.

Lansky, Melvin R. *Essential Papers on Dreams*. New York: New York University Press, 1992.

Lavie, Peretz. *The Enchanted World of Sleep*. New Haven, CT: Yale University Press, 1996.

Lewis, Lisa L. *The Sleep-Deprived Teen: Why Our Teenagers Are So Tired, and How Parents and Schools Can Help Them Thrive*. Coral Gables, FL: Mango Publishing, 2022.

Lewis, Penelope A. *The Secret World of Sleep: The Surprising Science of the Mind at Rest*. New York: Palgrave Macmillan, 2014.

Lowen, Rebecca S. *Creating the Cold War University: The Transformation of Stanford*. Berkeley: University of California Press, 1997.

Luce, Gay Gaer. *Current Research on Sleep and Dreams*. Bethesda, MD: US Department of Health, Education, and Welfare, Public Health Service, 1966.

Macnish, Robert. *The Philosophy of Sleep*. Glasgow: W. R. McPhun, 1830.

Manacéïne, Marie de. *Sleep: Its Physiology, Pathology, Hygiene and Psychology*. London: Walter Scot, 1897.

Mendelson, Wallace B. *Nepenthe's Children: The History of the Discoveries of Medicines for Sleep and Anesthesia*. New York: Pythagoras Press, 2020.

Miller, Richard Lawrence. *The Encyclopedia of Addictive Drugs*. Westport, CT: Greenwood Press, 2002.

Montgomery-Downs, Hawley, ed. *Sleep Science*. New York: Oxford University Press, 2020.

Moore-Ede, Martin. *The Twenty-Four-Hour Society: Understanding Human Limits in a World That Never Stops*. New York: Addison-Wesley, 1993.

Morin, Charles. *Insomnia: Psychological Assessment and Management*. New York: Guilford Press, 1993.

Okrent, Daniel. *The Guarded Gate: Bigotry, Eugenics, and the Law That Kept Two Generations of Jews, Italians, and Other European Immigrants Out of America*. New York: Scribner, 2019.

Packard, Vance. *The Hidden Persuaders*. New York: Pocket Books, 1958.

Penfield, Wilder. *The Difficult Art of Giving: The Epic of Alan Gregg*. Boston: Little, Brown, 1967.

Perlis, Michael, Mark Aloia, and Brett Kuhn. *Behavioral Treatments for Sleep Disorders*. Amsterdam: Academic Press, 2011.

Piéron, Henri. *Le Problème Physiologique du Sommeil*. Paris: Masson, 1913.

Pollak, Charles, et al. *The Encyclopedia of Sleep and Sleep Disorders*. New York: Facts on File, 2010.

Popov, Y., and L. Rokhin, eds. *I. P. Pavlov: Psychopathology and Psychiatry, Selected Works*. Moscow: Foreign Languages Publishing House, n.d.

Pressman, Mark R., and William C. Orr. *Understanding Sleep: The Evaluation and Treatment of Sleep Disorders*. Washington, DC: American Psychological Association, 1997.

Randall, David K. *Dreamland: Adventures in the Strange Science of Sleep*. New York: W. W. Norton, 2013.

Raphael, Marc Lee, ed. *The Columbia History of Jews & Judaism in America*. New York: Columbia University Press, 2008.

Rechtschaffen, Allan, and Anthony Kales. *A Manual of Standardized Technology, Techniques and Scoring System for Sleep Stages of Human Subjects*. Washington, DC: Public Health Service, US Government Printing Office, 1968.

Reiss, Benjamin. *Wild Nights: How Taming Sleep Created Our Restless World*. New York: Basic Books, 2017.

Richter, Curt. *Biological Clocks in Medicine and Psychiatry*. Springfield, IL: Charles C. Thomas, 1965.

Rock, Andrea. *The Mind at Night: The New Science of How and Why We Dream*. New York: Basic Books, 2004.

Roenneberg, Till. *Internal Time: Chronotypes, Social Jet Lag, and Why You're So Tired*. Cambridge, MA: Harvard University Press, 2012.

Rolls, Geoff. *Classic Case Studies in Psychology*, 2nd edition. London: Hodder Education, 2010.

Rowan, Andrew N., Franklin M. Loew, with Joan Weer. *The Animal Research Controversy: Protest, Process & Public Policy: An Analysis of Strategic Issues*. North Grafton, MA: Center for Animals and Public Policy, Tufts University School of Veterinary Medicine, 1995.

Sacks, Oliver. *Awakenings*. New York: Vintage Books, 1999.

Samuel, Lawrence R. *Shrink: A Cultural History of Psychoanalysis in America*. Lincoln: University of Nebraska Press, 2013.

Sanei, Saied, and J. A. Chambers. *EEG Signal Processing*. Chichester, England: John Wiley & Sons, 2007.

Schultz, Stanley. *The Great Depression: A Primary Source History*. Milwaukee: Gareth Stevens, 2006.

Scrivner, Lee. *Becoming Insomniac: How Sleeplessness Alarmed Modernity*. Basingstoke, England: Palgrave Macmillan, 2014.

Shermer, Michael, ed. *The Skeptic Encyclopedia of Pseudoscience*. Santa Barbara, CA: ABC-CLIO, 2002.

Shure, Caitlin. "Brain Waves, a Cultural History." Unpublished PhD thesis. Columbia University Graduate School of Arts and Sciences, 2018.

Siegel, Jerome. *The Neural Control of Sleep and Waking*. New York: Springer-Verlag, 2002.

Singer, Isidore. *Russia at the Bar of the American People*. New York: Funk & Wagnall's, 1914.

Smolensky, Michael, and Lynne Lamberg. *The Body Clock Guide to Better Health: How to Use Your Body's Natural Clock to Fight Illness and Achieve Maximum Health*. New York: Owl Books, 2001.

Strogatz, Steven. *The Mathematical Structure of the Human Sleep-Wake Cycle*. Berlin: Springer-Verlag, 1986.

Strogatz, Steven. *Sync: The Emerging Science of Spontaneous Order*. New York: Hyperion, 2003.

Todes, Daniel P. *Ivan Pavlov: A Russian Life in Science*. Oxford: Oxford University Press, 2014.

Walker, Matthew. *Why We Sleep: Unlocking the Power of Sleep and Dream*. New York: Scribner, 2017.

Weitzman, Elliot, ed. *Advances in Sleep Research, Volume 1*. Flushing, NY: Spectrum Publications, 1974.

Wever, Rütger A. *The Circadian Systems of Man: Results of Experiments Under Temporal Isolation*. New York: Springer-Verlag, 1979.

Wolf-Meyer, Matthew J. *The Slumbering Masses: Sleep, Medicine, and Modern American Life*. Minneapolis: University of Minnesota Press, 2016.

Wolstenholme, G. E. W., and Maeve O'Connor, eds. *Ciba Foundation Symposium on the Nature of Sleep*. Boston: Little, Brown, 1961.

Worden, Frederic G., Judith P. Swazey, and George Adeleman, eds. *The Neurosciences: Paths of Discovery, Volume 1.* Boston: Birkhäuser, 1992.

Zipperstein, Steven. J. *Pogrom: Kishinev and the Tilt of History.* New York: Liveright, 2018.

Articles

See notes

Author's Interviews

Sonia Ancoli-Israel

Thomas Anders

Armond Aserinsky

Donald Bliwise

Joe Borelli

Jill Aserinsky Buckley

Mary Carskadon

Stephanie Crowley

Nick Dement

William Dement

Dale Dirks

Richard Ellman

Bill Gonda

H. Craig Heller

Grant Hoyt

Sharon Keenan

Meir Kryger

Lynne Lamberg

Bruce Levitt

Mike Levitt

Marc Miller

Merrill Mitler

Debra Myers

Woody Myers

Rafael Pelayo

Stuart Rawlings

Arnold Richards

Cathy Dement Roos

Mark Rosekind

Tom Roth

Jon Sassin

Barbara Scavullo

Jerome Siegel

Terra Ziporyn Snider

Hortense Kleitman Snower

Harry Trosman

James Walsh

Images

Page 6: Nathaniel Kleitman Papers, Box 33, Folder 6, Hanna Holborn Gray Special Collections Research Center, University of Chicago Library.

Page 40: Nathaniel Kleitman Papers, Box 33, Folder 5.

Page 57: Nathaniel Kleitman Papers, Box 33, Folder 4.

Page 91: Courtesy of Jill Buckley.

Page 106: From Eugene Aserinsky and Nathaniel Kleitman, "Regularly Occurring Periods of Eye Motility, and Concomitant Phenomena, During Sleep," *Science* 118, no. 3062 (September 4, 1953): 273–274. Reprinted with permission from AAAS.

Page 111: Courtesy of Catherine Dement Roos.

Page 116: Reprinted from *Sleep Medicine Reviews* 7, no. 4, William C. Dement, "Knocking on Kleitman's Door," 289–292, copyright 2003, with permission from Elsevier.

Page 165: © Richard Meek/The LIFE Picture Collection/Shutterstock.com.

Pages 190–191: From William C. Dement and Christian Guilleminault, "Sleep Disorders: The State of the Art," *Hospital Practice* 8, no. 11 (November 1, 1973). Reprinted by permission of the publisher (Taylor & Francis Ltd., https://www.tandfonline.com).

Page 238: From Simona Sher, Amit Green, et al., "The Possible Role of Endozepines in Sleep Regulation and Biomarker of Process S of the Borbély Sleep Model," *Chronobiology*

International 38, no. 1 (January 2, 2021). Reprinted by permission of the publisher (Taylor & Francis Ltd., https://www.tandfonline.com).

Page 275: Courtesy of Mary A. Carskadon.

Page 278: Nathaniel Kleitman Papers, Box 33, Folder 1.

Page 285: Courtesy of Mary A. Carskadon.

Notes

Introduction

1. *At first, no one noticed:* Interview with Joe Borelli, October 23, 2013. Note: Parts of this chapter were adapted from my article "Wake-Up Call," *Discover*, April 2015, 54–59, published online as "Getting Enough Sleep Can Be a Matter of Life and Death," accessed December 6, 2022, https://www.discovermagazine.com/mind/getting-enough -sleep-can-be-a-matter-of-life-and-death.
2. *He settled on:* Interviews with Borelli, October 23, 2013, and June 25, 2020.
3. *By coincidence, Guilleminault:* Luciana Palombini et al., "Upper Airway Resistance Syndrome: Still Not Recognized and Not Treated," *Sleep Science* 4, no. 2 (2011): 72–78; Luciana B. M. de Godoy et al., "Treatment of Upper Airway Resistance Syndrome in Adults: Where Do We Stand?," *Sleep Science* 8, no. 1 (2015): 42–48.
4. *"You've had this disorder":* Interview with Borelli, October 23, 2013.
5. *Borelli was reluctant:* Ibid.
6. *I interviewed Borelli:* Miller, "Wake-Up Call."
7. *Just a century:* "Nathaniel Kleitman, PhD, 1895–1999," UChicago Medicine, August 1, 1999, accessed May 31, 2021, https://www.uchicagomedicine.org/forefront/news /nathaniel-kleitman-phd-1895-1999.
8. *Over 2,500:* "Sleep Medicine in America," American Academy of Sleep Medicine, accessed August 18, 2022, http://sleepeducation.org/news/2014/11/25/sleep-medicine -in-america-infographic. *4,000 more in other countries:* No one has systematically tallied the number of sleep centers around the world. But the forty-plus member organizations of the World Sleep Society collectively report about 3,500 centers—3,000 of them in China—and web searches for individual member countries turn up hundreds more. In addition, a 2012 report by the European Sleep Research Society tallied over 1,037 sleep clinics on that continent alone. See "Current Associate Society Members," World Sleep Society, https://worldsleepsociety.org/membership/societymembership /membersocieties/; "OSA Risk Across the UK," Asthma and Lung UK, https://www .blf.org.uk/support-for-you/obstructive-sleep-apnoea-osa/health-care-professionals/risk -and-sleep-services-map; "Find Your Ideal Sleep Health at a Clinic Near You," Australian Sleep Clinics, accessed August 20, 2022, https://australiansleepclinics.com .au; Claudio L. Bassetti, ed., *European Sleep Research Society 1972–2012: 40th Anniversary of the ESRS* (Hildesheim, Germany: Wecom Gesellschaft für Kommunikation

mbH & Co., 2012), 83–123; Jessica Evans et al., "Sleep Laboratory Test Referrals in Canada: Sleep Apnea Rapid Response Survey," *Canadian Respiratory Journal* 21, no. 1 (January/February 2014): e4–e10.

9. *The World Sleep Society:* "2020 Annual Report," World Sleep Society, accessed August 21, 2022, https://worldsleepsociety.org/about/annualreport.

10. *In a 2020 study:* "Size of the Sleep Economy Worldwide in 2019, by Product Category (in Billion U.S. Dollars)," Statista, https://www.statista.com/statistics/1119487 /size-of-the-sleep-economy-worldwide-by-product-category/; "Investor Presentation 2020," Casper, https://s24.q4cdn.com/768080344/files/doc_financials/2020/q2/v2/CSPR-Investor -Presentation_August-2020.pdf; "Investing in the Growing Sleep-Health Economy," McKinsey & Company, accessed August 21, 2022, https://www.mckinsey.com/-/media /mckinsey/industries/private%20equity%20and%20principal%20investors/our %20insights/investing%20in%20the%20growing%20sleep%20health%20economy /investing-in-the-growing-sleep-health-economy.ashx.

11. *Although a session:* "The Sleep-Tech Industry Is Waking Up," *Economist*, February 13, 2022, https://www.economist.com/business/2022/02/12/the-sleep-tech-industry-is-waking -up; Bertrand Beauté, "The $500 Billion Business of Sleep," Swissquote, https://en .swissquote.lu/international-investing/investing-ideas/500-billion-business-sleep; Casper Editorial Team, "15 Podcasts That Will Help You Doze Off Easier," Casper, accessed August 21, 2022, https://casper.com/blog/sleep-podcasts/.

12. *In the United States:* "Data and Statistics," CDC, https://www.cdc.gov/sleep/data _statistics.html.

13. *Diagnosticians now recognize:* "Exploding Head Syndrome," American Academy of Sleep Medicine, http://sleepeducation.org/sleep-disorders-by-category/parasomnias/exploding -head-syndrome/overview-facts; "REM Sleep Behavior Disorder," American Academy of Sleep Medicine, http://sleepeducation.org/sleep-disorders-by-category/parasomnias /rem-sleep-behavior-disorder/overview-facts.

14. *And the sense:* Figures for Germany, Brazil, Australia, United States, United Kingdom, and France: "Percentage of Adults Around the World Completely or Somewhat Satisfied with Their Sleep as of 2020, by Country," Statista, https://www.statista.com /statistics/1233775/adults-worldwide-satisfied-sleep-country/. Figures for global sleep satisfaction, and for Japan and South Korea: "How the World Sleeps," Raconteur, accessed August 19, 2022, https://www.raconteur.net/infographics/how-the-world-sleeps/.

15. *Beyond the effects:* William C. Rempel, "Tanker Inquiry Reflects Blame Beyond Captain," *Los Angeles Times,* May 21, 1989, 35; "Details About the Accident," final report, Alaska Oil Spill Commission, 1990, accessed July 6, 2022, https://evostc.state.ak.us /oil-spill-facts/details-about-the-accident/; Doug Struck, "Twenty Years Later, Impacts of the *Exxon Valdez* Linger," *Yale Environment* 360, March 24, 2009, accessed July 6, 2022, https://e360.yale.edu/features/twenty_years_later_impacts__of_the_exxon _valdez_linger. Space Shuttle *Challenger:* Andrew Dunbar and Stephen Waring, *The Power to Explore: A History of Marshall Space Flight Center, 1960–1990* (Washington, DC: US Government Printing Office, 1999), 339–387; "Report of the Presidential Commission on the Space Shuttle Challenger Accident, Volume 2, Appendix

G—Human Factor Analysis," NASA, https://history.nasa.gov/rogersrep/v2appg.htm; Leo Hickman, "The Hidden Dangers of Sleep Deprivation," *The Guardian*, February 9, 2011, accessed June 13, 2022, https://www.theguardian.com/lifeandstyle/2011/feb/09 /dangers-sleep-deprivation; Merrill Mitler et al., "Catastrophes, Sleep, and Public Policy: Consensus Report," *Sleep* 11, no. 1 (February 1988): 100–109.

16. *The invasion of:* Jessica C. Levenson et al., "The Association Between Social Media Use and Sleep Disturbance in Young Adults," *Preventive Medicine* 85 (April 2016): 36–41; Lisbeth Lund et al., "Electronic Media Use and Sleep in Children and Adolescents in Western Countries: A Systematic Review," *BMC Public Health* 21, no. 1598 (2021), accessed August 30, 2022, https://doi.org/10.1186/s12889-021-11640-9.

17. *A recent study:* RAND Corporation, "Why Sleep Matters: Quantifying the Economic Costs of Insufficient Sleep," Rand Europe, accessed August 28, 2022, https://www .rand.org/randeurope/research/projects/the-value-of-the-sleep-economy.html.

18. *They've won recognition:* J. W. Shepard et al., "History of the Development of Sleep Medicine in the United States," *Journal of Clinical Sleep Medicine* 2, no. 1 (2005): 61–82; Markus Helmut Schmidt, "In Memoriam: A Tribute to an Unsung Hero," *Sleep* 34, no. 3 (March 1, 2011): 403–404; "S-1—National Institutes of Health Revitalization Act of 1993," US Congress, accessed July 26, 2022, https://www.congress.gov /bill/103rd-congress/senate-bill/1/actions.

19. *As Borelli began:* interview with Borelli, October 23, 2013.

20. *"If sleep does not serve":* Michael E. Long, "What Is This Thing Called Sleep?," *National Geographic,* December 1987, 821.

21. *Although virtually every species:* Jerome Siegel, *The Neural Control of Sleep and Waking* (New York: Springer-Verlag, 2002), 129–131; Jerome Siegel, "Do All Animals Sleep?," *Trends in Neuroscience* 31, no. 4 (April 2008): 208–213.

22. *That's because sleep:* Siegel, *The Neural Control*, 130.

23. *There are gray areas:* Elizabeth Pennisi, "The Simplest of Slumbers," *Science* 374, no. 6567 (October 29, 2021): 526–529.

24. *Many observers have:* Siegel, *The Neural Control*, 129–131. Debate over purpose of sleep: Beth Azar, "Wild Findings on Animal Sleep," *American Psychological Association* 37, no. 1 (January 2006): 54–55; Pennisi, "The Simplest of Slumbers"; "Why Do We Sleep, Anyway?," Healthy Sleep, accessed August 29, 2022, https://healthysleep .med.harvard.edu/healthy/matters/benefits-of-sleep/why-do-we-sleep.

25. *They all started:* "Nathaniel Kleitman, PhD, 1895–1999," UChicago Medicine.

26. *The second explorer:* Eugene Aserinsky, "Memories of Famous Neuropsychologists: The Discovery of REM Sleep," *Journal of the History of the Neurosciences* 5, no. 3 (1996): 213–227. Date of Aserinsky's entry into the field: Email to author from Catherine Uecker, University of Chicago Library Special Collections, November 4, 2021.

27. *Explorer number three:* Rosanne Spector, "William Dement, Giant in Sleep Medicine, Dies at 91," Stanford Medicine, June 18, 2020, https://med.stanford.edu/news /all-news/2020/06/william-dement-giant-in-field-of-sleep-medicine-dies-at-91.html; Harrison Smith, "William Dement, Known as 'the Father of Seep Medicine,' Dies at 91," *Washington Post,* June 20, 2020, accessed December 2, 2022, https://www

.washingtonpost.com/local/obituaries/william-dement-psychiatrist-known-as-the
-father-of-sleep-medicine-dies-at-91/2020/06/22/ca70e21a-b3c8-11ea-a510
-55bf26485c93_story.html.

28. *The fourth explorer:* "2020 Prize Recipient: Mary A. Carskadon, PhD," Division of Sleep Medicine, Harvard Medical School, accessed December 2, 2022, https://sleep.hms.harvard.edu/news-events/hms-division-sleep-medicine-prize/prize
-recipients/2020-prize-recipient-mary-carskadon; interview with Mary Carskadon, July 18, 2020.

Chapter 1

1. *By the time:* Nathaniel Kleitman autobiography, dictated to Paulena Kleitman circa 1970s, Nathaniel Kleitman Papers, Box 1, Folder 7, Hanna Holborn Gray Special Collections Research Center, University of Chicago Library. *The young cloth merchant:* I've been unable to find any record of the elder Kleitman's occupation, but circumstantial evidence exists. The couple's relative prosperity—evidenced by their ability to travel to Egypt—suggests that they belonged to the local merchant class. The *Vsia Rossiia 1895 Business Directory,* accessed through Jewishgen.org (https://www.jewishgen.org/databases/jgdetail_2.php), lists only one enterprise in the couple's home district of Akkerman under the surname Kleitman: a fabric store. Because no given name is listed, it's impossible to be certain of the owner's identity. However, Pesea's second husband was known to be in the textile trade, and it seems reasonable to infer that she met him through family business connections. *a week's journey:* Nathaniel Knight (professor of history, Seton Hall University), email message to author, October 23, 2020.

2. *Soon after their arrival:* Kleitman autobiography.

3. *From an early age:* Kleitman autobiography. *Pesea remarried two years later:* Kleitman autobiography; Bessarabia marriage records, accessed via Jewishgen.org, https://www.jewishgen.org/databases/jgdetail_2.php. Her new husband, Avram Roitberg, was a fabric store owner, per the Bessarabia Business Directory of 1924–1925, https://www.jewishgen.org/bessarabia/files/projects/5BusinessDirectory/tighina-District.htm. *who may also have been:* Leyb's occupation is unknown, but the fact that Pesea's second husband was a cloth merchant (as her first may have been) suggests that her father might have shared the same trade. *When Leyb died two years after that:* Kleitman autobiography; Bessarabia Death Records, 1858–1914, https://www.jewishgen.org/databases/jgdetail_2.php (accessed September 1, 2022).

4. *Although their house:* Kleitman autobiography. *like all but the grandest:* Steven J. Zipperstein, *Pogrom: Kishinev and the Tilt of History* (New York: Liveright, 2018), 41–42. *Kishinev's Jewish community:* "Moldova Virtual Jewish History Tour," Jewish Virtual Library, https://www.jewishvirtuallibrary.org/moldova-virtual-jewish-history-tour; Herman Rosenthal and Max Rosenthal, "Kishinef (Kishinev)," *Jewish Encyclopedia*, accessed August 31, 2022, https://jewishencyclopedia.com/articles/9350-kishinef-kishinev. *These Ashkenazim, whose ancestors:* Paul Johnson, *A History of the Jews* (New York: Harper Perennial, 1987), 230–231, 356; "Moldova Virtual Jewish History Tour," Jewish Virtual Library; Rosenthal and Rosenthal, "Kishinef (Kishinev)."

5. *Anti-Semitism was:* Johnson, *A History of the Jews,* 357–365. *Young males faced:* Yohanan Petrovsky-Shtern, "Military Service in Russia," Yivo Encyclopedia of Jews in Eastern Europe, accessed August 31, 2022, https://yivoencyclopedia.org/article.aspx /Military_Service_in_Russia. *According to family lore:* Kleitman autobiography; interview with Hortense Kleitman Snower, September 10, 2016. *But conditions worsened:* Johnson, *A History of the Jews,* 359–365; "Russia Jewish Virtual History Tour," Jewish Virtual Library, accessed September 1, 2022, https://www.jewishvirtuallibrary.org /russia-virtual-jewish-history-tour.

6. *In Kishinev, the atmosphere:* Rosenthal and Rosenthal, "Kishinef (Kishinev); Zipperstein, *Pogrom,* 45–57; S. M. Dubnow, *History of the Jews in Russia and Poland, Volume 3* (Philadelphia: Jewish Publication Society of America, 1920), 71.

7. *The two-day rampage:* "Kishinev," *Yivo Encyclopedia of Jews in Eastern Europe,* accessed August 31, 2022, https://yivoencyclopedia.org/article.aspx/kishinev; Zipperstein, *Pogrom,* 70, 73. *What is known:* Hortense Kleitman Snower interview; Isidore Singer, *Russia at the Bar of the American People* (New York: Funk & Wagnall's, 1914), 20, 27.

8. *As an adult:* Hortense Kleitman Snower interview.

9. *The massacre sparked:* Zipperstein, *Pogrom,* 11–14, 187–189. *Between 1880 and 1914:* Johnson, *A History of the Jews,* 365, 370.

10. *Kleitman's trajectory:* Kleitman autobiography.

11. *In 1914:* Ibid.

12. *A few hours later:* Ibid.

13. *Kleitman spent:* Ibid.

14. *The last hope:* Ibid.; "Profile: Henry Morgenthau Sr., the US Ambassador to Constantinople," Chatham House, https://www.chathamhouse.org/2015/02/profile-henry -morgenthau-sr-us-envoy-constantinople; Steve Lipman, "Armenia's Jewish Hero," Jewish Telegraphic Agency, April 22, 2005, accessed August 31, 2022, https://www.jta .org/2005/04/22/ny/armenias-jewish-hero; Lindsey Cummings, "Economic Warfare and the Evolution of the Allied Blockade of the Eastern Mediterranean: August 1914– April 1917," unpublished master's thesis, Georgetown University, 2015. *Morgenthau arranged for:* Kleitman autobiography.

15. *In July 1915:* Kleitman autobiography.

16. *Kleitman landed:* Ibid.; Daniel Okrent, *The Guarded Gate: Bigotry, Eugenics, and the Law That Kept Two Generations of Jews, Italians, and Other European Immigrants Out of America* (New York: Scribner, 2019), 1–4.

17. *The restrictionists often regarded:* Okrent, *Guarded Gate,* frontispiece, 204–208.

18. *Although ethnically targeted:* Cannato, *American Passage: The History of Ellis Island* (New York: Harper Perennial, 2010), 76, 127–128, 182–183, 195, 242, 261.

19. *Still, the admissions:* Kleitman autobiography. Note: In Kleitman's recollection, he also had to take a literacy test, though this requirement did not actually become law until 1917 (Cannato, *American Passage,* 308; Vincent Cannato, email message to author, November 24, 2020).

20. *After a ferry ride:* Kleitman autobiography.

Chapter 2

1. *When Kleitman landed:* Nathaniel Kleitman autobiography, dictated to Paulena Kleitman circa 1970s, Nathaniel Kleitman Papers, Box 1, Folder 7, Hanna Holborn Gray Special Collections Research Center, University of Chicago Library.
2. *With over half a million:* Paul Johnson, *A History of the Jews* (New York: Harper Perennial, 1987), 372–373; Deborah Dwork, "Health Conditions of Immigrant Jews on the Lower East Side of New York: 1880–1914," *Medical History* 25 (1981): 1–40; "New York (Manhattan) Sectors Population & Density 1800–1910," Demographia, http://demographia .com/db-nyc-sector1800.htm; Alex M. Mcleese, "Yesterday and Today," *Harvard Crimson,* August 4, 2009, https://www.thecrimson.com/article/2009/8/4/yesterday-and-today-new -york-ny/; Geraldine Murphy, "The People of New York City: Summary of 'Jews and Italians in Greater New York City, 1880 to World War I' Part I by Binder and Reimers," Macaulay Honors College at CUNY, https://eportfolios.macaulay.cuny.edu/murphy16/2016/03/09 /summary-of-jews-and-italians-in-greater-new-york-city-1880-to-world-war-i-by-binder -and-reimers/; Christopher D. Brazee et al., "East Village/Lower East Side Historic District Designation Report," October 9, 2012, New York City Landmarks Preservation Commission, accessed September 12, 2022, http://s-media.nyc.gov/agencies/lpc/lp/2491.pdf.
3. *At first, he hoped:* Kleitman autobiography.
4. *Founded in 1847:* "Our History," CUNY, https://www.ccny.cuny.edu/about/history; "Today: The City University of New York Is the Nation's Leading Urban Public University," *CUNY,* https://www.cuny.edu/about/history/today/; "Henry Morgenthau: 1956–1946," *Lapham's Quarterly,* https://www.laphamsquarterly.org/contributors/morgenthau; "George Washington Goethals," Grove School of Engineering, https://www.ccny.cuny.edu /engineering/george-washington-goethals; "Bernard Baruch: Private Life of a Public Man," CUNY, accessed September 12, 2022, https://blogs.baruch.cuny.edu/bernardbaruch/.
5. *Thanks to courses:* Kleitman autobiography.
6. *As a resident alien:* Ibid.; *The City College Quarterly, Volumes 18–19,* March 1922, 37, https:// books.google.com/books?id=P1s9AAAAYAAJ&pg=RA1-PA37&lpg=RA1-PA37 &dq=%22city+college%22+%22new+york%22+%22student+army+training+corps%22 &source=bl&ots=y8Ca8nov6L&sig=r70vA2rYkjdbZ0hKAZujhC9EDak&hl=en &sa=X&ved=0ahUKEwiK2e2f97XcAhVp44MKHdN7Aps4ChDoAQgsMAI#v =onepage&q=%22student%20army%20training%20corps%22&f=false; Betsy B. Creekmore,"World War I—Students' Army Training Corps, 'A' Section," Volopedia, accessed September 12, 2022, https://volopedia.lib.utk.edu/entries/world-war-i-students -army-training-corps-a-section/.
7. *Kleitman graduated in 1919:* Kleitman autobiography.
8. *His master's thesis:* "Sugar in the Blood of the Frog," CLIO, https://clio.columbia.edu /catalog/4301956.
9. *After receiving his MA:* Kleitman autobiography.
10. *His options in both areas:* Stephen Steinberg, "How Jewish Quotas Began," *Commentary,* September 1971; Edward C. Halperin, "The Jewish Problem in U.S. Medical Education, 1920–1955," *Journal of the History of Medicine* 56 (April 2001): 140–167.

11. *In academia, a tidal wave:* Steinberg, "How Jewish Quotas Began."

12. *Several leading universities:* Ibid.; Halperin, "The Jewish Problem."

13. *Likewise, Jews who sought:* Lewis S. Feuer, "The Stages in the Social History of Jewish Professors in American Colleges and Universities," *American Jewish History* 71, no. 4 (June 1982): 437–438.

14. *Yet the barriers:* Ibid.

15. *Women professors:* Patricia Albjerg Graham, "Expansion and Exclusion: A History of Women in American Higher Education," *Signs* 3, no. 4 (Summer 1978): 759–773; Patsy Parker, "The Historical Role of Women in Higher Education," *Administrative Issues Journal* 5, no. 1 (Spring 2015): 3–14.

16. *African Americans, by contrast:* "Key Events in Black Higher Education," *Journal of Blacks in Higher Education,* https://www.jbhe.com/chronology/.

17. *Kleitman's first sponsor:* Kleitman autobiography; "Dr. Abraham J. Goldfarb," *New York Daily News,* April 19, 1962; Abraham J. Goldfarb naturalization record, ancestry .com; Abraham J. Goldfarb Columbia degree, Annual Registry of the College of the City of New York, 1918, 11, https://archive.org/details/annualregisterof191213191415 city; "Generation 2—Salant Family," Isaacs-Salant Family, accessed September 12, 2022, https://isaacssalantfamilytree.com/?s=william+salant.

18. *Now, he told Goldfarb:* Kleitman autobiography. Note: A close reading of this passage in Kleitman's manuscript suggests that Salant had something more than academic qualifications in mind in his search for a candidate to fill the position of instructor. According to Kleitman, Salant had first approached Professor Frank Pike, at Columbia University, Department of Physiology, "who could not recommend anyone." After striking out with Pike (whose obituary describes him a native-born Gentile), Salant turned to Goldfarb at City College, where 90 percent of the student body was Jewish. When Goldfarb recommended Kleitman, Salant asked Pike why he had not done so. Pike replied that "NK was 'no good.'" Nonetheless, Kleitman wrote, "Dr. Salant engaged NK, in desperation." The trajectory of Salant's search, and his willingness to hire Kleitman despite a poor review from Pike, are difficult to explain without considering the era's academic anti-Semitism.

19. *Soon afterward:* Kleitman autobiography.

20. *That June:* Kleitman autobiography.

21. *Anton Julius Carlson:* Lester R. Dragstedt, "Anton Julius Carlson: 1875–1956," National Academy of Sciences, accessed September 11, 2022, http://www.nasonline .org/publications/biographical-memoirs/memoir-pdfs/carlson-anton-j.pdf.

22. *At the end:* Kleitman autobiography.

23. *Before Kleitman could:* Feuer, "The Stages in the Social History," 433–434.

24. *It's also possible:* Kleitman autobiography.

25. *Published in 1913:* Henri Piéron, *Le Problème Physiologique du Sommeil* (Paris: Masson, 1913), passim.

26. *Although Western thinkers:* Sudhansu Chokroverty and Michel Billiard, eds., *Sleep Medicine: A Comprehensive Guide to Its Development, Clinical Milestones, and Advances in Treatment* (New York: Springer, 2015), 50. *The first was the drive:* Kenton

Kroker, *The Sleep of Others and the Transformation of Sleep Research* (Toronto: University of Toronto Press, 2015), 88–90, 96–107, 110–111. *The second impetus:* Ibid., 74–79.

27. *Piéron's book cataloged:* Piéron, *Le Problème Physiologique*, 370, 403–415; Michael J. Thorpy, "History of Sleep Medicine," *Handbook of Clinical Neurology* 98 (2011): 3–25; Kroker, *The Sleep of Others*, 79–83; Henry Hubbard Foster, "The Necessity for a New Standpoint in Sleep Theories," *American Journal of Psychology* 12, no. 2 (January 1901): 145–177.

28. *Thus arose various:* Piéron, *Le Problème Physiologique*, 390–398; Thorpy, "History of Sleep Medicine"; Foster, "The Necessity."

29. *In 1862:* Piéron, *Le Problème Physiologique*, 25, 401, 435; Mathias Basner, "Arousal Threshold Determination in 1862: Kohlschütter's Measurements on the Firmness of Sleep," *Sleep Medicine* 11 (2010): 417–422.

30. *Kohlschütter's study suffered:* Ibid.

31. *In 1888:* Piéron, *Le Problème Physiologique*, 25, 225, 434, 435, 436; Matthias M. Weber and Wolfgang Burgmair, "'The Assistant's Bedroom Served as a Laboratory': Documentation in 1888 of Within Sleep Periodicity by the Psychiatrist Eduard Robert Michelson," *Sleep Medicine* 10 (2009): 378–384.

32. *When Michelson crunched:* Ibid.

33. *Le Problème Physiologique du Sommeil described:* Piéron, *Le Problème Physiologique,* 41, 50, 54, 103, 141, 270, 431; Kroker, *The Sleep of Others,* 153, 158–161; Michel Huteau, "Un Météore de la Psychology Française: Nicolae Vaschide (1874–1907)," *Bulletin de psychologie* 2, no. 494 (2008): 173–199.

34. *Other researchers had:* Piéron, *Le Problème Physiologique*, 23, 51, 105; Antonio Zadra and Robert Stickgold, *When Brains Dream: Understanding the Science and Mystery of Our Dreaming Minds* (New York: W. W. Norton, 2022), https://books .google.com/books?id=ADPxDwAAQBAJ&pg=PT29&lpg=PT29&dq=%22sante +de+sanctis%22+dogs&source=bl&ots=4fODl3nLlb&sig=ACfU3U35bVVQgZuk41 RRHTs4ANUHeDvOng&hl=en&sa=X&ved=2ahUKEwjgjOC_gIT6AhXcL0Q IHTP5BHIQ6AF6BAgfEAM#v=onepage&q=%22sante%20de%20sanctis%22 %20dogs&f=false; Chiara Bartolucci et al., "Sante De Sanctis' Contribution to the Study of Dreams Between '800 and '900 Century: The Originality of the Integrated Method," *International Journal of Dream Research* 9, no. 1 (2016): 22–33.

35. *as Charles Dickens suggested:* Charles Dickens, *Works of Charles Dickens, Volume 5* (New York: Hurd and Houghton, 1872), 27.

36. *The book also:* Piéron, *Le Problème Physiologque*, 170, 375, 380; Kroker, *The Sleep of Others*, 194, 196; James E. Lebensohn, "The Eye and Sleep," *Archives of Ophthalmology* 25, no. 3 (1941): 401–411.

37. *Yet Mauthner's theory:* Cobb, *The Idea of the Brain: The Past and Future of Neuroscience* (New York: Basic Books, 2020): 80–102; Kleitman, *Sleep and Wakefulness as Alternating Phases in the Cycle of Existence.*(Chicago: University of Chicago Press, 1939), 337, 492–501; Thorpy, "History of Sleep Medicine."

38. *The relationship between sleep:* Piéron, *Le Problème Physiologqgue*, 375, 404; Cobb, *The Idea of the Brain,* 134–156; Thorpy, "History of Sleep Medicine"; Foster, "The Necessity."

39. *One thing that:* Piéron, *Le Problème Physiologique*, 261–262; M. Bentivoglio and G. Grassi-Zucconi, "The Pioneering Experimental Studies on Sleep Deprivation," *Sleep* 20, no. 7 (July 1997): 570–576, doi: 10.1093/sleep/20.7.570; Marie de Manacéïne, "Quelques Observations Expérimentales sure l'Influence de l'Insomnie Absolue," *Archives Italiennes de Biologie* 21 (1894): 322–325.

40. *Other investigators replicated:* Piéron, *Le Problème Physiologique*, 259, 270; Bentivoglio and Grassi-Zucconi, "Pioneering Experimental Studies"; G. T. W. Patrick and J. Allen Gilbert, "Studies from the Psychological Laboratory of the University of Iowa," *Psychological Review* 3, no. 5 (September 1896): 469–483; Kerry Grens, "Go to Bed!," *The Scientist,* March 2016, accessed September 8, 2022, https://www.the-scientist.com/features/go-to-bed-33974.

41. *Such results lent credence:* The first researcher to uncover evidence of fatigue toxins was Japanese physiologist Kuniomi Ishimori. In a study published in a Tokyo medical journal in 1909, Ishimori extracted alcohol-soluble substances from the brains of five sleep-deprived dogs and injected them into seven normal dogs; the recipients fell into a deep sleep within sixty minutes. His work, however, drew no attention outside Japan and was quickly forgotten even there. See Kisou Kubota, "Kuniomi Ishimori and the First Discovery of Sleep-Inducing Substances in the Brain," *Neuroscience Research* 6 (1989): 497–518.

42. *The most extensive study:* Piéron, *Le Problème Physiologique*, 294–306; Claude Gottesmann, *Henri Piéron and Nathaniel Kleitman: Two Major Figures of 20th Century Sleep Research* (New York: Nova Science Publishers, 2013), 15–17, 26.

43. *Yet Piéron was:* Gottesmann, *Henri Piéron,* 31; Kroker, *The Sleep of Others,* 205.

44. *Kleitman found these:* Interview by Mel Stuart, questions and correspondence, questionnaire, 1991. Nathaniel Kleitman Papers, Box 1, Folder 4.

45. *Two more elements:* Kroker, *The Sleep of Others,* 193–196; Lazaros Triarhou, "The Percipient Observations of Constantin von Economo on Encephalitis Lethargica and Sleep Disruption and Their Lasting Impact on Contemporary Sleep Research," *Brain Research Bulletin* 69 (2006): 244–258; Morris S. Dickman, "Von Economo Encephalitis," *Archives of Neurology* 58 (October 2001): 1696–1698.

46. *With no known:* Kroker, *The Sleep of Others,* 195–196; Triarhou, "The Percipient Observations"; Dickman, "Von Economo"; Oliver Sacks, *Awakenings* (New York: Vintage Books, 1999), passim.

47. *The other factor:* Daniel P. Todes, *Ivan Pavlov: A Russian Life in Science* (Oxford: Oxford University Press, 2014), 306, 735.

48. *Pavlov believed this:* Nathaniel Kleitman, "Studies on the Physiology of Sleep: I. The Effects of Prolonged Sleeplessness on Man," *American Journal of Physiology* 66 (1923): 67–92; Y. Popov and L. Rokhin, eds., *I. P. Pavlov: Psychopathology and Psychiatry, Selected Works* (Moscow: Foreign Languages Publishing House, n.d.), 109–125, 513–514.

49. *His mentor, as:* Dwight J. Ingle, "Anton J. Carlson: A Biographical Sketch," *Perspectives in Biology and Medicine* 22, no. 2 (Winter 1979): S114–S137; Lester R. Dragstedt, "Anton Julius Carlson: 1875–1956," National Academy of Sciences, 1961, accessed September 10, 2022, http://www.nasonline.org/publications/biographical-memoirs /memoir-pdfs/carlson-anton-j.pdf.

50. *Early in his career:* Kroker, *The Sleep of Others,* 212–215; Ingle, "Anton J. Carlson"; Dragstedt, "Anton Julius Carlson."

51. *Later, he'd turned:* "Scientist's Scientist," *Time,* February 10, 1941, 44–48; Ingle, "Anton J. Carlson"; Anton J. Carlson, *The Control of Hunger in Health and Disease* (Chicago: University of Chicago Press, 1916), 30, 161. Note: Vlcek's name is often misspelled as "Vleck" in secondary sources.

52. *As a boy:* Kroker, *The Sleep of Others,* 212–215; Carlson, 30, 161.

53. *Carlson's research had:* Kroker, *The Sleep of Others,* 212–215; Carlson, *The Control of Hunger,* 38, 153–155; Mark A. Andrews, "Why Does Your Stomach Growl When You Are Hungry?," *Scientific American,* January 21, 2022, accessed September 11, 2022, https://www.scientificamerican.com/article/why-does-your-stomach-gro/#.

54. *From Carlson, Kleitman:* Kroker, *The Sleep of Others,* 212–215.

55. *The cycle Kleitman:* Interview by Mel Stuart, 1991; Kleitman, "The Effects of Prolonged Sleeplessness on Man."

56. *Sometime in the:* Kleitman autobiography; Kleitman letter to Edward J. Murray, February 22, 1964, Nathaniel Kleitman Papers, Box 1, Folder 5.

57. *As warm-up exercises:* Nathaniel Kleitman, "Studies on the Physiology of Sleep: V. Some Experiments on Puppies," *American Journal of Physiology* 84 (1928): 386–395; Kleitman lab notes, Nathaniel Kleitman Papers, Box 1, Folder 8.

58. *Then he embarked:* Kleitman, "The Effects of Prolonged Sleeplessness on Man."

59. *Kleitman's avowed goal:* Ibid.

60. *The fledgling researcher:* Ibid.

61. *Kleitman's approach to:* Ibid.

62. *The most striking:* Ibid.

63. *Although participants invariably:* Ibid.

64. *There were other:* Ibid.

65. *Kleitman titled:* Ibid.

66. *He began with:* Ibid.

67. *Kleitman surveyed:* Ibid.

68. *Lastly, he turned:* Ibid.

69. *He then offered:* Ibid.

70. *This blockade led:* Ibid.

71. *As for sleep's:* Ibid.

72. *After cautioning that:* Ibid.

73. *Kleitman had stated:* Kleitman autobiography; Gottesmann, *Henri Piéron,* 114.

74. *During the summer:* Kleitman autobiography; Franklin McLean, "Arno B. Luckhardt, Physiologist," *Science* 127, no. 3297 (March 7, 1958): 509.

75. *His hometown had:* Kleitman autobiography; Yitzchak Koren, *The Jews of Kishinev* (New York: JewishGen, 2019), 173–175; Eli Wiesel et al., eds., *Final Report of the International Commission on the Holocaust in Romania,* 2005, accessed September 12, 2022, https://www.jewishvirtuallibrary.org/jsource/Holocaust/Romania/one.pdf.

76. *Kleitman spent most:* Kleitman autobiography.

Chapter 3

1. *The Physiology Building:* Postcard (c. 1916) from author's collection; Anonymous, *The University of Chicago: An Official Guide* (Chicago: University of Chicago Press, 1916), 42–44. The building, renamed Culver Hall, now houses a variety of biology labs.

2. *At the start:* Kleitman to J. B. Hoskisson, April 10, 1973; Kleitman to Gérard Lemaine, July 1, 1974, Nathaniel Kleitman Papers, Box 12, Folder 9, Hanna Holborn Gray Special Collections Research Center, University of Chicago Library.

3. *Nor did he confine:* Lists of Kleitman publications, 1924–1925, 1925–1926, 1926–1927, 1927–1928, 1928–1929,1930–1931, Nathaniel Kleitman Papers, Box 15, Folder 3.

4. *Nonetheless, he set:* Kenton Kroker, *The Sleep of Others and the Transformation of Sleep Research* (Toronto: University of Toronto Press, 2015), 205–256.

5. *For a few hours:* Nathaniel Kleitman autobiography, dictated to Paulena Kleitman circa 1970s, Nathaniel Kleitman Papers, Box 1, Folder 7.

6. *whether due to*: For possible causes of eyebrow loss, see Annapurna Kumar and Kaliaperumal Karthikeyan, "Maderosis: A Marker of Many Maladies," *International Journal of Trichology* 4, no. 1 (January–March 2012): 3–18.

7. *Kleitman wasn't the:* Kroker, *The Sleep of Others,* 205–206.

8. *Aided by a rotating:* Kleitman, *Sleep and Wakefulness,* passim.

9. *When Kleitman began:* Alan Derickson, *Dangerously Sleepy: Overworked Americans and the Cult of Manly Wakefulness* (Philadelphia: University of Pennsylvania Press, 2014), 15–17.

10. *A native of Angola:* "Another Angola Boy Winning His Way," *Angola Herald,* September 10, 1920, 1.

11. *Laird was two:* "Dr. Donald Laird, Popularizer of Psychology, Dies," *New York Times,* December 23, 1969, 32; "Donald A. Laird: 1897–1969," Colgate University Biographical Files, Special Collections and University Archives, Colgate University Libraries.

12. *Sleep, however, was not:* "Dr. Donald Laird Declares Effect of Liquor Is Undesirable One," *Garrett Clipper,* January 8, 1925, 2; "Criminals Not Mentally Normal, Says Colgate University Professor," *Brooklyn Daily Eagle,* March 1, 1925, 14; "Tests Show Women Students to Be More Unstable Than Men," *Journal* (Logan, UT), September 30, 1925, 6; "Long Day Due to Men in the Wrong Jobs," *Rutland Daily Herald,* March 11, 1925, 9.

13. *Laird's sleep studies:* "The Sleep Lab at Colgate," Colgate University Biographical Files; "Read It and Sleep—a Book for Insomnia," *Wisconsin State Journal,* March 30, 1930, 36.

14. *He set up:* Donald A. Laird, "Human Guinea Pigs Help Science Find New Facts About Sleep," *McAllen Daily Press,* May 27, 1928; Donald A. Laird and Charles G. Muller,

"Sleep: Science Explains How to Make the Most of Your Slumbers," *Maclean's*, December 1, 1932, 10, 42–45; "What Science Is Doing to Make You Sleep Better," *El Paso Times*, December 16, 1934, 6; Donald Anderson Laird and Charles Geoffrey Muller, *Sleep: Why We Need It and How to Get It* (New York: John Day, 1930), 99.

15. *But Laird seemed:* Laird and Muller, "Sleep: Science Explains."

16. *Laird had a:* Donald A. Laird, "Authority on Sleep Explains How to Reduce the Hours of Repose," *Platteville Journal*, April 15, 1925, 8.

17. *Laird later urged:* Laird, "Human Guinea Pigs."

18. *He also flip-flopped:* Laird, "Authority on Sleep Explains."

19. *In 1932, however:* Laird and Muller, "Sleep: Science Explains."

20. *By then, Laird had:* Laird and Muller, *Sleep*; Donald A. Laird, "Finding Out About Human Sleep: Announcement of an Unusual Lecture Illustrated Throughout with Original Motion Pictures," undated promotional flyer, Colgate University Biographical Files.

21. *Such manuals, which:* Examples include Robert Macnish's *The Philosophy of Sleep* (Glasgow: W. R. McPhun, 1830) and Marie de Manacéïne's *Sleep: Its Physiology, Pathology, Hygiene, and Psychology* (London: Walter Scot, 1897).

22. *But he left:* "Ayer Sets Up Body for Consumer Study: Dr. Laird to Direct Analyzing of Public's Reactions," *New York Times*, January 23, 1939, 20.

23. *his work is largely forgotten:* Although Laird is cited several times in Kleitman's 1939 textbook *Sleep and Wakefulness*, none of the studies listed could be described as seminal. And he goes unmentioned in Kenton Kroker's highly detailed *The Sleep of Others and the Transformations of Sleep Research*, or in the condensed histories of sleep science found in sleep medicine journals and textbooks.

24. *At least Laird:* Carl Shoup, "Now Science Proposes to Abolish Sleep," *St. Louis Post-Dispatch*, August 30, 1925.

25. *Kleitman, by contrast:* This assessment is based on searches I conducted on Newspapers .com (https://www.newspapers.com), NewspaperArchive (https://newspaperarchive .com), and the *New York Times* (https://www.nytimes.com) for articles quoting Kleitman from 1923 through 1935.

26. *Kleitman began cautiously:* Claude Gottesmann, *Henri Piéron and Nathaniel Kleitman: Two Major Figures of 20th Century Sleep Research* (New York: Nova Science Publishers, 2013), 45–54.

27. *He also continued:* "Studies Sleep by Going Without It," *Ogden Standard-Examiner*, February 21, 1927, 2; "Stay Awake 115 Hours in Test: Chicago Scientists Suffer Agonies in Study of Effects of Wakefulness," *Courier-Journal* (Louisville, KY), June 11, 1925, 17.

28. *The stories usually:* "Are Humans Vegetables; and Why Do They Sleep?," *Chino Champion*, August 24, 1928, 7, syndicated from *New York Herald-Tribune*.

29. *Yet Kleitman was:* Nathaniel Kleitman and George Crisler, "A Quantitative Study of a Salivary Conditioned Reflex," *American Journal of Physiology* 79, no. 3 (February 1927): 571.

30. *Pavlov's theory seemed:* Y. Popov and L. Rokhin, eds., *I. P. Pavlov: Psychopathology and Psychiatry, Selected Works* (Moscow: Foreign Languages Publishing House, n.d.), 53–59, 109–125, 513.

31. *Pavlov had arrived:* Ibid., 53–59, 109–125, 513; Daniel P. Todes, *Pavlov: A Russian Life in Science* (Oxford: Oxford University Press, 2014), 306, 343, 402–3, 522–523.

32. *But Kleitman and:* "Dr. Crisler Dies at 64 at His Home," *Orlando Sentinel,* September 27, 1966, 25; Kleitman and Crisler, "A Quantitative Study," 571–614.

33. *The study involved:* Kleitman and Crisler, "A Quantitative Study."

34. *The pair's first:* Ibid.

35. *"[Our] results differ":* Ibid., 602.

36. *Paulena Schweizer was:* Hortense Kleitman Snower interview; "Paulena Kleitman," FamilySearch, https://www.familysearch.org/tree/person/details/GMNH-3P4; "Canadian County," Encyclopedia of Oklahoma History and Culture, https://www.okhistory .org/publications/enc/entry.php?entry=CA038&l=; "Goldberg Cohen Family Tree," http://goldbergcohenfamily.info/gedcom/p199.html#I9949; John Alley, "Oklahomans in the Spanish-American War," This Land Press, December 30, 2014, accessed September 17, 2022, https://thislandpress.com/2014/12/30/oklahomans-in-the-spanish-american -war/; "Happily Wedded: Jacob Schweizer and Minnie Joseph United in Marriage at the Home of the Latter's Parents Last Evening," *Wichita Daily Eagle,* January 22, 1891, 5.

37. *Dark-haired and:* "A War Pageant Star," *Kansas City Star,* September 18, 1917, 7.

38. *She'd gone on:* Paulena Schweizer academic transcript, September 29, 1920, University of Chicago Library Special Collections. Her major was not specified, but her coursework toward a PhB centered on the social sciences.

39. *and to volunteer:* Hortense Kleitman Snower interview; "About," Jane Addams Hull-House Museum, accessed September 17, 2022, https://www.hullhousemuseum.org /about-jane-addams.

40. *When the thirty-two-year-old:* Kleitman autobiography.

41. *In the fall:* Ibid.

42. *By then, construction:* "Abbott Memorial Hall," Phorio Building Database, https:// en.phorio.com/abbott_memorial_hall,_chicago,_united_states; "Abbott Memorial Hall," University of Chicago Photographic Archive, http://photoarchive.lib.uchicago.edu /db.xqy?one=apf2-00033.xml; John Easton, "UChicago Medicine Celebrates 90 Years of Patient Care," UChicago Medicine, October 20, 2017, accessed June 1, 2021, https://www.uchicagomedicine.org/forefront/patient-care-articles/uchicago-medicine -celebrates-90-years-of-patient-care.

43. *Thanks to his:* William C. Dement, *The Sleepwatchers* (Menlo Park, CA: Nychthemeron Press, 1995), 28; interview with Armond Aserinsky, October 11, 2021.

44. *The young scientist:* Jack A. Rall, "The XIIIth International Physiological Congress in Boston in 1929: American Physiology Comes of Age," *Advances in Physiology Education* 40 (2016): 5–16, accessed June 1, 2021, https://journals.physiology.org/doi /full/10.1152/advan.00126.2015.

45. *The white-bearded icon:* Ibid., 11.

46. *Meanwhile, Kleitman had examined:* Kroker, *The Sleep of Others,* 217.

47. *According to Pavlov:* Ibid., 218–219.

48. *Whether it shocked:* Kleitman autobiography.

49. *The lecture also brought:* "Physiology Wonders Why We Go to Sleep," *Honolulu Advertiser*, September 17, 1929, 12.

50. *The* Advertiser's *pundit:* Kroker, *The Sleep of Others*, 349–350; Laird and Muller, *Sleep*, 165–171; Edmund Jacobson, *You Can Sleep Well: The ABC's of Restful Sleep for the Average Person* (New York: Whittlesey House, 1938), 3–130.

51. *By 1930, an estimated:* "Sleep: A Historical Perspective," Healthy Sleep, accessed February 20, 2020, http://healthysleep.med.harvard.edu/interactive/timeline.

52. *Like the bromides:* Francisco Lopez-Muñoz, Ronaldo Ucha-Adabe, and Cecilio Alamo, "The History of Barbiturates a Century After Their Clinical Introduction," *Neuropsychiatric Disease and Treatment* 1, no. 4 (2005): 329–343; Peter Davis, ed., *Contested Ground: Public Purpose and Private Interest in the Regulation of Prescription Drugs* (New York: Oxford University Press, 1996), 43.

53. *There was narcolepsy:* Kroker, *The Sleep of Others*, 338–343.

54. *There was shift work:* Benjamin Reiss, *Wild Nights: How Taming Sleep Created Our Restless World* (New York: Basic Books, 2017), 39–40.

55. *Researchers were making:* Lazaros Triarhou, "The Percipient Observations of Constantin von Economo on Encephalitis Lethargica and Sleep Disruption and Their Lasting Impact on Contemporary Sleep Research," *Brain Research Bulletin* 69 (2006): 244–258.

56. *Von Economo never:* Yasemin Kaya et al., "Constantin von Economo (1876–1931) and His Legacy to Neuroscience," *Child's Nervous System* 21 (February 24, 2015): 217–220.

57. *That year, however:* C. W. Hess, "Walter R. Hess," *Swiss Archives of Neurology, Psychiatry and Psychotherapy* 158, no. 4 (2008): 255–261.

58. *Kleitman wasn't convinced:* Nathaniel Kleitman and Nelaton Camille, "The Behavior of Decorticated Dogs," *American Journal of Physiology* 100, no. 3 (April 1932): 474–480.

59. *So Kleitman turned:* Ibid.

60. *This was, in fact:* Róbert Bódizs et al., "Sleep in the Dog: Comparative, Behavioral and Translational Relevance," *Current Opinion in Behavioral Sciences* 33 (2020): 25–33; Austin Meadows, "How Much Sleep Do Dogs Need?," Sleep.org, March 12, 2021, accessed June 1, 2021, https://www.sleep.org/how-much-do-dogs-sleep/. *cultures that practiced biphasic sleep:* A. Roger Ekirch, *At Day's Close: Night in Times Past* (New York: W. W. Norton, 2006), 300–311.

61. *Previous experiments had:* Kleitman and Camille, "The Behavior of Decorticated Dogs."

62. *With a young Haitian:* Achille Aristide, *Memoire sur la municipalité en Haïti*, Imprimerie de L'Etat, Port-au-Prince, 1952, accessed June 1, 2021, https://ufdc.ufl.edu/AA00000876 /00001/3?search=nelaton; https://ufdc.ufl.edu/AA00000876/00001/48?search=nelaton; Rockefeller Foundation Directory of Fellowships and Scholarships, 1917–1970, 50, https://books.google.de/books?id=FmxBAAAAIAAJ&q=%22CAMILLE,+NELATON +(Haiti)+b.+1895+Haiti.+M.D.+Natl.%22&dq=%22CAMILLE,+NELATON+(Haiti) +b.+1895+Haiti.+M.D.+Natl.%22&hl=en&sa=X&redir_esc=y; George S. Schuyler, "Views and Reviews," *Pittsburgh Courier*, May 17, 1952, 7.

63. *Kleitman performed:* Kleitman and Camille, "The Behavior of Decorticated Dogs."

64. *To Kleitman, these:* Ibid.

65. *Decades later:* Triarhou, "The Percipient Observations of Constantin von Economo."

66. *The likeliest explanation:* Interview with Donald Bliwise (sleep researcher and professor of neurology, Emory University School of Medicine), April 8, 2022.

67. *Kleitman had guessed:* Triarhou, "The Percipient Observations of Constantin von Economo"; Anna Aulinas, "Physiology of the Pineal Gland and Melatonin," Endotext, accessed June 1, 2021, https://www.ncbi.nlm.nih.gov/books/NBK550972/.

68. *Kleitman's work in the early 1930s:* Kroker, *The Sleep of Others*, 222–223.

69. *Across the country:* Gene Smiley, "Great Depression," Econlib, https://www.econlib .org/library/Enc/GreatDepression.html; Robert Longley, "Hoovervilles: Homeless Camps of the Great Depression," ThoughtCo, https://www.thoughtco.com/hoover villes-homeless-camps-of-the-great-depression-4845996; Stanley Schultz, *The Great Depression: A Primary Source History* (Milwaukee: Gareth Stevens, 2006), 14.

70. *At the University:* Kroker, *The Sleep of Others*, 222–223.

71. *A food conglomerate:* Dwight J. Ingle, "Anton J. Carlson: A Biographical Sketch," *Perspectives in Biology and Medicine* 22, part 2 (Winter 1979): S126; cartoon viewed on eBay.com, June 1, 2021.

72. *Despite these distractions:* These studies are detailed extensively in Kleitman's 1939 textbook, *Sleep and Wakefulness as Alternating Phases in the Cycle of Existence.*

73. *Of all the:* Kleitman, *Sleep and Wakefulness*, 191–210.

74. *He began his:* Kleitman autobiography.

75. *Family members would:* Interview with Hortense Kleitman Snower, June 10, 2016; interview with Bruce Levitt (nephew of Nathaniel Kleitman), June 9, 2021.

76. *Although it was:* Snower interview; Kroker, *The Sleep of Others*, 228–230; Matthew J. Wolf-Meyer, *The Slumbering Masses: Sleep, Medicine, and Modern American Life* (Minneapolis: University of Minnesota Press, 2016), 133–134.

77. *Kleitman did:* Gottesmann, *Henri Piéron*, 60–64; Kroker, *The Sleep of Others*, 228–230.

78. *The motility experiments:* Gottesmann, *Henri Piéron*, 60–64; Kleitman, *Sleep and Wakefulness*, 138–140.

79. *Kleitman then turned:* Gottesmann, *Henri Piéron*, 60–61.

80. *Next, he studied:* Ibid., 65–66.

81. *In April 1934:* "Dr. Kleitman Advocates Snoozes in Large Quantities as He Gives Latest Facts on New Discoveries," *Ogden Standard-Examiner*, April 18, 1934, 1.

82. *Nor could most:* Ibid.

83. *Kleitman was beginning:* Kroker, *The Sleep of Others*, 223–226; Kleitman autobiography.

84. *With Paulena's help:* Key for color-coded system used on bibliographic cards, undated, Nathaniel Kleitman Papers, Box 15, Folder 8.

85. *For his investigations:* Kroker, *The Sleep of Others*, 223.

86. *In March 1934:* Ibid., 223, 225–226.

87. *It's easy to imagine:* Wilder Penfield, *The Difficult Art of Giving: The Epic of Alan Gregg* (Boston: Little, Brown, 1967), 6–15; "Alan Gregg: Biographical Overview," US

National Library of Medicine Profiles in Science, accessed June 1, 2021, https://profiles
.nlm.nih.gov/spotlight/fs/feature/biographical-overview.

88. *In any case:* Kroker, *The Sleep of Others*, 226–227.

89. *"I told him we":* Alan Gregg diary, March 28, 1934, "Prof. Kleitman," Rockefeller Foundation Records, Rockefeller Archive Center, Record Group 12.

90. *Months of silence:* Kroker, *The Sleep of Others*, 226–227.

91. *In May 1934:* Ibid.

92. *Still skeptical, but:* Ibid.

93. *With Gregg's support:* Ibid., 229–231.

94. *Around the same:* Frederic Bremer, "Cerveau 'Isolé' et Physiologie du Sommeil," *Comptes Rendus des Séances de la Société de Biologie* 118 (1935): 1235–1241; Peretz Lavie and Myriam Kerkhofs, "Frederic Bremer 1892–1982." *Newsletter, World Federation of Sleep Research Societies* 6, no. 2 (1999), accessed September 18, 2022, http://www.edumed .org.br/cursos/neurociencia/cdrom/Biblioteca/HistorySleepResearch.htm; Claude Gottesmann, "What the Cerveau Isolé Preparation Tells Us Nowadays About Sleep-Wake Mechanisms," *Neuroscience and Biobehavioral Reviews* 12 (1988): 39–48.

95. *To Bremer's bewilderment:* Bremer, "Cerveau 'Isolé.'"

96. *The cerveau isolé study:* Kleitman, *Sleep and Wakefulness*, 50–51.

97. *But a formidable:* Kroker, *The Sleep of Others*, 231–234.

98. *In the years:* Ibid., 231–232.

99. *Only one was:* Gottesmann, *Henri Piéron*, 68–69.

100. *In 1937, Kleitman began:* Kleitman, *Sleep and Wakefulness*, 252–259.

101. *This was not:* Ibid.

102. *For a month, Kleitman tried:* Ibid.

103. *To determine where:* Ibid.

104. *Why couldn't he:* Ibid.

105. *Hoping to eliminate those variables:* Kroker, *The Sleep of Others*, 232–233.

106. *The rejection must:* Ibid.

107. *Then a colleague:* Interview with Colleen Olson (Mammoth Cave National Park ranger-historian), September 2, 2016; Kleitman to J Harlan Bretz, March 1, 1938, Nathaniel Kleitman Papers, Box 12, Folder 12; Kleitman, *Sleep and Wakefulness*, 259.

Chapter 4

1. *Between ten and fifteen million:* "How Mammoth Cave Formed," National Park Service, https://www.nps.gov/maca/learn/nature/how-mammoth-cave-formed.htm; "Rocks of Mammoth Cave," National Park Service, https://www.nps.gov/maca/learn/nature /rocks-of-mammoth-cave.htm; "Exploring the World's Longest Known Cave," National Park Service, accessed September 22, 2022, https://www.nps.gov/articles/000/exploring -the-worlds-longest-known-cave.htm.

2. *Indigenous people first:* "Early Native Americans," National Park Service, https://www .nps.gov/maca/learn/historyculture/native-americans.htm; "History of Mammoth Cave," OhRanger.com, http://www.ohranger.com/mammoth-cave/history-mammoth-cave; Rick Olson, "The Lamps That Lit Their Way," October 9, 2008, Mammoth Cave Research

Symposia, Paper 6, http://digitalcommons.wku.edu/mc_reserch_symp/9th_Research
_Symposium_2008/Day_two/6.

3. *The earliest commercial:* "Kentucky: Mammoth Cave National Park," National Park
Service, accessed September 22, 2022, https://www.nps.gov/articles/mammothcave
.htm.

4. *One of the latter:* "Steven Bishop," National Park Service, https://www.nps.gov/people
/stephen-bishop.htm; Austyn Gaffney, "This U.S. National Park Has the World's
Longest Cave System—and an Unusual History," *National Geographic,* April 5, 2021,
https://www.nationalgeographic.com/travel/article/unusual-history-of-mammoth-cave
-national-park-worlds-longest-cave-system; "Mammoth Cave: Timeline," National Park
Service, accessed September 22, 2022, https://www.nps.gov/maca/learn/historyculture
/timeline.htm.

5. *As hotels sprang up:* "The Kentucky Cave Wars," National Park Service, https://www
.nps.gov/articles/000/the-kentucky-cave-wars.htm; Dave Tabler, "The Kentucky Cave
Wars," Appalachian History.Net, https://www.appalachianhistory.net/2017/04/kentucky
-cave-wars.html; "Mammoth Cave National Park," *New World Encyclopedia,* accessed
September 22, 2022, https://www.newworldencyclopedia.org/entry/Mammoth_Cave
_National_Park.

6. *By 1938:* Email messages from Colleen Olson (Mammoth Cave National Park ranger-
historian) and Terry Langford (Mammoth Cave National Park curator) to author,
September 6, 2016.

7. *When Kleitman told:* John Sonnichsen, "Legacy: J Harlen Bretz (1882–1982)," *University of Chicago Magazine,* November–December 2009, http://magazine.uchicago
.edu/0912/features/legacy.shtml; Nathaniel Kleitman, *Sleep and Wakefulness as Alternating Phases in the Cycle of Existence* (Chicago: University of Chicago Press, 1939), 259;
Cassandra Tate, "Bretz, J Harlen (1882–1981)," HistoryLink.org, November 29, 2007,
accessed September 22, 2022, https://www.historylink.org/File/8382; interview with
Colleen Olson, September 2, 2016.

8. *That February, Kleitman:* Matthew Wolf-Meyer, "Where Have All Our Naps Gone?
Or Nathaniel Kleitman, the Consolidation of Sleep, and the Historiography of
Emergence," *Anthropology of Consciousness* 24, no. 2 (2013): 101–102; Kleitman to
W. W. Thompson, May 5, 1938, Nathaniel Kleitman Papers, Box 12, Folder 12, Hanna
Holborn Gray Special Collections Research Center, University of Chicago Library.
Note: In his letter to Thompson, Kleitman placed a comma after "at least." This could
be read as meaning that Kleitman was anxious to avoid unnecessary publicity but
was less concerned about avoiding *necessary* publicity. However, the sentence that fol-
lows makes it clear that he wanted to avoid *any* publicity: "We believe that all prepara-
tions can be made unobtrusively and that the necessary communications with us while
we are in the cave can be made in a routine manner without attracting the attention
of the numerous visitors to the cave." In this context, it appears that the comma was
inserted erroneously, and I've taken the liberty of removing it.

9. *Thompson agreed:* Colleen Olson interview; email message from Olson to author, September 3, 2016; Joseph Mader, "Captions for Picture Series on Scientists in Mammoth

Cave," Nathaniel Kleitman Papers, Box 12, Folder 12; Wolf-Meyer, "Where Have All Our Naps Gone?"; Bill Hayes, *Sleep Demons: An Insomniac's Memoir* (Chicago: University of Chicago Press, 2018), 11; Matthew Walker, *Why We Sleep: Unlocking the Power of Sleep and Dream* (New York: Scribner, 2017), 16; Colleen Olson interview; "Scientists Deep in Cave Phone for Prize Fight News," *St. Louis Post-Dispatch*, June 22, 1938, 2; S. V. Stiles, "Twenty-Four Hours Is Habit! Two Chicago Psychologists Find It a Hard One to Break, Even in Mammoth Cave," *Cincinnati Enquirer*, June 12, 1938, 36.

10. *On June 4:* Kleitman, *Sleep and Wakefulness*, 259; Stiles, "Sleep Is a Habit"; "Science: Cave Men," *Time*, July 18, 1938, http://content.time.com/time/subscriber/article /0,33009,760040,00.html; "Bruce Harrison Richardson," Ancestry.com, accessed September 22, 2022, https://www.ancestry.com/family-tree/person/tree/26481178/person /13764018302/facts?_phsrc=FCR2&_phstart=successSource; Colleen Olson interview. Note: Contemporary newspaper accounts give the depth of Rafinesque Hall as 119 feet, but Olson informed me that the correct figure is 140 feet.

11. *Kleitman and Richardson kept:* Wolf-Meyer, "Where Have All Our Naps Gone?"

12. *A hotel employee:* Mader, "Captions for Picture Series on Scientists in Mammoth Cave"; "2 Try Living in State Cave to Break 24-Hour Day Habit," *Courier-Journal*, June 12, 1938, 13; email message from Mammoth Cave National Park curator Terry Langford, July 28, 2021; Kleitman to H. S. Sanborn, May 3, 1938, Nathaniel Kleitman Papers, Box 12, Folder 12.

13. *To pass the time:* "2 Try Living in State Cave"; "Physiologists Report on Life Underground," *Escanaba Daily Press*, July 3, 1938, 1; "Their Test of 28-Hour Day Indecisive," *Oregon Statesman*, July 8, 1938, 1; Marder, "Captions for Picture Series on Scientists in Mammoth Cave."

14. *For five twenty-eight-hour cycles:* W. W. Thompson to Kleitman, June 10, 1938, Nathaniel Kleitman Papers, Box 12, Folder 12.

15. *Kleitman was distressed:* Statement by Nathaniel Kleitman and Bruce Richardson, July 6, 1938, Nathaniel Kleitman Papers, Box 12, Folder 12.

16. *He feared that:* Kleitman to Thompson, May 5, 1938, Nathaniel Kleitman Papers, Box 12, Folder 12.

17. *Yet there was danger, too:* Thompson to Kleitman, June 10, 1938; statement by Kleitman and Richardson; Stiles, "Twenty-Four Hours Is Habit!"

18. *Stiles's seven-hundred-word dispatch:* "2 Try Living in State Cave."

19. *The story was picked up:* "Two Men Conducting '28-Hour Day' Experiment Deep in Mammoth Cave," *New Castle News*, June 14, 1938, 5; "28-Hour Day Test in Depths of Mammoth Cave," *St. Louis Post-Dispatch*, June 12, 1938, 8; "Scientists Deep in Cave Phone for Prize Fight News"; "American Experience: The Louis-Schmeling Fights, 1936 and 1938," PBS, https://www.pbs.org/wgbh/americanexperience/features /fight-louis-schmeling-fights-1936-and-1938/; "Physiologists Report on Life Underground," *Escanaba Daily Press*, July 3, 1938, 1; Thompson to Kleitman, June 13, 1938, Nathaniel Kleitman Papers, Box 12, Folder 12.

20. *Then Kleitman received:* Wolf-Meyer, "Where Have All Our Naps Gone?"; to Kleitman from Jack H. Lieb, July 1, 1938, Nathaniel Kleitman Papers, Box 12, Folder 12.

21. *The clip, which was shown:* "Sleep Research," YouTube video, 1:24, posted by "Yogi242424," July 1, 2008, https://www.youtube.com/watch?v=xMH9eF5Bq70; "Hearst Metrotone News Collection," UCLA Film and Television Archive, accessed September 22, 2022, https://www.cinema.ucla.edu/sites/default/files/HearstNewsreel.pdf.

22. *Then comes a close-up:* "Sleep Research," YouTube.

23. *After Richardson adds:* Ibid.

24. *The newsreel is in some ways:* Colleen Olson interview.

25. *The announcer mispronounces:* Ibid.

26. *And contrary to the scientists' declaration:* "Physiologists Report on Life Underground"; Kleitman, *Sleep and Wakefulness,* 260–264.

27. *Yet such details:* "Old World Looks Strange After a Month in Cavern," *Washington, C.H. Record-Herald,* July 7, 1938, 3; "Scientists Check Data on 33-Day Cave Story," *Democrat and Chronicle,* July 8, 1938, 1; "Bearded Scientists Near End of 33 Days' Stay in Mammoth Cave," *Courier-Journal,* July 5, 1938, 1; Mader, "Captions for Picture Series on Scientists in Mammoth Cave"; Colleen Olson interview.

28. *What struck Kleitman most sharply:* "Old World Looks Strange After a Month in Cavern"; Colleen Olson interview; Mader, "Captions for Picture Series on Scientists in Mammoth Cave"; Nathaniel Kleitman autobiography, dictated to Paulena Kleitman circa 1970s, Nathaniel Kleitman Papers, Box 1, Folder 7.

29. *The end of the expedition:* "Scientific Cavemen Emerge" and "Scientists Quit Playing Moles in Sleep Study," *Decatur Herald,* July 8, 1938, 1; "Old World Looks Strange"; "Welcome Caveman-Savant Home," *New Castle News,* July 12, 1938, 13.

30. *That week, there were other headlines: Decatur Herald,* July 8, 1938, 1.

31. *Kleitman was now a national celebrity:* Kenton Kroker, *The Sleep of Others and the Transformation of Sleep Research* (Toronto: University of Toronto Press, 2015), 233–234.

32. *Although Kleitman never:* Ibid.

33. *Yet the impact:* Dale Carnegie, "Dale Carnegie Says," *Evening Standard* (Uniontown, PA), April 5, 1939, 6.

34. *Newspaper features described:* "Ten Thousand Nights Upset Many Popular Ideas on Sleeping," *Palm Beach Post,* December 11, 1938, 20.

35. *Kleitman cemented his:* Kleitman, *Sleep and Wakefulness,* passim; Vicki Cheng, "Nathaniel Kleitman, Sleep Expert, Dies at 104," *New York Times,* August 19, 1999, B8; University of Chicago Medicine press release, "Nathaniel Kleitman, PhD, 1895–1999," accessed September 22, 2022, https://www.uchicagomedicine.org/forefront/news/nathaniel-kleitman-phd-1895-1999.

36. *Kleitman had spent:* Kleitman Autobiography; Kleitman, *Sleep and Wakefulness,* passim. In the acknowledgments, he noted apologetically that his reading ability was "limited to French, German, Italian, and Russian" and credited a graduate student in anthropology for translating source material from Spanish, Dutch, and the Scandinavian languages.

He also thanked Paulena for correcting the first draft and final proofs, as well as compiling the index and 1,400-item bibliography.

37. *The book was:* Donald Anderson Laird and Charles Geoffrey Muller, *Sleep: Why We Need It and How to Get It* (New York: John Day, 1930); Edmund Jacobson, *You Can Sleep Well: The ABC's of Restful Sleep for the Average Person* (New York: Whittlesey House, 1938).

38. *"Among a mass":* "Book Notices," *Journal of the American Medical Association* 113, no. 23 (December 2, 1939), 2086.

39. *Besides its scope:* Kleitman, *Sleep and Wakefulness*, 3–4.

40. Sleep and Wakefulness *was divided:* These sections covered the functional differences between wakefulness and sleep; the course of events during the sleep phase; the periodicity (or cyclic structure) of sleep; interference with the sleep-wake cycle (as in sleep deprivation experiments); spontaneous changes in the sleep-wake cycle (as in narcolepsy, encephalitis lethargica, insomnia, and other diseases and disorders); means of influencing the sleep-wakefulness cycle (pharmaceutical or otherwise); states resembling sleep (including hibernation and hypnosis); and theories of sleep.

41. *These set the stage:* Kleitman, *Sleep and Wakefulness*, 502.

42. *Individuals as well as species:* Ibid., 504–505, 509.

43. *Just as there:* Ibid., 516.

44. *Building on this:* Ibid., 521. Kleitman admits such "general properties of protoplasm" may exist and suggests that "[t]he cause of the persistence of the several rhythms in animal activity is a fascinating problem and should attract more interest than it does at present." But his theory holds that in individual animals and across species, the evolution of the sleep-wake cycle reflected that of the nervous system as a whole and was thus a story of increasing dominance by the cerebral cortex.

45. *What makes people:* Ibid., 510–513.

46. *Why does sleep depth:* Ibid., 517–518.

47. *Still, Kleitman was:* Ibid., 527–528.

48. *He turned out:* Mallory Stallings, "Animals That Need Hardly Any Sleep," Sleep.org, Feb. 25, 2021, accessed September 22, 2022, https://www.sleep.org/animals-that-sleep-the-least/.

49. *That summer:* Kleitman autobiography.

50. *Kleitman's stepfather:* Ibid.; "The Central Database of Jewish Victims' Names," Yad Vashem, accessed September 22, 2022, https://yvng.yadvashem.org/index.html?language=en&s_id=&s_lastName=roitberg&s_firstName=avraam&s_place=&s_dateOfBirth=&cluster=true and https://yvng.yadvashem.org/index.html?language=en&s_id=&s_lastName=torban&s_firstName=esther&s_place=&s_dateOfBirth=.

51. *Most of the city's seventy thousand Jews:* "Kishinev," Yivo Encyclopedia of Jews in Eastern Europe, https://yivoencyclopedia.org/article.aspx/Kishinev; Samuel Aroni, "Ghettos: Memories of the Holocaust: Kishinev(Chisinau) (1941–1944)," Jewish Virtual Library, accessed September 22, 2022, https://www.jewishvirtuallibrary.org/memories-of-the-holocaust-kishinev-introduction-and-acknowledgements; Miriam

Chakin, *A Nightmare in History: The Holocaust, 1933–1945* (New York: Clarion Books, 1987), 60.

52. *With nonmilitary research:* Jerome Siegel, *The Neural Control of Sleep and Waking* (New York: Springer-Verlag, 2002), 31.

53. *He became a special consultant:* Alan Derickson, *Dangerously Sleepy: Overworked Americans and the Cult of Manly Wakefulness* (Philadelphia: University of Pennsylvania Press, 2014), 42–43.

54. *Based on his:* Ibid.

55. *In June 1942:* "Arranging Shifts for Maximum Production," United States Department of Labor, Division of Labor Standards, 1942, accessed September 20, 2022, https://books.google.com/books?id=F0oRpO_ej8IC&pg=PA1&lpg=PA1&dq=%22department+of+labor%22+%22arranging+shifts+for+maximum+production%22&source=bl&ots=xspUwtdhMM&sig=ACfU3U110P98EjqUm0sytDzNa5uT9ok9jA&hl=en&sa=X&ved=2ahUKEwiCpfjgnaT6AhVcMEQIHSrQAC8Q6AF6BAgeEAM#v=onepage&q=%22department%20of%20labor%22%20%22arranging%20shifts%20for%20maximum%20production%22&f=false.

56. *The idea stirred:* E. L. Bishop to Kleitman, July 27, 1942; J. L. Tucker to Kleitman, July 20, 1942; F. J. Burt to Kleitman, July 22, 1942; A. C. Clifford to Kleitman, July 15, 1942; Norma Hendren to Kleitman, July 13, 1942; C. J. Simpson to Kleiman, July 13, 1942; C. J. Miller to Kleitman, July 13, 1942; Harry E. Baker to Kleitman, July 28, 1942; H. M. Miller to Kleitman, etc., Nathaniel Kleitman Papers, Box 10, Folder 6.

57. *The American Can Company:* V. A. Zimmer to Kleitman, August 31, 1942; *Industrial Relations News Letter,* July 31, 1942, Nathaniel Kleitman Papers, Box 10, Folder 7.

58. *Kleitman was invited:* Derickson, *Dangerously Sleepy,* 44; Nathaniel Kleitman, "A Scientific Solution to the Multiple Shift Problem," *Mining Congress Journal,* January 1943, 15–16.

59. *Yet no record exists:* Kleitman to R. L. Mason, September 29, 1942, Nathaniel Kleitman Papers, Box 10, Folder 6; Nathaniel Kleitman to F. C. Smith, April 12, 1947; Kleitman to V. A. Zimmer, August 3, 1942; V. A. Zimmer to Kleitman, July 31, 1942; Nathaniel Kleitman Papers, Box 10, Folder 7; Derickson, *Dangerously Sleepy,* 43–44.

60. *After the war:* Wolf-Meyer, "Where Have All Our Naps Gone?"

Chapter 5

1. *On a spring morning:* T. J. La Vaque, "The History of EEG: Hans Berger," *Journal of Neurotherapy: Investigations in Neuromodulation, Neurofeedback and Applied Neuroscience* 3, no. 2 (1999): 1–9; "Wavecraft: Messages from an Unknown World," *UCL Institute of Neurology Library and Archive,* May 1–August 1, 2018, http://www.kenbarrettstudio.co.uk/wp-content/uploads/2018/01/wavecraft-artwork-and-text-1.pdf (accessed September 6, 2021).

2. *That evening:* La Vaque, "The History of EEG."

3. *At the time:* Michael Shermer, ed., *The Skeptic Encyclopedia of Pseudoscience* (Santa Barbara, CA: ABC-CLIO, 2002), 217.

4. *In many fields:* David Millett, "Hans Berger: From Psychic Energy to the EEG," *Perspectives in Biology and Medicine* 44, no. 4 (Autumn 2001): 522–542.

5. *Berger embraced such:* Cornelius Borck, *Brainwaves: A Cultural History of Electroencephalography, Science, Technology and Culture, 1700–1945* (London: Routledge / Taylor & Francis Group, 2018), 33–35, 37–39; Millett, "Hans Berger."

6. *After earning his:* La Vaque, "The History of EEG"; Raphael Ginzberg, "Three Years with Hans Berger: A Contribution to His Biography," *Journal of the History of Medicine and Allied Sciences* 4, no. 4 (1949): 361–371.

7. *In secret, however:* Millett, "Hans Berger."

8. *Since existing devices:* La Vaque, "The History of EEG."

9. *In 1870, German physicians:* Charles G. Gross, "The Discovery of Motor Cortex and Its Background," *Journal of the History of the Neurosciences* 16 (2007): 320–331.

10. *In 1875, British physician:* Lord Cohen of Birkenhead, "Richard Caton (1842–1926) Pioneer Electrophysiologist," *Proceedings of the Royal Society of Medicine* 52 (March 4, 1959): 645–651; Omar J. Ahmed and Sydney S. Cash, "Finding Synchrony in the Desynchronized EEG: The History and Interpretation of Gamma Rhythms," *Frontiers in Integrative Neuroscience* 7 (August 2013), accessed September 24, 2022, https://www.frontiersin.org/articles/10.3389/fnint.2013.00058/full.

11. *Caton's findings suggested:* Cohen, "Richard Caton."

12. *In follow-up reports:* Richard Caton, "IV.—Interim Report on Investigation of the Electric Currents of the Brain," *American Journal of EEG Technology* 11, no. 1 (1971): 23–24.

13. *Perhaps most intriguingly:* Cohen, "Richard Caton."

14. *In 1890, Polish:* Omar J. Ahmed and Sydney S. Cash, "Finding Synchrony in the Desynchronized EEG: The History and Interpretation of Gamma Rhythms," *Frontiers in Integrative Neuroscience* 7, no. 56 (August 12, 2013), accessed September 13, 2021, https://www.ncbi.nlm.nih.gov/pmc/articles/PMC3740477/; "Adolf Beck: A Pioneer in Electroencephalography in Between Richard Caton and Hans Berger," *Advances in Cognitive Psychology* 9, no. 4 (2013): 216–221; Anton Coenen and Oksana Zayachkivska, *Adolf Beck, Co-Founder of the EEG: An Essay in Honour of His 150th Birthday* (Utrecht, Netherlands: Digitalis/Biblioscope, 2013), 7–12.

15. *Many hoped it:* Caitlin Shure, "Brain Waves, a Cultural History," unpublished PhD thesis, Columbia University Graduate School of Arts and Sciences, 2018.

16. *Yet by 1902:* Thomas F. Collura, "History and Evolution of Electroencephalographic Instruments and Techniques," *Journal of Clinical Neurophysiology* 10, no. 4 (1993): 476–504; Cohen, "Richard Caton"; Neal Stephenson, "Mother Earth Board," *Wired,* December 1, 1996, accessed September 12, 2021, https://www.wired.com/1996/12/ffglass/; John Munro, *Heroes of the Telegraph* (London: Religious Tract Society, 1891), 99–100; "Mirror Galvanometer," Magnet Academy, National High Magnetic Field Laboratory, accessed September 12, 2021, https://nationalmaglab.org/education/magnet-academy/watch-play/interactive/mirror-galvanometer. Caton had adapted a Thomson galvanometer (originally patented as a telegraph receiver), in which magnets

were attached to a small mirror hanging from a thread within a coil of copper wire. A light was trained on the mirror, casting a bright dot on a graduated scale; when a pulse of current flowed through the coil, the mirror twisted back and forth, and the dot traced its oscillations. Beck had used a slightly more advanced model, called a *d'Arsonval galvanometer*. Neither device could create a record of the dot's rapid motions, making it impossible to track the current's frequency or amplitude in detail.

17. *Berger began his quest:* La Vaque, "The History of EEG"; Millett, "Hans Berger."

18. *Discouraged, he turned:* Millett, "Hans Berger"; Borck, *Brainwaves*, 32–35, 39–41.

19. *In 1907, he made:* Borck, *Brainwaves*, 40–41, 25; Millett, "Hans Berger," 529.

20. *In 1910, Berger:* W. Bruce Frye, "History of the Origin, Evolution, and Impact of Electrocardiography," *American Journal of Cardiology* 73, no. 13 (May 15, 1994): 937–949; Collura, "History of Electroencephalographic Instruments"; Arthur Douglas Hirschfelder, *Diseases of the Heart and Aorta* (Philadelphia: J. B. Lippincott, 1918), 85–91.

21. *Yet this technique:* Millett, "Hans Berger," 530–531; Borck, *Brainwaves*, 41–44.

22. *Soon afterward, Berger:* Millett, "Hans Berger."

23. *Meanwhile, a few:* Collura, "History of Electroencephalographic Instruments"; Ahmed and Cash, "Finding Synchrony"; La Vaque, "The History of EEG"; Ari Ben-Menahem, *Historical Encyclopedia of Natural and Mathematical Sciences* (Berlin: Springer-Verlag, 2009), 3606; Coenen and Zayachkivska, *Adolf Beck*.

24. *The next incursion:* Ahmed and Cash, "Finding Synchrony"; Anton Coenen, Edward Fine, and Oksana Zayachkivska, "Adolf Beck: A Forgotten Pioneer in Electroencephalography," *Journal of the History of the Neurosciences* 23, no. 3 (April 2014): 276–286.

25. *Later that year:* Millett, "Hans Berger."

26. *After the armistice:* Ibid.

27. *In 1924, a new:* Ibid.; Borck, *Brainwaves*, 51–53.

28. *Berger came up:* Millett, "Hans Berger."

29. *But the new:* Borck, *Brainwaves*, 60–66.

30. *At the same time:* Collura, "History and Evolution of Encephalographic Instruments"; Robert M. Kaplan, "The Mind Reader: The Forgotten Life of Hans Berger, Discoverer of the EEG," *Australasian Psychiatry* 19, no. 2 (April 2011): 168–169.

31. *Perhaps unsurprisingly:* Collura, "History and Evolution of Encephalographic Instruments"; Borck, *Brainwaves*, 54–56.

32. *A turning point:* Borck, *Brainwaves*, 55–60; "The Moving Coil Galvanometer," Toppr, accessed October 2, 2021, https://www.toppr.com/guides/physics/moving-charges-and -magnetism/moving-coil-galvanometer/.

33. *These qualities enabled:* Borck, *Brainwaves*, 25, 56–64.

34. *By late 1928:* Ibid., 60–66, 127–128; Collura, "History and Evolution of Encephalographic Techniques"; La Vaque, "The History of EEG."

35. *In 1929, Berger:* Borck, *Brainwaves*, 63, 111–112, 122–125, 127–129, 140–146.

36. *During the five years:* La Vaque, "History of EEG."

37. *In 1933, Edgar Douglas Adrian:* Stanley Finger, *The Minds Behind the Brain: A History of the Pioneers and Their Discoveries* (New York: Oxford University Press, 1999), 239–255; Borck, *Brainwaves,* 153–157.

38. *Reviewing the literature:* Borck, *Brainwaves,* 157; E. D. Adrian and B. H. C. Matthews, "The Berger Rhythm: Potential Changes from the Occipital Lobes in Man," *Brain* 57 (1934): 355–385.

39. *In this endeavor:* Finger, *The Minds Behind the Brain,* 153–157.

40. *Berger was unable:* Ibid.; Borck, *Brainwaves,* 56, 126.

41. *Matthews was a brilliant engineer:* Collura, "History and Evolution of Electroencephalographic Instruments"; Bryan H. C. Matthews, "A New Electrical Recording System for Physiological Work," *Journal of Scientific Instruments* 6, no. 7 (1929): 225–242.

42. *In the spring:* Borck, *Brainwaves,* 158; Adrian and Matthews, "The Berger Rhythm"; Collura, "History and Evolution of Electroencephalographic Instruments."

43. *"We found Berger's alpha rhythms":* Borck, *Brainwaves,* 158; Jacob Empson, *Human Brainwaves: The Psychological Significance of the Electroencephalogram* (New York: Stockton Press, 1986), 22.

44. *That May:* Borck, *Brainwaves,* 158, 160–161.

45. *In December:* Adrian and Matthews, "The Berger Rhythm."

46. *At year's end:* Borck, *Brainwaves,* 124, 159–160, 163, 170.

47. *If Edgar Adrian lent:* Kenton Kroker, *The Sleep of Others and the Transformation of Sleep Research* (Toronto: University of Toronto Press, 2015), 272–273, 275, 280–282; Robert Galambos, "Hallowell Davis: 1896–1992," in *Biographical Memoirs, Volume 75* (Washington, DC: National Academies Press, 1988), 10, accessed December 16, 2022, http://www.nasonline.org/publications/biographical-memoirs/memoir-pdfs/davis-hallowell.pdf.

48. *Davis, too, started:* Kroker, *The Sleep of Others,* 282; Frederic G. Worden, Judith P. Swazey, and George Adeleman, eds., *The Neurosciences: Paths of Discovery, Volume 1* (Boston: Birkhäuser, 1992), 316–317.

49. *To do so:* James L. Stone and John R. Hughes, "Early History of Electroencephalography and Establishment of the American Clinical Neurophysiology Society," *Journal of Clinical Neurophysiology* 30, no. 1 (2013): 28–44.

50. *The two students:* Ibid.

51. *Davis quickly recognized:* Kroker, *The Sleep of Others,* 282–284; Stone and Hughes, "Early History of Electroencephalography"; Dominic Hall, "BackStory: The Development of New Medical Devices Requires Many Hands and Moments of Inspiration," *Harvard Medicine,* Spring 2021, accessed October 4, 2021, https://hms.harvard.edu/magazine/business-medicine/backstory; Galambos, "Hallowell Davis," 9.

52. *Soon after his conversion:* Kroker, *The Sleep of Others,* 286

53. *That December:* Collura, "History and Development of Electroencephalographic Instruments"; Stone and Hughes, "Early History of Electroencephalography"; Kroker, *The Sleep of Others,* 284–285; F. A. Gibbs, H. Davis, and W. D. Lennox, "The

Electroencephalogram in Epilepsy and in Conditions of Impaired Consciousness," *Archives of Neuropsychiatry* 34, no. 6 (December 1935): 1133–1148.

54. *The study also:* "EEG (electroencephalogram)," Mayo Clinic, accessed September 24, 2022, https://www.mayoclinic.org/tests-procedures/eeg/about/pac-20393875.

55. *By the end of 1936:* Borck, *Brainwaves*, 166; Kroker, *The Sleep of Others*, 288–290.

56. *Before the advent:* Nathaniel Kleitman, *Sleep and Wakefulness as Alternating Phases in the Cycle of Existence* (Chicago: University of Chicago Press, 1939), 16.

57. *Or they could measure:* Kroker, *The Sleep of Others*, 237–254.

58. *Edmund Jacobson:* Ibid.; Edmund Jacobson, *You Can Sleep Well: The ABC's of Restful Sleep for the Average Person* (New York: Whittlesey House, 1938), 183–203.

59. *The EEG offered:* Jennet Conant, *Tuxedo Park: A Wall Street Tycoon and the Secret Palace of Science That Changed the Course of World War II* (New York: Simon & Schuster, 2003), 18–19, 25–28, 37–42, 47–73, 106.

60. *In 1934, Loomis:* Kroker, *The Sleep of Others*, 268, 287–288.

61. *Soon afterward, Loomis:* Ibid., 288–289.

62. *Among the custom-made:* Ibid., 287–290; Conant, *Tuxedo Park*, 108–110; Collura, "History and Evolution of Electroencephalographic Instruments"; Alfred L. Loomis, E. Newton Harvey, and Garret Hobart, "Potential Rhythms of the Cerebral Cortex During Sleep," *Science* 81, no. 2111 (June 14, 1935): 597–598.

63. *When everything was:* Conant, *Tuxedo Park*, 110–111; Hallowell Davis, P. Davis, A. L. Loomis, et al., "Human Brain Potentials During the Onset of Sleep," *Journal of Neurophysiology* 1 (1938): 24–38; Alfred L. Loomis, E. Newton Harvey, and Garret Hobart, "Electrical Potentials of the Human Brain," *Journal of Experimental Psychology* 14, no. 3 (June 1936): 249–279.

64. *On one occasion:* Borck, *Brainwaves*, 174.

65. *Einstein's EEG:* Alfred L. Loomis, E. Newton Harvey, and Garret Hobart, "Further Observations on the Potential Rhythms of the Cerebral Cortex During Sleep," *Science* 82, no. 2122 (August 30, 1935): 198–200; Loomis et al., "Electrical Potentials."

66. *It took more:* Kroker, *The Sleep of Others*, 293–296.

67. *A—Alpha:* Hallowell Davis, P. Davis, A. L. Loomis, et al., "Human Brain Potentials During the Onset of Sleep," *Journal of Neurophysiology* 1 (1938): 24–38.

68. *This list represented:* Kroker, *The Sleep of Others*, 295–297, 303–304; Kleitman, *Sleep and Wakefulness*, 42–44.

69. *The dozen papers:* Stone and Hughes, "Early History of Electroencephalography"; Conant, *Tuxedo Park*, 301–302.

70. *At the University of Chicago:* H. Bake and R. W. Gerard, "Brain Potentials During Sleep," *American Journal of Physiology* 119, no. 4 (July 31, 1937): 692–703.

71. *In 1938, Kleitman:* Helen Blake, R. W. Gerard, and N. Kleitman, "Factors Influencing Brain Potentials During Sleep," *Journal of Neurophysiology* 2, no. 1 (Jan. 1939): 48–60.

72. *All these studies:* Stone and Hughes, "Early History of Electroencephalography."

73. *Davis focused his:* Galambos, "Hallowell Davis."

74. *Gerard went on:* Ben Libet and Orr E. Reynolds, "R.W. Gerard: Born October 7, 1900—Died February 17, 1974," *Physiologist* 17, no. 2 (May 1974): 165–168.

75. *Blake married:* "George R. Carlson in the Cook County, Illinois Marriage Index, 1930–1960," marriage license, January 23, 1939, Ancestry.com; "George Raymond Carlson, in the 1940 U.S. Federal Census," Ancestry.com; Helen's married name is per *Announcements of the University of Chicago: The Medical Schools* (1941), 122, accessed September 24, 2022, https://archive.org/details/annualannounceme98unse /page/n121/mode/2up?q=carlson; "Helen Jean Carlson, in the California, US, Death Index, 1940–1997," Ancestry.com.

76. *Berger's research career:* Lawrence A. Zeidman, James Stone, and Daniel Kondziella, "New Revelations About Hans Berger, Father of the Electroencephalogram (EEG), and His Ties to the Third Reich," *Journal of Child Neurology* 29, no. 7 (2013): 1002–1010; Borck, *Brainwaves*, 139–140; Millett, "Hans Berger"; Robert M. Kaplan, "The Mind Reader: The Forgotten Life of Hans Berger, Discoverer of the EEG," *Australasian Psychiatry* 19, no. 2 (December 2010): 168–169.

77. *Kleitman, for his part:* Eugene Aserinsky, "Memories of Famous Neuropsychologists: The Discovery of REM Sleep," *Journal of the History of the Neurosciences* 5, no. 3 (1996): 213–227.

Chapter 6

1. *Eugene Aserinsky would:* Eugene Aserinsky, "Memories of Famous Neuropsychologists: The Discovery of REM Sleep," *Journal of the History of the Neurosciences* 5, no. 3 (1996): 213–227.

2. *a slight, black-haired:* Lynne Lamberg, "The Student, the Professor, and the Birth of Modern Sleep Research," *Medicine on the Midway,* Spring 2004.

3. *had a problem with patriarchs:* Interview with Armond Aserinsky, October 11, 2021.

4. *Aserinsky grew up:* Armond Aserinsky interview.

5. *His mother died:* "Sonya Hectin: Facts," Ancestry.com, www.ancestry.com/family-tree /person/tree/73413864/person/38268594587/facts; Lillian Aserinsky marriage license, April 11, 1931, Ancestry.com, accessed September 27, 2022, https://www.ancestry.com /discoveryui-content/view/7621935:61406?tid=&pid=&queryId=b01efba0f592c3b308 edda94765ea8c3&_phsrc=FCR223&_phstart=successSource.

6. *Eugene was marooned:* Armond Aserinsky interview; interview with Jill Aserinsky Buckley, February 24, 2022; J. D. White et al., *The Dental Cosmos, Volume 58* (Philadelphia: The S.S. White Dental Manufacturing Co., 1916), 848; Chip Brown, "The Stubborn Scientist Who Unraveled a Mystery of the Night," *Smithsonian,* October 2003.

7. *Although his nocturnal:* Armond Aserinsky interview; Brown, "The Stubborn Scientist."

8. *During his studies:* Armond Aserinsky interview.

9. *Sylvia gave birth:* Ibid.

10. *Though he would:* Aserinsky, "Memories of Famous Neuropsychologists"; videotaped

speech to Associated Professional Sleep Societies, 1995 (collection of Jill Aserinsky Buckley).

11. *records list his rank:* "Aserinsky, Eugene," Enlisted Record and Report of Separation, honorable discharge, November 22, 1945, National Personnel Records Center. This document lists Aserinsky's occupational specialty as "Classification Specialist (275)"—a position whose responsibilities included interviewing, testing, and recommending assignments of officers and enlisted personnel, as well as general record-keeping, according to the Military Yearbook Project, accessed May 25, 2022, https://militaryyearbookproject.org/references/old-mos-codes/wwii-era/usmc-wwii-codes/administrative-and-clerical/275-classification-specialist.

12. *Sylvia took a job:* Armond Aserinsky interview.

13. *In his late twenties:* Ibid.; Brown, "The Stubborn Scientist."

14. *After enrolling in September 1948:* Email message to author from Catherine Uecker (head of research and instruction, University of Chicago Library Special Collections), November 4, 2021.

15. *he was dismayed:* Aserinsky, "Memories of Famous Neuropsychologists."

16. *The meeting that launched:* No record exists of the date of this meeting. However, the item on blinking that inspired Kleitman's assignment for Aserinsky appeared in the January 14, 1950, issue of *Nature* (see note 22 below).

17. *At this point:* Kenton Kroker, *The Sleep of Others and the Transformation of Sleep Research* (Toronto: University of Toronto Press, 2015), 307–309, 313–314; Giuseppe Moruzzi and Horace Magoun, "Brain Stem Reticular Formation and Activation of the EEG," *EEG and Clinical Neurophysiology* 1 (1949): 455–473.

18. *Kleitman could not:* Moruzzi and Magoun, "Brain Stem Reticular Formation"; Peretz Lavie, "Historical Perspective: History of Sleep Research," *World Federation of Sleep Research Societies Newsletter* 6, no. 2 (1999), accessed November 6, 2021, http://www.edumed.org.br/cursos/neurociencia/cdrom/Biblioteca/HistorySleepResearch.htm.

19. *As for his own:* Kroker, *The Sleep of Others,* 307–309, 313–314; Lawrence Lader, "Why Can't You Sleep?," *Cincinnati Enquirer,* June 15, 1947, 99; "Maintain Sleep Cycle," *Tampa Bay Times,* February 28, 1948, 9; A. E. Hotchner, "When Your Temperature Goes Up...," *St. Louis Globe-Democrat,* December 17, 1950, 110.

20. *After a quarter century:* Guide to the Nathaniel Kleitman Papers, 1896–2001, Hanna Holborn Gray Special Collections Research Center, University of Chicago Library; *"Probably the longest period of 'probation'":* Nathaniel Kleitman autobiography, dictated to Paulena Kleitman circa 1970s, Nathaniel Kleitman Papers, Box 1, Folder 7, Hanna Holborn Gray Special Collections Research Center, University of Chicago Library.

21. *But he hadn't:* Kroker, *The Sleep of Others,* 306–308.

22. *Still, he was:* Aserinsky, "Memories of Famous Neurophysiologists"; Robert Lawson, "Blinking and Sleep," *Nature* 165, no. 4185 (January 14, 1950): 81–82.

23. *Although Kleitman's orders:* Claude Gottesmann, *Henri Piéron and Nathaniel Kleitman: Two Major Figures of 20th Century Sleep Research* (New York: Nova Science Publishers, 2013), 84–85; Nathaniel Kleitman and Jonas E. X. Schreider, "Diurnal Variations

in Oculomotor Performance," *L'Année Psychologique* 50 (1951): 201–215. Note: In several sources, this volume is mistakenly listed as having been published in 1949, but it appeared in the journal's 1951 *Volume jubilaire* in honor of Piéron's seventieth birthday.

24. *Aserinsky regarded:* Aserinsky, "Memories of Famous Neuropsychologists"; Andrea Rock, *The Mind at Night: The New Science of How and Why We Dream* (New York: Basic Books, 2004), 4.

25. *Kleitman then suggested:* Aserinsky, "Memories of Famous Neuropsychologists"; Armond Aserinsky interview.

26. *As the younger:* Aserinsky, "Memories of Famous Neuropsychologists."

27. *After a few months:* Ibid.

28. *In early 1951:* Ibid.; Armond Aserinsky interview.

29. *For his part:* Armond Aserinsky interview; Ingrid Gonçalves, "Forgotten History: Postwar Prefabs," *Core*, Winter 2017, accessed February 18, 2023, http://thecore.uchicago.edu/Winter2017/departments/postwar-prefabs.shtml.

30. *Even the rent:* Brown, "The Stubborn Scientist."

31. *But in his:* Aserinsky, "Memories of Famous Neuropsychologists."

32. *To avoid remaining:* Ibid.

33. *No funds were:* Ibid.

34. *Since the 1930s:* Thomas F. Collura, "History and Evolution of Electroencephalographic Instruments and Techniques," *Journal of Clinical Neurophysiology* 10, no. 4 (1993): 476–504.

35. *The machine that:* The origin of this machine has never been firmly established, beyond that it was a prototype Offner Dynograph. But evidence suggests it was first used for experiments carried out by Blake and Gerard (eventually joined by Kleitman) in 1937 and 1938. The ink-writer portion of the prototype was designed in 1936, according to Franklin Offner and Ralph W. Gerard, "A High Speed Crystal Ink Writer," *Science* 84, no. 2174 (August 28, 1936): 209–210. The amplifier portion came shortly afterward, per Alvin M. Weinberg and Peter J. Dallos, "Franklin F. Offner: 1911–1999," *Memorial Tributes, Volume 10* (Washington, D.C.: National Academies Press, 2002), 189–192, https://www.nae.edu/188090/FRANKLIN-FOFFNER-19111999. See also: Albert M. Grass, *The Electroencephalographic Heritage* (Quincy, MA: Grass Instrument Company, 1984), 36–37, https://grassfoundation.org/wp-content/uploads/2021/05/Electroencephalographic-Heritage.pdf. *It appears to have had only three channels:* Aserinsky, "Memories of Famous Neuropsychologists"; email message to author from EEG expert Tom Collura, December 28, 2022; email message to author from EEG expert Sharon Keenan, February 27, 2023. I've found no record of the number of channels the prototype Offner possessed, but Aserinsky's description of his method of tracking EOG and EEG simultaneously suggests that the machine (like many others of its day) had no more than three.

36. *And as Aserinsky:* Aserinsky, "Memories of Famous Neuropsychologists"; Armond Aserinsky interview.

37. *Aserinsky asked:* Aserinsky, "Memories of Famous Neuropsychologists."

38. *Taking readings from:* Ibid.

39. *At Kleitman's suggestion:* Gottesmann, *Henri Piéron*, 92.

40. *To hone his skills:* Armond Aserinsky interview; Rock, *The Mind at Night*, 4.

41. *The lab where:* William Dement, *The Sleepwatchers* (Menlo Park, CA: Nychthemeron Press, 1995), 28; Armond Aserinsky interview.

42. *Each practice run:* Brown, "The Stubborn Scientist."

43. *After squirting:* Mallika Rao, "Third State," *HiberNation* (podcast), July 15, 2021, accessed November 7, 2021, https://hibernation.simplecast.com/episodes/third-state.

44. *He plugged the leads:* Dement, *The Sleepwatchers*, 28.

45. *Then, as Armond:* Armond Aserinsky interview.

46. *To avoid interruptions:* Brown, "The Stubborn Scientist."

47. *"I knew my dad needed help":* Rock, *The Mind At Night*, 4. While Aserinsky strove to master the machine, Kleitman embarked with his family on his long-awaited Arctic expedition, spending May through July 1951 in the Norwegian outpost of Tromsø. In one study, he recorded his own and his daughters' bodily rhythms as they adopted eighteen- and twenty-eight-hour sleep-wake cycles for three-week periods. The perpetual summer daylight had no discernible effect on their ability to adapt to these artificial schedules. As in Mammoth Cave, Kleitman's temperature rhythms stuck to a twenty-four-hour cycle; the young women adjusted quickly. (Because his wife, Paulena, fell ill and was unable to participate, it remained unclear whether the difference in flexibility was related to age or individual factors.) In a second study, Kleitman and his daughters interviewed one hundred local people on their seasonal sleep patterns. Contrary to travelers' accounts, which claimed that the populace slept for just three or four hours a night in summer, the researchers found that residents changed their habits only slightly. See: Nathaniel Kleitman and Esther Kleitman, "Effect of Non-Twenty-Four-Hour Routines of Living on Oral Temperature and Heart Rate," *Journal of Applied Physiology* 6, no. 5 (November 1953): 283–291; Nathaniel Kleitman and Hortense Kleitman, "The Sleep-Wakefulness Pattern in the Arctic," *Scientific Monthly* 76, no. 6 (June 1953): 349–356.

48. *By December, Aserinsky felt ready:* Aserinsky, "Memories of Famous Neuropsychologists"; Brown, "The Stubborn Scientist."

49. *Aserinsky consulted:* Aserinsky, "Memories of Famous Neuropsychologists."

50. *His wife was pregnant:* Armond Aserinsky interview.

51. *For weeks, Aserinsky:* Aserinsky, "Memories of Famous Neuropsychologists"; Armond Aserinsky interview.

52. *Aserinsky was beginning:* Aserinsky, "Memories of Famous Neuropsychologists."

53. *This was not:* Ibid.; Eugene Aserinsky and Nathaniel Kleitman, "Regularly Occurring Periods of Eye Motility, and Concomitant Phenomena, During Sleep," *Science* 118, no. 3062 (September 4, 1953): 273–274.

54. *In early 1952:* Aserinsky, "Memories of Famous Neuropsychologists"; Lynne Lamberg, "The Student, the Professor and the Birth of Modern Sleep Research," *Medicine on the Midway,* Spring 2004, 16–25; Lynne Lamberg, "Scientists Never Dreamed Finding Would Shape a Half-Century of Sleep Research," *JAMA* 290, no. 20 (November 26, 2003): 2652–2654.

55. *Indeed, further investigation:* John Peever and Patrick M. Fuller, "The Biology of REM Sleep," *Current Biology* 26, no. 1 (November 20, 2017): R1237–R1248.

56. *But why?:* Aserinsky, "Memories of Famous Neuropsychologists."

57. *This was strange:* Ibid.; Nathaniel Kleitman, *Sleep and Wakefulness as Alternating Phases in the Cycle of Existence* (Chicago: University of Chicago Press, 1939), 158–165.

58. *Aserinsky wasn't sure:* Aserinsky, "Memories of Famous Neuropsychologists."

59. *Scientists would eventually:* Julie A. E. Christensen et al., "Rapid Eye Movements Are Reduced in Blind Individuals," *Journal of Sleep Research* 28, no. 6 (April 1, 2019), accessed September 27, 2022, https://doi.org/10.1111/jsr.12866.

60. *As the evidence for REM:* Aserinsky, "Memories of Famous Neuropsychologists"; Grass, *The Electroencephalographic Heritage,* 26, 32; W. C. Dement, "Remembering Nathaniel Kleitman," *Archives Italiennes de Biologie* 139 (2001): 11–17.

61. *This polygraph probably:* Collura, "History of EEG Instruments and Techniques."

62. *Aserinsky began sampling:* Aserinsky, "Memories of Famous Neuropsychologists"; Lamberg, "Scientists Never Dreamed."

63. *Kleitman soon showed:* Aserinsky, "Memories of Famous Neuropsychologists."

64. *For Aserinsky, who preferred:* Ibid.

65. *This must have been:* Armond Aserinsky interview.

66. *On the downside:* Aserinsky, "Memories of Famous Neuropsychologists."

67. *Aserinsky, of course:* Ibid.

68. *In addition to:* Aserinsky and Kleitman, "Regularly Occurring Periods of Eye Motility."

69. *To Aserinsky, these:* Aserinsky, "Memories of Famous Neuropsychologists."

70. *At this point:* Ibid.

71. *The demonstration must:* Ibid.

72. *It is hard to judge:* Ibid.

73. *Kleitman's own accounts:* Nathaniel Kleitman, *Sleep and Wakefulness, Revised and Enlarged Edition* (Chicago: University of Chicago Press, 1963), 92–107, 379.

74. *What's clear is:* Aserinsky, "Memories of Famous Neuropsychologists."

75. *In his telling:* Ibid. Note: Sleep scientist Claude Gottesmann, who investigated Aserinsky's claims for a 2013 book, concluded that "based on a misunderstanding… Aserinsky somewhat underestimated Kleitman's contribution." See Gottesmann, *Henri Piéron,* 92.

76. *That concession, however:* Aserinsky, "Memories of Famous Neuropsychologists."

77. *Outraged by this:* Ibid.; "Dreams Set Your Eyes to Dancing," *Boston Globe,* April 7, 1953, 7; "Notes on Science: Dreams," *New York Times,* May 17, 1953, E7.

78. *Meanwhile, he'd grudgingly:* Aserinsky and Kleitman, "Regularly Occurring Periods of Eye Motility"; Aserinsky, "Memories of Famous Neuropsychologists"; Armond Aserinsky interview.

79. *Aserinsky and Kleitman's:* Aserinsky and Kleitman, "Regularly Occurring Periods of Eye Motility."

80. *Within a few:* Brown, "The Stubborn Scientist."

81. *Yet the report:* Aserinsky, "Memories of Famous Neuropsychologists."

82. *Aserinsky and his:* Eugene Aserinsky and Nathaniel Kleitman, "Two Types of Ocular Motility Occurring in Sleep," *Journal of Applied Physiology* 8, no. 1 (July 1955): 1–10.

83. *By then, he'd become:* Brown, "The Stubborn Scientist."

84. *His wife, meanwhile:* Armond Aserinsky interview.

85. *The French sleep scientist:* Brown, "The Stubborn Scientist."

Chapter 7

1. *For all their differences:* Interview with William Dement, September 16, 2016; Kathryn Severyns Dement, *My Life and Times,* 72–75.

2. *Kathryn Severyns Dement, thirty-seven:* William Dement interview; Kathryn Dement, *My Life and Times* (Auburn, CA: Sierra Dreams Press, 1990), 9–75.

3. *From an early age:* William Dement interview.

4. *He joined:* William C. Dement, with Stuart Rawlings. *Dreamer: The Memoir of a Sleep Researcher* (unpublished manuscript, 2018), 16–18.

5. *Another would be:* Dement interview; Dement and Rawlings, *Dreamer,* 20–21; Dement, *My Life and Times,* 76–79.

6. *"I couldn't have a conversation":* William Dement interview.

7. *Those qualities began:* Ibid.; Dement and Rawlings, *Dreamer,* 21–33.

8. *Freud's central idea:* Stephen P. Thornton, "Sigmund Freud (1856–1939)," Internet Encyclopedia of Philosophy, accessed December 26, 2021, https://iep.utm.edu/freud/; Erwin, *The Freud Encyclopedia,* 18, 179, 212–214, 321–324, 363–368.

9. *By the late '40s:* Lawrence R. Samuel, *Shrink: A Cultural History of Psychoanalysis in America* (Lincoln: University of Nebraska Press, 2013), 95.

10. *As he later:* William C. Dement and Christopher Vaughan, *The Promise of Sleep: A Pioneer in Sleep Medicine Explores the Vital Connection Between Health, Happiness, and a Good Night's Sleep* (New York: Delacorte, 1999), 35.

11. *One factor that:* William Dement interview.

12. *He also played:* Dement and Rawlings, *Dreamer,* 21–22, 29–33.

13. *When the president:* William Dement interview.

14. *Yet for all his:* Ibid.

15. *Nevertheless, by his:* Ibid.; Dement and Rawlings, *Dreamer,* 29–33; "Robert M. Hutchins, Long a Leader in Educational Change, Dies at 78," *New York Times,* May 16, 1977, accessed December 26, 2021, https://www.nytimes.com/1977/05/16/archives/robert-m-hutchins-long-a-leader-in-educational-change-dies-at-78.html.

16. *Dement was in:* Dement and Vaughan, *The Promise of Sleep,* 27.

17. *The realization:* William Dement interview.

18. *After class:* Dement and Vaughan, *The Promise of Sleep,* 28.

19. *There was no answer:* Ibid.

20. *Dement bought a copy:* Ibid., 28, 32.

21. *When Dement arrived:* Ibid., 34–35.

22. *In his books:* Ibid., 39–40.

23. *Three other female:* Chip Brown, "The Stubborn Scientist Who Unraveled a Mystery of the Night," *Smithsonian,* October 2003.

24. *Dement also recalled:* William C. Dement, "Dr. Aserinsky Remembered," *SRS Bulletin* 4, no. 2 (September 1998): 14.

25. *In later years, Dement:* Brown, "The Stubborn Scientist."

26. *Aserinsky, for his part:* Eugene Aserinsky, "Memories of Famous Neuropsychologists: The Discovery of REM Sleep," *Journal of the History of the Neurosciences* 5, no. 3 (1996): 213–227. Note: I've been unable to verify Aserinsky's assertion that Dement oversaw only five sleep sessions. Aserinsky's lab records have been donated to UCLA but were not yet accessible to researchers when this book went to press.

27. *After Aserinsky's unhappy:* William C. Dement, "Knocking on Kleitman's Door: The View from 50 Years Later," *Sleep Medicine Reviews* 7, no. 4 (2003): 289–292.

28. *Dement thought:* Sigmund Freud and James Strachey, *The Interpretation of Dreams* (New York: Basic Books, 2010), 105–106, 577, 588.

29. *REM, the young researcher:* William C. Dement, *The Sleepwatchers* (Menlo Park, CA: Nychthemeron Press, 1995), 29.

30. *Manteno's polygraph department:* William Dement interview; Dement and Rawlings, *Dreamer,* 40–41; Prasad Vannemreddy et al., "Frederic Gibbs and his Contributions to Epilepsy Surgery and Electroencephalography," *Neurosurgery* 70 (2012): 774–782.

31. *The readings, it turned:* Dement, *The Sleepwatchers,* 30.

32. *The only difference:* William C. Dement, "Dream Recall and Eye Movement During Sleep in Schizophrenics and Normals," *Journal of Nervous and Mental Disease* 122 (1955): 263–269.

33. *Disappointed but undaunted:* Dement, *The Sleepwatchers,* 30; Dement and Vaughan, *The Promise of Sleep,* 37–38. Note: Although Aserinsky later claimed his own landmark REM studies were based on "continuous all-night recording" (Aserinsky, "Memories of Famous Neuropsychologists"), his 1953 and 1955 papers make no such stipulation. And the assertion in those papers that the first REM cycle occurs, on average, three hours after sleep onset—rather than 90 minutes, as later studies showed—indicates that readings were taken only intermittently during many, if not most, of the reported sleep runs.

34. *"In blatant violation":* Dement, *The Sleepwatchers,* 30.

35. *That year, Dement:* Claude Gottesmann, *Henri Piéron and Nathaniel Kleitman: Two Major Figures of 20th Century Sleep Research* (New York: Nova Science Publishers, 2013), 95–104, 118.

36. *He also began dating:* Dement and Rawlings, *Dreamer,* 46–52; Kathryn Dement, *My Life and Times,* 102.

37. *who was working as a secretary:* Dement and Rawlings, *Dreamer,* 47; Andrea Rock, *The Mind at Night: The New Science of How and Why We Dream* (New York: Basic Books, 2004), 102; Lynne Lamberg, "Night Pilot," *Psychology Today,* July 1988, 34–42; Stephanie Crowley and Charmaine Eastman, "The Queen of Dreams: Remembering Rosalind D. Cartwright, PhD," Sleep Research Society, https://www.sleepresearchsociety.org/the-queen-of-dreams-remembering-rosalind-d-cartwright-phd/; Mark Kelland, "Personality Theory in a Cultural Context," OpenStax CNX, November 4, 2015, accessed Jan. 2, 2022, http://cnx.org/contents/9484b2cb-a393-45aa-96bf-e9ae9380dd3e@1.1.

38. *A native of Charleston:* Dement and Rawlings, *Dreamer,* 46–52; interview with Cathy Dement Roos, November 2, 2020; interview with Barbara Scavullo, April 13, 2022.

39. *Despite these rough:* Cathy Dement Roos interview; interview with Nick Dement, October 19, 2021.

40. *They lived together:* Dement and Rawlings, *Dreamer,* 46–52.

41. *After that, Kleitman:* William Dement interview; Dement and Vaughan, *The Promise of Sleep,* 40.

42. *Pat helped deepen:* W. C. Dement, "Remembering Nathaniel Kleitman," *Archives Italiennes de Biologie* 139 (2001): 11–17.

43. *Over the next four:* Ibid.

44. *Kleitman would refer:* Lynne Lamberg, "The Student, the Professor and the Birth of Modern Sleep Research," *Medicine on the Midway,* Spring 2004, 16–25.

45. *For Dement, the older:* Dement, "Remembering Nathaniel Kleitman."

46. *Dement's early papers:* Arthur J. Snider, "Ever See a Dream Talking?," *Minneapolis Star,* March 30, 1957, 37; William Dement and Nathaniel Kleitman, "The Relation of Eye Movements During Sleep to Dream Activity: An Objective Method for the Study of Dreaming," *Journal of Experimental Psychology* 33, no. 5 (1957): 339–346.

47. *Another finding that:* Dement and Kleitman, "The Relation of Eye Movements"; "In Dreams You're Star and Audience," *Arizona Republic,* April 19, 1956, 1.

48. *But the paper:* William Dement and Nathaniel Kleitman, "Cyclic Variations in EEG During Sleep and Their Relation to Eye Movements, Body Motility, and Dreaming," *Electroencephalography and Clinical Neurophysiology* 9 (1957): 673–690.

49. *The paper was:* Dement and Kleitman, "Cyclic Variations."

50. *When Dement and Kleitman:* Ibid.

51. *Although "Cyclic Variations":* Dement, *The Sleepwatchers,* 30.

52. *As the first:* Dement and Kleitman, "Cyclic Variations."

53. *As the authors:* Ibid.

54. *Yet for all:* Dement, "Remembering Nathaniel Kleitman."

55. *By June 1957:* Dement and Rawlings, *Dreamer,* 52. It's interesting to note that Dement's choice of internship was also affected by his tendency to procrastinate over tasks that didn't excite him. Ambivalent about pursuing a career in medicine, he put off applying for the national matching program until it was too late. Having missed the deadline, he turned to Charles Fisher, a psychiatrist at Mount Sinai with whom he'd been corresponding over their shared interest in the psychophysiology of dreaming. With Fisher's sponsorship, Dement was able to secure an internship at the hospital (see Chapter 8).

56. *On his last:* Dement, "Remembering Nathaniel Kleitman."

57. *At Kleitman's urging:* Ibid.

58. *Dement and his spouse:* Dement and Rawlings, *Dreamer,* 52–53.

59. *Later that summer:* Dement, "Remembering Nathaniel Kleitman."

60. *Dement began:* Dement and Rawlings, *Dreamer,* 52; Dement and Kleitman, "Cyclic Variations."

Chapter 8

1. *The beacon that:* Arthur H. Aufses Jr. and Barbara J. Niss, *This House of Noble Deeds: The Mount Sinai Hospital, 1852–2002* (New York: New York University Press, 2002), 8.

2. *There, a psychiatrist:* Howard Shevrin, ed., *Subliminal Explorations of Perceptions, Dreams, and Fantasies: The Pioneering Contributions of Charles Fisher* (Madison, CT: International Universities Press, 2003), 1–27.

3. *Small, dapper:* Charles Fisher, interview with Arnold D. Richards, New York Psychoanalytic Institute and Society Oral History Project (undated). Fisher's appearance is per author interview with psychologist Steven Ellman, a former colleague, October 14, 2018.

4. *Although Fisher earned:* Ibid.; Shevrin, *Subliminal Explorations*, 6; "Charles Fisher," Cook County, Illinois Marriage Index, 1930–1960, Ancestry.com, accessed March 9, 2022, https://search.ancestry.com/cgi-bin/sse.dll?dbid=1500&h=6037923&indiv=try&o_vc=Record:OtherRecord&rhSource=60525.

5. *During the war:* Fisher interview with Richards.

6. *Fisher's obsession grew:* Shevrin, *Subliminal Explorations*, 3–6; Sigmund Freud and James Strachey, *The Interpretation of Dreams* (New York: Basic Books, 2010), 204.

7. *Fisher saw Pötzl's findings:* Shevrin, *Subliminal Explorations*, 3–6; Freud, *The Interpretation of Dreams*, 188.

8. *In the early 1950s:* Fisher interview with Richards; Shevrin, *Subliminal Explorations*, 10–41, 145.

9. *Fisher determined that:* Shevrin, *Subliminal Explorations*, 17–19, 162–168.

10. *Fisher's first paper:* Ibid., vii.

11. *Dement read it:* Fisher interview with Richards.

12. *The timing of Dement's:* Charles R. Acland, *Swift Viewing: The Popular Life of Subliminal Influence* (Durham, NC: Duke University Press, 2011), 66–82.

13. *In April 1957:* Vance Packard, *The Hidden Persuaders* (New York: Pocket Books, 1958), 35–36; "New Book Tells Why People Buy What They Do," *Wichita Eagle*, April 11, 1957, 39.

14. The Hidden Persuaders *was:* Acland, *Swift Viewing*, 105–107.

15. *The researcher, James M. Vicary:* Ibid., 92–105, 111; Kelly B. Crandall, "Invisible Commercials and Hidden Persuaders: James M. Vicary and the Subliminal Advertising Controversy of 1957," Unpublished undergraduate thesis, University of Florida, 2006, http://plaza.ufl.edu/cyllek/docs/KCrandall_Thesis2006.pdf.

16. *Vicary's announcement sparked:* "You'll Be Brainwashed and Not Even Know It," *Lincoln Journal*, September 13, 1957, 1.

17. *"Welcome to 1984":* William O'Barr, "'Subliminal' Advertising," *Advertising & Society Review* 13, no. 4 (2013), accessed Feb. 7, 2022, https://muse.jhu.edu/article/193862.

18. *Even* Advertising Age *signaled:* Ibid.; Crandall, "Invisible Commercials."

19. *Within a few months:* Crandall, "Invisible Commercials."

20. *Years later, Vicary:* O'Barr, "'Subliminal' Advertising"; Anthony R. Pratkanis, "The Cargo-Cult Science of Subliminal Persuasion," *Skeptical Inquirer* 16 (Spring 1992): 260–272.

21. *By 1958, however:* Fisher interview with Richards.

22. *"The subliminal stuff":* Ibid.

23. *Dement's sojourn:* William C. Dement with Stuart Rawlings, *Dreamer: The Memoir of a Sleep Researcher* (unpublished manuscript, 2018), 52–54.

24. *That December, Fisher:* Ibid., 55–59.

25. *Once again, however:* Ibid., 58–59.

26. *Soon afterward, Dement and Fisher:* Fisher interview with Richards.

27. *One of the first:* Ibid.; William C. Dement, "Thirty Years of Narcolepsy Research (1953–1983)," William C. Dement papers (SC0633), Box 5, Folder 4.1.2.1, Department of Special Collections and University Archives, Stanford University Libraries, Stanford, CA.

28. *It was now known:* D. Todman, "Narcolepsy: A Historical Review," *Internet Journal of Neurology* 9, no. 2 (2007), accessed Feb. 7, 2022, https://print.ispub.com/api/0/ispub-article/7361; Emmanuel Mignot, "History of Narcolepsy," *Archives Italiennes de Biologie* 139, no. 3 (2001): 207–220. *or in the waking center:* R. B. Aird, N. S. Gordon, and H. C. Gregg, "Use of Phenacemide (Phenerone) in Treatment of Narcolepsy and Catalepsy," *AMA Archives of Neurology and Psychiatry* 70, no. 4 (1953): 510–515.

29. *Others insisted:* H. A. Droogleever Fortuyn et al., "Narcolepsy and Psychiatry: An Evolving Association of Increasing Interest," *Sleep Medicine* 12 (2011): 714–719.

30. *The latter explanation:* Nathan G. Hale, *The Rise and Crisis of Psychoanalysis in the United States: Freud and the Americans, 1917–1985* (New York: Oxford University Press, 1995), 178–184, 282–283.

31. *In some cases:* Fisher interview with Richards; Shevrin, *Subliminal Explorations*, 39.

32. *Fisher, like many:* William C. Dement and Christopher Vaughan, *The Promise of Sleep: A Pioneer in Sleep Medicine Explores the Vital Connection Between Health, Happiness, and a Good Night's Sleep* (New York: Delacorte, 1999), 201; Charles Fisher, "Psychoanalytic Implications of Recent Research on Sleep and Dreaming: Part I: Empirical Findings," *Journal of the American Psychoanalytic Association* 13 (April 1965): 197–270.

33. *A smattering of:* Fortuyn et al., "Narcolepsy and Psychiatry."

34. *On the appointed night:* William C. Dement, *The Sleepwatchers* (Menlo Park, CA: Nychthemeron Press, 1995), 59; Dement and Vaughan, *The Promise of Sleep*, 196, 201.

35. *in a city of eight million:* US Department of Commerce, "Estimates of the Population of Selected Eastern Standard Metropolitan Areas," *Current Population Reports,* Series F-25, no. 181, August 14, 1958, 2.

36. *they found only four:* Dement and Vaughan, *The Promise of Sleep,* 201–202.

37. *In the fall of 1959:* Nathaniel Kleitman autobiography, dictated to Paulena Kleitman circa 1970s, Nathaniel Kleitman Papers, Box 1, Folder 7, Hanna Holborn Gray Special Collections Research Center, University of Chicago Library; Nathaniel Kleitman, *Sleep and Wakefulness, Revised and Enlarged Edition* (Chicago: University of Chicago Press, 1963), vii.

38. *Kleitman's successor:* Richard Sandomir, "Allan Rechtschaffen, Eminent Sleep Researcher, Dies at 93," *New York Times*, December 19, 2021, A3; *Golden Gloves boxer:* Interview with Donald Bliwise, December 30, 2021.

39. "Al was a tough son of a bitch": Donald Bliwise interview.

40. *Like Dement, Rechtschaffen:* Sandomir, "Allan Rechtschaffen."

41. *Among Rechtschaffen's apprentices:* David Foulkes, Allan Rechtschaffen, and Thomas Roth, "Obituary for Gerald W. Vogel, MD," *Sleep* 35, no. 6 (June 1, 2012): 889; Gerald Vogel, "Studies in the Psychophysiology of Dreams, III: The Dream of Narcolepsy," *Archives of General Psychiatry* 3, no. 4 (October 1960): 421–482.

42. *Vogel conducted:* Vogel, "Studies in the Psychophysiology."

43. *The paper, published:* Ibid.

44. *The subject had found:* Ibid.

45. *Vogel—true to his time:* Ibid.

46. *Vogel hypothesized:* Ibid.

47. *One afternoon:* Ibid.

48. *Vogel's paper:* Ibid.

49. *Soon after Dement:* Dement and Vaughan, *The Promise of Sleep*, 42.

50. *The previous record:* "D.J. to Go After Record," *Times* (San Mateo, CA), January 1, 1959, 610.

51. *The thirty-two-year-old:* Gay Gaer Luce, *Current Research on Sleep and Dreams* (Bethesda, MD: US Department of Health, Education, and Welfare, Public Health Service, 1966), 19–20.

52. *After two days:* Ibid.; Geoff Rolls, *Classic Case Studies in Psychology*, 2nd edition (London: Hodder Education, 2010), 161–162.

53. *To Dement, Tripp's:* Dement and Vaughan, *The Promise of Sleep*, 42–43.

54. *The hint of:* Luce, *Current Research on Sleep*, 20. Note: Although Tripp recovered quickly from his psychotic symptoms, he suffered what the psychologist's report called a "mild depression" for the next three months. This may have been an aftereffect of his ordeal— but it could also have been connected with his indictment shortly after the stunt on charges of having accepted $36,000 in payola. (See Nick Ravo, "Peter Tripp, 73, Popular Disc Jockey," *New York Times*, February 13, 2000, 44.)

55. *"It was as if":* Dement and Vaughan, *The Promise of Sleep*, 43.

56. *Dement set out:* Ibid., 44; Fisher interview with Richards.

57. *They repeated the process:* Calvin Trillin, "A Third State of Existence," *New Yorker*, September 18, 1965, 58–125.

58. *Early results confirmed:* William Dement, "The Effect of Dream Deprivation," *Science* 131, no. 3415 (June 10, 1960): 1705–1707.

59. *Yet to the researchers':* Dement and Vaughan, *The Promise of Sleep*, 45; Rolls, *Classic Case Studies*, 161–162.

60. *Hoping that further:* Dement, *The Sleepwatchers*, 139; W. C. Dement, "Remembering Nathaniel Kleitman," *Archives Italiennes de Biologie* 139 (2001): 11–17.

61. *One Barnard grad student:* Dement, *The Sleepwatchers*, 139–140.

62. *Dement also conducted:* Ibid., 140–143.

63. *Meanwhile, over several:* Dement and Vaughan, *The Promise of Sleep*, 44.

64. *A native of France's:* M. Jouvet, "How Sleep Was Dissociated into Two States: Telen-cephalic and Rhomboencephalic Sleep?," *Archives Italiennes de Biologie,* 142 (2004): 317–326.

65. *Awarded a Fulbright:* Ibid.

66. *In 1957, Jouvet won:* Ibid.

67. *Jouvet wanted to see:* Ibid.

68. *To get around this:* Ibid.

69. *Now Jouvet was ready:* Ibid.

70. *Besides monitoring:* Ibid.

71. *Jouvet noticed something else:* Ibid.

72. *What on earth:* Ibid.

73. *The French researcher:* Ibid.

74. *First, REM was triggered:* M. Jouvet, F. Michel, and J. Courjon, "Sur un Stade d'Activité Électrique Cérébrale Rapide au Cours du Sommeil Physiologique," *Comptes Rendus des Séances de l'Académie de la Société Biologique et de Ses Filiales* 153 (1959): 1024–1028; M. Jouvet and F. Michel, "Corrélations Électromyographiques du Sommeil Chez le Chat Décortiqué et Mésencephalique Chronique," *Comptes Rendus des Séances de la Société de Biologie et de Ses Filiales* 153 (1959): 422–425; M. Jouvet, F. Michel, and J. Courion; Jouvet, "How Sleep Was Dissociated into Two States."

75. *Together, as Jouvet later:* Michel Jouvet, *The Paradox of Sleep: The Story of Dreaming* (Cambridge, MA: MIT Press, 1999), 5.

76. *Eager to follow:* W. C. Dement, "The Paradox of Sleep: The Early Years," *Archives Ital-iennes de Biologie* 142 (2004): 333–345.

77. *As the new decade:* Howard Roffwarg interview with Milton Erman, 2010, "Conversa-tions with Our Founders," Sleep Research Society, accessed February 8, 2022, https://www.sleepresearchsociety.org/about/conversations-our-founders/; Peter Wortsman, "Howard Roffwarg: A Scientific Champion of Sleep," *Journal of the College of Physicians and Surgeons of Columbia University* 19, no. 2 (1999), accessed August 24, 2016, http://www.cumc.columbia.edu/psjournal/archive/archives/jour_v19no2/profile.html.

78. *Roffwarg started helping:* Dement, *The Sleepwatchers*, 34–35; Howard P. Roffwarg, Joseph N. Muzio, and William C. Dement, "Ontogenic Development of the Human Sleep-Dream Cycle," *Science* 152, no. 3722 (April 29, 1966): 604–619.

79. *A growing number:* Rock, *The Mind at Night*, 11; Donald Bliwise interview.

80. *Psychologist Joe Kamiya:* Cynthia Kerson, Tom Collura, and Joanne Kamiya. *Joe Kamiya: Thinking Inside the Box* (Corpus Christi: BMED Press, 2020), 22–23, 28; Sonia Ancoli-Israel, "In Memoriam: Joe Kamiya, PhD," *Society News,* Sleep Research Society, August 6, 2021, accessed February 11, 2022, https://www.sleepresearchsociety.org/in-memoriam-joe-kamiya-phd/.

81. *At Downstate Medical:* Luce, *Current Research on Sleep*, 28; Donald R. Good-enough et al., "A Comparison of 'Dreamers' and 'Nondreamers': Eye Movements,

Electroencephalograms, and the Recall of Dreams," *Journal of Abnormal and Social Psychology* 59, no. 3 (1959): 295–302.

82. *The pair sometimes:* Robert O'Brien, "Almost Everybody Has Dreams on Every Night," *Life*, May 5, 1958, 120–128.

83. *Polygraph-equipped sleep labs:* David Foulkes, "Dream Research: 1953–1993," *Sleep* 19, 8 (1996): 609–624; "The Collected Abstracts of Research Reports Presented at the First Annual Meeting of the Sleep Research Society at the University of Chicago, 25–26 March 1961," Sleep Research Society, accessed January 16, 2022, https://www.sleepresearchsociety.org/publications/historical-papers/.

84. *Thanks largely to:* Adrian R. Morrison, "Coming to Grips with a 'New' State of Consciousness: The Study of Rapid-Eye-Movement Sleep in the 1960s," *Journal of the History of the Neurosciences* 22, no. 4 (2013): 392–407; G. E. W. Wolstenhome and Maeve O'Connor, eds., *Ciba Foundation Symposium on the Nature of Sleep* (Boston: Little, Brown, 1961), VII–XII; "Sir John Eccles: Biographical," Nobel Prize, accessed October 1, 2022, https://www.nobelprize.org/prizes/medicine/1963/eccles/biographical/; Dement and Rawlings, *Dreamer*, 57.

85. *REM was still new:* Morrison, "Coming to Grips."

86. *One branch of that:* Ibid., 396.

87. *For a quarter century:* Vinod Kumar, ed., *Biological Timekeeping: Clocks, Rhythms and Behavior* (New Delhi: Springer, 2017), 12.

88. *Although sleep-wake cycles:* Cold Spring Harbor Symposia on Quantitative Biology, *Volume XXV: Biological Clocks* (Cold Spring Harbor, NY: Biological Laboratory, 1961), IX–XIII, accessed December 16, 2022, http://library.cshl.edu/symposia/1960/participants.html.

89. *Chronobiology did not yet:* Franz Halberg, "Chronobiology," *Scholarpedia* 6, no. 3 (2011): 3008, accessed December 4, 2022, http://www.scholarpedia.org/article/Chronobiology.

90. *This field had long:* Wilse B. Webb, "Sleep as a Biological Rhythm: A Historical Review," *Sleep* 17, no. 2 (1994): 188–194.

91. *In part, this was a legacy:* Kumar, *Biological Timekeeping*, 8.

92. *In* Sleep and Wakefulness: Kleitman, *Sleep and Wakefulness*, passim. The book contains references to the work of Maynard Johnson, J. S. Szymanski, Curt Richter, Sutherland Simpson, and others.

93. *Yet his theory:* Webb, "Sleep as a Biological Rhythm."

94. *The Cold Spring Harbor symposium was:* Serge Daan, "Colin Pittendrigh, Jürgen Aschoff, and the Natural Entrainment of Circadian Systems," *Journal of Biological Rhythms* 15, no. 3 (June 2000): 195–207.

Chapter 9

1. *In March 1961:* "Historical Note," Association for the Psychophysiological Study of Sleep Records, Hanna Holborn Gray Special Collections Research Center, University of Chicago Library, accessed February 11, 2022, https://www.lib.uchicago.edu/e/scrc/findingaids/view.php?eadid=ICU.SPCL.SLEEP; "Some Selected Details from the Earliest

Sleep Conferences (1961–1964)," Hanna Holborn Gray Special Collections Research Center, University of Chicago Library, courtesy Donald Bliwise; Kleitman autobiography.

2. *In contrast to:* "Some Selected Details"; "The Collected Abstracts of Research Reports Presented at the First Annual Meeting of the Sleep Research Society at the University of Chicago, 25–26 March 1961," Sleep Research Society, accessed December 6, 2021, https://www.sleepresearchsociety.org/publications/historical-papers/. The psychologically oriented talks strayed far beyond Freudian orthodoxy. Topics included "The Application of the REM Technique to the Study of Telepathy and Dreaming" (reporting ambiguous results from an experiment measuring "telepathic influence" on dream content) and "Drugs, Dreams, and the Experimental Subject" (describing effects such as "a tendency for prochlorperazine to increase the expression of heterosexuality and phenobarbital to increase the expression of homosexuality" in dreams).

3. *David Foulkes:* "Some Selected Details"; "The Collected Abstracts"; David Foulkes, "Dream Research: 1953–1993," *Sleep* 19, no. 8 (1996): 609–624; "William David Foulkes," Prabook, accessed January 26, 2022, https://prabook.com/web/william_david .foulkes/1701518.

4. *The conference's success:* "Historical Note," Association for the Psychophysiological Study of Sleep Records; Kenton Kroker, *The Sleep of Others and the Transformation of Sleep Research* (Toronto: University of Toronto Press, 2015), 330–331; Cynthia Kerson, "In Memoriam: Joe Kamiya, 1926–2021," *Biofeedback* 49, no. 4 (2021): 103–105, accessed January 26, 2022, https://meridian.allenpress.com/biofeedback /article-abstract/49/4/103/475843/In-Memoriam-Joe-Kamiya-1926-2021.

5. *Dement published nothing:* Willam C. Dement CV, c. 1988, William C. Dement Papers, Box 2, Folder 1.10.4.

6. *He was working on:* William Dement and Allan Rechtschaffen interview with Sonia Ancoli-Israel, "Conversations with Our Founders," Sleep Research Society, 2010, accessed March 5, 2022, https://www.sleepresearchsociety.org/about/conversations -our-founders/?_ga=2.136795888.111096961.1646506615-566757450.1619465445; W. C. Dement, "The History of Narcolepsy and Other Sleep Disorders," *Journal of the History of the Neurosciences* 2, no. 2 (1993): 121–134.

7. *Unlike Rechtschaffen's former:* William C. Dement, *The Sleepwatchers* (Menlo Park, CA: Nychthemeron Press, 1995), 59; Allan Rechtschaffen et al., "Nocturnal Sleep of Narcoleptics," *Electroencephalography and Clinical Neurophysiology* 15, no. 4 (August 1963): 599–609.

8. *Meanwhile, the still-officially-nameless:* "Some Selected Details"; Michael Thorpy, "Elliot D. Weitzman MD and Early Sleep Research Medicine in New York," *Sleep Medicine* 16 (2015): 1295–1300.

9. *For Dement:* W. C. Dement, "The Paradox of Sleep: The Early Years," *Archives Italiennes de Biologie* 142 (2004): 333–345; Michel Jouvet, "Recherches sir les Structures Nerveless et les Mécanismes Responsables des Différentes Phases du Sommeil Physiologique," *Archives Italiennes de Biologie* 100 (1962): 125–206.

10. *Back in New York:* William C. Dement with Stuart Rawlings, *Dreamer: The Memoir of a Sleep Researcher* (unpublished manuscript, 2018), 59; Dement CV; E. Kahn, W.

Dement, and C. Barmack, Jr., "Incidence of Color in Immediately Recalled Dreams," *Science* 137 (1962): 1054–1055; C. Fisher and W. Dement, "Manipulation Experimentale du Cycle Rêve-Sommeil par Rapport aux Etats Psychopathalogiques," *Revue de Médecine Psychosomatique* 4 (1962): 5–12; H. Roffwarg, W. Dement, J. Muzio, and C. Fisher, "Dream Imagery: Relationship to Rapid Eye Movements of Sleep," *Archives of General Psychiatry* 7 (1962): 235–258.

11. *He'd begun to feel:* Dement and Rawlings, *Dreamer*, 59.

12. *Dement was losing:* Interview with William Dement, October 19, 2016.

13. *Fisher is still:* "Perchance to Dream," *Newsweek*, November 30, 1959, 104. Note: This quote is sometimes mistakenly attributed to Dement, perhaps because it appears at the end of an article about his research. In fact, Dement himself took credit for it in his 1995 book, *The Promise of Sleep* (page 44), perhaps having forgotten where he'd first seen the famous statement.

14. *Dement, however:* William Dement interview.

15. *Fisher, for his part:* Charles Fisher, interview with Arnold D. Richards, New York Psychoanalytic Institute and Society Oral History Project (undated).

16. *Dement took the hint:* Interview with William Dement, September 24, 2016; Harvey W. Fineberg, "David Hamburg (1925–2019)," *Science* 364, no. 6444 (June 7, 2019): 940.

17. *Not long afterward:* William Dement interview, September 24, 2016.

Chapter 10

1. *When the Dements:* Interview with William Dement, September 24, 2016; interview with Cathy Dement Roos, November 2, 2020; "796 Raimundo Way, Stanford, CA 94305," Zillow, accessed March 1, 2022, https://www.zillow.com/homedetails /796-Raimundo-Way-Stanford-CA-94305/2139534155_zpid/?.

2. *David Hamburg, the psychiatry:* Interview with William Dement, October 19, 2016.

3. *Dement received:* Calvin Trillin, "A Third State of Existence," *New Yorker*, September 18, 1965, 58–125.

4. *Despite the cramped:* William Dement interview.

5. *To start with:* Trillin, "A Third State of Existence."

6. *In his new lab:* Ibid.; W. C. Dement, "The Paradox of Sleep: The Early Years," *Archives Italiennes de Biologie* 142 (2004): 333–345.

7. *"I felt peculiarly uninhibited":* Trillin, "A Third State of Existence."

8. *The failure of these:* Ibid.

9. *Dement was not:* Kenton Kroker, *The Sleep of Others and the Transformation of Sleep Research* (Toronto: University of Toronto Press, 2015), 333; "Program for the Association for the Psychophysiological Study of Sleep Meeting, 1963," Association for the Psychophysiological Study of Sleep Records, Box 1, Folders 7–8, Hanna Holborn Gray Special Collections Research Center, University of Chicago Library.

10. *Nearly two hundred researchers:* "Program for the Association for the Psychophysiological Study."

11. *Also present was:* Chip Brown, "The Stubborn Scientist Who Unraveled a Mystery of the Night," *Smithsonian*, October 2003. Repayment of the loan is per a letter from Kleitman to Aserinsky, March 2, 1963, collection of Jill Aserinsky Buckley.

12. *At the meeting:* Ibid.

13. *Dement recalled that:* William C. Dement, "Dr. Aserinsky Remembered," *SRS Bulletin* 4, no. 2 (September 1998): 14. In this article, Dement mistakenly recalled the 1965 APSS meeting in Bethesda, Maryland, as the site of his first encounter with Aserinsky since 1953. According to Armond Aserinsky, however, his father attended the 1963 meeting and was shaken by his reception as a sort of living ghost.

14. *Aserinsky later set up:* Brown, "The Stubborn Scientist"; interviews with Armond Aserinsky and Jill Aserinsky Buckley.

15. *Scientifically speaking:* "Program for the Association for the Psychophysiological Study"; Kroker, *The Sleep of Others*, 333.

16. *Soon after the APSS meeting:* William C. Dement with Stuart Rawlings, *Dreamer: The Memoir of a Sleep Researcher* (unpublished manuscript, 2018), 70.

17. *When her bleary-eyed husband:* William C. Dement, *The Sleepwatchers* (Menlo Park, CA: Nychthemeron Press, 1995), 59–60.

18. *The impetus was:* Allan Rechtschaffen et al., "Nocturnal Sleep of Narcoleptics," *Electroencephalography and Clinical Neurophysiology* 15 (1963): 599–609; Dement, *The Sleepwatchers*, 59. The study also found other links between narcolepsy and REM sleep. For example, the cataplectic episodes that plagued many patients while awake resembled the loss of muscle tone seen during REM. Their hallucinations and bouts of paralysis during transitions between sleep and waking likewise suggested that aspects of REM were occurring at the wrong time.

19. *Citing Jouvet's discovery:* Rechtschaffen et al., "Nocturnal Sleep of Narcoleptics."

20. *To solve the mystery:* William C. Dement and Christopher Vaughan, *The Promise of Sleep: A Pioneer in Sleep Medicine Explores the Vital Connection Between Health, Happiness, and a Good Night's Sleep* (New York: Delacorte, 1999), 46; Dement, *The Sleepwatchers*, 59–60; W. C. Dement, "The History of Narcolepsy and Other Sleep Disorders," *Journal of the History of the Neurosciences* 2, no. 2 (1993): 121–134. Note: Dement also ran an ad in the *San Francisco Examiner* (June 10, 1963, 32), asking specifically for patients diagnosed with narcolepsy. He does not mention this ad in any of his writings, which suggests that it failed to generate a significant response.

21. *as epilepsy, perhaps:* These were common misdiagnoses of narcolepsy at the time. See: Sleep Review Staff, "Narcolepsy: Historical Perspectives, Epidemiology, Socioeconomic Impact, Symptoms & Treatment," *Sleep Review*, April 4, 2001, accessed March 22, 2022, https://sleepreviewmag.com/sleep-disorders/hypersomnias/narcolepsy/narcolepsy/.

22. *Other respondents had:* Dement does not mention this group in his writings, but the prevalence of psychosomatic medicine in the 1950s and early '60s would have made their existence inevitable.

23. *Those who were receiving:* Sleep Review Staff, "Narcolepsy."

24. *Dement was struck:* William C. Dement, *Some Must Watch While Some Must Sleep* (San Francisco: San Francisco Book Company, 1976), 76–79.

25. *He was also surprised:* Emmanuel J. M. Mignot, "History of Narcolepsy at Stanford University," *Immunology Research* 58 (2014): 315–339.

26. *Today, experts estimate:* "Narcolepsy Fact Sheet," National Institute of Neurological Disorders and Stroke, accessed April 16, 2022, https://www.ninds.nih.gov/Disorders/Patient-Caregiver-Education/Fact-Sheets/Narcolepsy-Fact-Sheet.

27. *If he could nail down:* Interview with William Dement, October 19, 2016.

28. *Yet he felt no:* Ibid.

29. *That September, Dement:* M. Jouvet, "Society Proceedings: International Symposium on the Anatomical and Physiological Basis of Sleep," *Electroencephalography and Clinical Neurophysiology* 17 (1964): 440–450.

30. *In its survey:* Ibid.

31. *In other brain regions:* Dement, *The Sleepwatchers*, 45.

32. *Several presentations focused:* Jouvet, "Society Proceedings."

33. *That development owed:* Francisco Lopez-Muñoz and Cecilio Alamo, "Historical Evolution of the Neurotransmission Concept," *Journal of Neural Transmission* 116 (2009): 515–533; Matthew Cobb, *The Idea of the Brain: The Past and Future of Neuroscience* (New York: Basic Books, 2020), 292–311.

34. *By the early 1950s:* Lopez-Muñoz and Alamo, "Historical Evolution"; Matthew Cobb, *The Idea of the Brain: The Past and Future of Neuroscience* (New York: Basic Books, 2020), 292–311.

35. *Dement suggested one:* Jouvet, "Society Proceedings"; Adrian R. Morrison, "Coming to Grips with a 'New' State of Consciousness: The Study of Rapid-Eye-Movement Sleep in the 1960s," *Journal of the History of the Neurosciences* 22, no. 4 (2013): 392–407.

36. *The meeting left Dement:* Dement, "The Paradox of Sleep."

37. *Back in Palo Alto:* Ibid.; Trillin, "A Third State of Existence."

38. *In his next study:* Dement, "The Paradox of Sleep"; Trillin, "A Third State of Existence."

39. *Dement designed a protocol:* Trillin, "A Third State of Existence."

40. *Still, Dement:* Ibid.

41. *And after completing:* William Dement to Maurice B. Visscher, June 20, 1968, William C. Dement Papers (SC0633), Box 8, Folder 4.27.

42. *Constrained by his tiny lab:* Trillin, "A Third State of Existence."

43. *To determine whether:* Ibid.

44. *After more than:* Ibid.

45. *For that, more space:* Ibid.

46. *In the new location:* Ibid.; Seymour S. Kety, Edward V. Evarts, and Harold L. Williams, eds., *Sleep and Altered States of Consciousness* (Baltimore: Williams and Wilkins, 1967), 458. Note: The twenty-second time period, which is shorter than the sixty to ninety seconds described in Trillin's article, is from a book chapter written by Dement and his lab assistants.

47. *After a few weeks:* Dement, "The Paradox of Sleep."

48. *This study ran:* Ibid.; Trillin, "A Third State of Existence"; Gay Gaer Luce, *Current Research on Sleep and Dreams* (Bethesda, MD: US Department of Health, Education, and Welfare, Public Health Service, 1966), 87.

49. *One striking change:* Trillin, "A Third State of Existence."

50. *That kind of heightened:* Dement, "The Paradox of Sleep."

51. *Dement began to:* Trillin, "A Third State of Existence."

52. *So far, though:* Ibid.

53. *In early January:* Geoff Rolls, *Classic Case Studies in Psychology*, 2nd edition (London: Hodder Education, 2010), 163.

54. *A Honolulu deejay:* Parth Shah et al., "Eleven Days Without Sleep: The Haunting Effect of a Record-Breaking Stunt," NPR News, accessed March 12, 2022, https://www.wbur.org /npr/562305141/eleven-days-without-sleep-the-haunting-effects-of-a-record-breaking -stunt.

55. *Although Gardner was:* Rolls, *Classic Case Studies*, 163.

56. *When Dement arrived:* George Gulevich, Willam Dement, and LaVerne Johnson, "Psychiatric and EEG Observations on a Case of Prolonged (264 Hours) Wakefulness," *Archives of General Psychiatry* 15 (July 1966): 29–35; Rolls, *Classic Case Studies*, 164–166; Dement and Vaughan, *The Promise of Sleep*, 242–247.

57. *Yet what impressed:* Gulevich et al., "Psychiatric and EEG Observations"; "17 Year Old Journalism Major Becomes King of Insomniacs," *Sacramento Bee*, January 8, 1964, 14; Rolls, *Classic Case Studies*, 164–166.

58. *What accounted for:* Gulevich et al., "Psychiatric and EEG Observations."

59. *Tripp, who would be:* Nick Ravo, "Peter Tripp, 73, Popular Disc Jockey," *New York Times*, February 13, 2000, 44; Esther Inglis-Arkell, "The Sleep Deprivation Publicity Stunt That Drove One Man Crazy," Gizmodo, March 24, 2014, accessed March 12, 2022, https:// gizmodo.com/the-sleep-deprivation-publicity-stunt-that-drove-one-ma-1550084876.

60. *Dement still believed:* Dement and Vaughan, *The Promise of Sleep*, 245–247.

61. *That March, Dement:* "Some Selected Details from the Earliest Sleep Conferences (1961–1964)," Association for the Psychophysiological Study of Sleep Records; Howard Roffwarg, "History and Description of the SRS," Association for the Psychophysiological Study of Sleep Records.

62. *One of those:* "Some Selected Details"; Conference Program, Fourth Annual Meeting, Association for the Psychophysiological Study of Sleep Records.

63. *Cartwright, who'd moved on:* Penelope Green, "Rosalind Cartwright, Psychologist and 'Queen of Dreams,' Dies at 98," *New York Times*, March 15, 2021, 22.

64. *The APSS had fewer:* "Some Selected Details."

65. *Her presentation concerned:* Rosalind Dymond Cartwright, "Similarities and Differences Between Drug-Induced Hallucinations and Dreams," private communication to APSS, March 1964, Association for the Psychophysiological Study of Sleep Records.

66. *Dement himself gave:* Conference Program, Fourth Annual Meeting.

67. *Besides moving Dement:* Ibid.; email message from Cathy Dement Roos to author, March 25, 2022.

68. Meet the Beatles *had just come out:* "US album release: Meet The Beatles!," Beatles Bible, accessed March 22, 2022, https://www.beatlesbible.com/1964/01/20/us-lp-meet-the-beatles/; interview with Cathy Dement Roos, November 2, 2020.

69. *One manifestation of:* Dement, "The History of Narcolepsy"; Dement and Vaughan, *The Promise of Sleep*, 202.

70. *Dement's playfulness was:* Interview with Nick Dement, October 5, 2022; Cathy Dement Roos interview.

71. *Pat was a coconspirator:* Interviews with Nick Dement and Cathy Dement Roos; interview with Woody Myers, September 29, 2021.

72. *Dement fostered:* Trillin, "A Third State of Existence"; Kety et al., *Sleep and Altered States*, 457–458.

73. *Dement's hut had:* Ibid.

74. *One morning, Trillin:* Trillin, "A Third State of Existence."

75. *The conversations in the lab:* Ibid.

76. *Dement's combination of:* Interview with Grant Hoyt, February 14, 2022; interview with Bill Gonda, February 15, 2022.

77. *A former lab technician:* Grant Hoyt interview.

78. *It wasn't that:* Ibid.

Chapter 11

1. *The Menlo Park VA:* James Dinwiddie (blogging as bdbuddha), "Hippies and High Tech," November 12, 2012, accessed March 22, 2022, https://bdbuddha.com/2012/11/10/hippies-and-high-tech/.

2. *Shortly before* Cuckoo's Nest: Ibid.; Dave Swanson, "How the Warlocks Became the Grateful Dead," UCR: Classic Rock and Culture, accessed March 23, 2022, https://ultimateclassicrock.com/warlocks-grateful-dead/.

3. *In the neighborhood:* Dinwiddie, "Hippies and High Tech."

4. *During Dement's stint:* "Activism@Stanford: 1960s," Stanford Libraries, accessed March 23, 2022, https://exhibits.stanford.edu/activism/feature/1960s; Richard Lyman, "At the Hands of the Radicals," *Stanford Magazine*, January/February 2009, accessed March 23, 2022, https://stanfordmag.org/contents/at-the-hands-of-the-radicals; interview with Cathy Dement Roos, November 2, 2020; Ana Leorne, "The Story of the Beatles' Last Official Concert, Which Took Place in San Francisco," SFGate, August 20, 2021, accessed October 5, 2022, https://www.sfgate.com/music/article/beatles-final-concert-san-francisco-candlestick-16411474.php.

5. *In 1967, they:* Robin Wander, "1967: The Summer of Love at Stanford," Stanford News, July 23, 2017, accessed March 18, 2022, https://news.stanford.edu/2017/07/23/1967-year-summer-love-stanford.

6. *In 1968, protesters:* "Activism@Stanford: 1960s."

7. *Later that year:* Sean L. Malloy, "Uptight in Babylon: Eldridge Cleaver's Cold War," *Diplomatic History* 37, no. 3 (June 2013): 538–571; John Blake, "The Black Panthers Are Back—and Never Really Went Away," CNN, February 17, 2016, accessed March 18.2022, https://www.cnn.com/2016/02/16/us/black-panthers/index.html; Cathy Dement Roos interview.

8. *In 1969, activists:* "Activism@Stanford: 1960s"; Lyman, "At the Hands of the Radicals."

9. *Although Dement was:* Interview with Grant Hoyt, February 14, 2022; interview with Bill Gonda, February 15, 2022.

10. *"There was a supportive environment":* Interview with Bill Gonda.

11. *Dement also acted:* Interview with William Dement, September 24, 2016; William C. Dement and Christopher Vaughan, *The Promise of Sleep: A Pioneer in Sleep Medicine Explores the Vital Connection Between Health, Happiness, and a Good Night's Sleep* (New York: Delacorte, 1999), 239.

12. *In the Dements':* Cathy Dement Roos interview, November 2, 2020; Michael Peña, "Resident Fellow Thom Massey, Champion of Multicultural Education, Dies at 61," Stanford Report, January 7, 2009, accessed March 23, 2022, https://news.stanford.edu/news/2009/january7/thom-010709.html; "Taking the Mic: Black Students Seek Representation in Education," Stanford Graduate School of Education: Stories, accessed March 22, 2023, https://gse100.stanford.edu/stories/taking-the-mic-black-students-seek-representation-in-education?src=title.

13. *Yet throughout this:* Interviews with Cathy Dement Roos and Bill Gonda.

14. *Jouvet unveiled one:* Calvin Trillin, "A Third State of Existence," *New Yorker*, September 18, 1965, 58–125; Michel Jouvet, "Behavioral and EEG Effects of Paradoxical Sleep Deprivation in the Cat," *Proceedings of the XXIIIrd International Congress of Physiological Sciences, Tokyo,* September 1965; Michel Jouvet, *The Paradox of Sleep: The Story of Dreaming* (Cambridge, MA: MIT Press, 1999), 82–87.

15. *His finding seemed:* J. A. Hobson, "Michel Jouvet: A Personal Tribute," *Archives Italiennes de Biologie* 142 (2004): 347–352; Jouvet, "Behavioral and EEG Effects"; Jouvet, *The Paradox of Sleep,* 81–93, 155.

16. *This led Jouvet:* Hobson, "Michel Jouvet"; Jouvet, *The Paradox of Sleep,* 155.

17. *Charles Fisher declared:* Charles Fisher, "Psychoanalytic Implications of Recent Research on Sleep and Dreaming, Part I: Empirical Findings," *Journal of the American Psychoanalytic Association* 13, no. 2 (April 1965): 197–270. *"plethysmographic bagel":* Charles Fisher, interview with Arnold D. Richards, New York Psychoanalytic Institute and Society Oral History Project (undated).

18. *Dement, however, had:* Howard Roffwarg, Joseph N. Muzio, and William C. Dement, "Ontogenic Development of the Human Sleep-Dream Cycle," *Science* 152 (April 29, 1966): 604–619.

19. *By 1969, Dement was:* James Ferguson et al., "'Hypersexuality' and Behavioral Changes in Cats Caused by Administration of p-Chlorphynalanine," *Science* 168 (April 1970): 499–501.

20. *Dement documented:* William C. Dement, *The Sleepwatchers* (Menlo Park, CA: Nychthemeron Press, 1995), 141.

21. *Truth be told:* Dement and Vaughan, *The Promise of Sleep,* 45–46.

22. *In the near term:* Ibid.

23. *At the time:* Nathaniel Kleitman, *Sleep and Wakefulness, Revised and Enlarged Edition* (Chicago: University of Chicago Press, 1963), 233–289; Nathaniel Kleitman, *Sleep and Wakefulness as Alternating Phases in the Cycle of Existence* (Chicago: University of Chicago Press, 1939), 325–397.

24. *In Germany, neurophysiologist:* Kenton Kroker, *The Sleep of Others and the Transforma-tion of Sleep Research* (Toronto: University of Toronto Press, 2015), 404–411; T. Young et al., "The Occurrence of Sleep-Disordered Breathing Among Middle-Aged Adults," *New England Journal of Medicine* 328, no. 17 (April 29, 1993): 1230–1235.

25. *Few treatments existed:* Kleitman, *Sleep and Wakefulness, Revised and Expanded Edi-tion,* 293–317; Sudhansu Chokroverty and Michel Billiard, eds., *Sleep Medicine: A Comprehensive Guide to Its Development, Clinical Milestones, and Advances in Treatment* (New York: Springer, 2015), 103–111; Meir H. Kryger, Thomas Roth, and William C. Dement, *Principles and Practice of Sleep Medicine,* 3rd edition (Philadelphia: Saunders, 2000), 10.

26. *And yet, several:* Kroker, *The Sleep of Others,* 359–361; "Kefauver-Harris Amendments Revolutionized Drug Development," FDA Consumer Health Information, accessed March 23, 2022, https://www.gvsu.edu/cms4/asset/F51281F0-00AF-E25A-5BF632E 8D4A243C7/kefauver-harris_amendments.fda.thalidomide.pdf; "The Hypnotic Poten-tial of Thalidomide," Science Daily, September 23, 2020, accessed March 23, 2022, https://www.sciencedaily.com/releases/2020/09/200923124626.htm.

27. *Sleep researchers, in response:* Kroker, *The Sleep of Others,* 362–363.

28. *Studies were beginning:* Ibid. In addition to sparking concerns about the health effect of sleeping pills, research was raising questions about the very definition of insomnia. In 1967, Rechtschaffen's PhD student Lawrence Munroe published a study showing that the difference between people who described themselves as "poor sleepers" or "good sleepers" was fuzzier than it seemed. In questionnaires, poor sleepers reported that it took them an average of fifty-nine minutes to fall asleep. Yet in the lab, their average time was fifteen minutes—only eight minutes longer than for self-reported good sleep-ers. Still, a follow-up study indicated that they weren't just imagining things: Poor sleepers tended to be more anxious and depressed than good sleepers. They also spent less time in REM and more in Stage 2, suggesting that their sense of unrest had some basis in reality. See: Lawrence J. Munroe, "Psychological and Physiological Differences Between Good and Poor Sleepers," *Journal of Abnormal Psychology* 72, no. 3 (1967): 255–264.

29. *In the process:* Allan Rechtschaffen and Anthony Kales, *A Manual of Standard-ized Technology, Techniques and Scoring System for Sleep Stages of Human Subjects* (Washington, DC: Public Health Service, US Government Printing Office, 1968); "Andres A. Gonzalez, "How has the AASM Manual for Sleep Scoring Evolved?," Medscape, August 19, 2019, https://www.medscape.com/answers/1188142-193006 /how-has-the-aasm-manual-for-sleep-stage-scoring-evolved.

30. *Kales, who'd launched:* Kroker, *The Sleep of Others,* 362–364; Anthony Kales, *Sleep: Physiology & Pathology* (Philadelphia: J. B. Lippincott, 1969), 5–13. *which would also fund much of Kales's:* Examples include Anthony Kales et al., "Sleep Laboratory Studies of Flurazepam: A Model for Evaluating Hypnotic Drugs," *Clinical Pharmacology and Therapeutics* 19, no. 5 (May 1976); Edward O. Bixler et al., "Flunitrazepam, an Inves-tigational Hypnotic Drug: Sleep Laboratory Evaluations," *Journal of Clinical Pharma-cology* 17 (October 1977): 569–578, 1379–1388; Martin B. Scharf et al., "Long-Term

Sleep Laboratory Evaluation of Flunitrazepam," *Pharmacology* 19 (1979): 173–181; Edward O. Bixley et al., "Effects of Hypnotic Drugs on Memory," *Life Sciences* 25, no. 16 (1979): 1379–1388.

31. *Such developments:* Dement, *The Sleepwatchers*, 71–72, 74–80.

32. *Since the 1960:* Lynne Lamberg, *Bodyrhythms: Chronobiology and Peak Performance* (New York: William Morrow, 1994), 24–25; William K. Stevens, "Curt Richter, Credited with Idea of Biological Clock, Is Dead at 94, *New York Times,* December 22, 1988, section D, 23.

33. *In 1965, he:* Lamberg, *Bodyrhythms*, 24–25; Curt Richter, *Biological Clocks in Medicine and Psychiatry* (Springfield, IL: Charles C. Thomas, 1965), 21; Lynne Lamberg, "Researchers Dissect the Tick and Tock of the Human Body's Master Clock," *JAMA* 278, no. 13 (October 1, 1997): 1049–1051.

34. *The master clock's:* Lamberg, *Bodyrhythms*, 25.

35. *At Montefiore Medical Center:* Michael Thorpy, "Elliot D. Weitzman MD and Early Sleep Research and Sleep Medicine in New York," *Sleep Medicine* 16 (2015): 1295–1300; Elliot D. Weitzman, Herbert Schaumburg, and William Fishbein, "Plasma 17-Hydroxycorticosteroid Levels During Sleep in Man," *Journal of Clinical Endocrinology and Metabolism* 26, no. 2 (February 1966): 121–127.

36. *Weitzman's paper caught:* Thorpy, "Elliot D. Weitzman MD."

37. *The answer:* Elliot D. Weitzman et al., "Acute Reversal of the Sleep-Waking Cycle in Man: Effect on Sleep Stage Patterns," *Archives of Neurology* 22 (June 1970): 483–489.

38. *Levels of cortisol:* E. D. Weitzman et al., "Reversal of Sleep-Waking Cycle: Effect on Sleep Stage Pattern and Certain Neuro-Endocrine Rhythms," *Psychophysiology* 4, no. 3 (1968): 366–367.

39. *With further studies:* Thorpy, "Elliot D. Weitzman MD."

40. *Across the Atlantic:* Till Roenneberg, *Internal Time* (Cambridge, MA: Harvard University Press, 2012), 3, 41.

41. *There, Aschoff:* Rütger A. Wever, *The Circadian System of Man: Results of Experiments Under Temporal Isolation* (New York: Springer-Verlag), 9–10.

42. *In some ways:* Roenneberg, *Internal Time*, 36–46; Wever, *The Circadian System*, 4–17. As in a handful of earlier temporal isolation experiments (all involving a single subject), most of Aschoff and Wever's subjects settled into a daily cycle lasting about twenty-five hours. A few, however, fell into wildly irregular sleep-wake patterns—and were shocked, at the end of the experiment, to find that it had lasted twenty-four days when they'd only experienced nineteen or twenty. Their cycles of rest and activity also fell out of sync with their other internal rhythms, such as daily fluctuations in temperature or hormone secretion.

43. *The bunker studies confirmed:* Roenneberg, *Internal Time*, 45–46.

44. *What spurred him:* W. C. Dement, "The Paradox of Sleep: The Early Years," *Archives Italiennes de Biologie* 142 (2004): 333–345.

45. *At that moment:* Gerald W. Vogel et al., "REM Sleep Reduction Effects on Depression Syndromes," *Archives of General Psychiatry* 32 (June 1975): 765–777; Gerald W. Vogel et al., "Endogenous Depression Improvement and REM Pressure," *Archives of General Psychiatry* 34, no. 1 (1977): 96–97; Gerald W. Vogel et al., "Improvement

of Depression by REM Sleep Deprivation: New Findings and a Theory," *Archives of General Psychiatry* 37, no. 3 (1980): 247–253.

46. *In the '80s:* Matt Wood, "Allan Rechtschaffen, PhD, Sleep Research Pioneer, 1927–2021," University of Chicago Biological Sciences Division News, December 9, 2021, accessed March 21, 2022, https://biologicalsciences.uchicago.edu/news/allan-rechtschaffen-obituary; Clete Kushida, Bernard M. Bergmann, and Allan Rechtschaffen, "Sleep Deprivation in the Rat: IV. Paradoxical Sleep Deprivation," *Sleep* 12, no. 1 (1989): 22–30; Allan Rechtschaffen et al., "Sleep Deprivation in the Rat: X. Integration and Discussion of the Findings," *Sleep* 12, no. 1 (1989): 68–87. In Rechtschaffen's study, rats deprived of REM sleep died after sixteen to fifty-four days, compared to eleven to thirty-two days in those subjected to total sleep deprivation.

47. *He was ready:* Dement, "The Paradox of Sleep."

Chapter 12

1. *Historians often date:* William C. Dement, "The Study of Human Sleep: A Historical Perspective," *Thorax* 53, supplement 3 (1998): S2–7; P. Montagna and S. Chokroverty, eds., *Handbook of Clinical Neurology, Volume 98 (3rd series): Sleep Disorders, Part I* (Amsterdam: Elsevier, 2011), accessed April 5, 2022, https://cdn2.hubspot.net/hubfs/4256583/Thorpy_2011_Handbook-of-Clinical-Neurology.pdf; Thomas Kilduff et al., "New Developments in Sleep Research," *Journal of Neuroscience* 28, no. 46 (November 12, 2008): 11814–11818.

2. *In reality, it was:* Sudhansu Chokroverty and Michel Billiard, eds., *Sleep Medicine: A Comprehensive Guide to Its Development, Clinical Milestones, and Advances in Treatment* (New York: Springer, 2015), 121.

3. *The Stanford clinic's mission:* William C. Dement, *The Sleepwatchers* (Menlo Park, CA: Nychthemeron Press, 1995), 136–138; William Dement interview with Tom Roth, 2010, "Conversations with Our Founders," Sleep Research Society, accessed April 5, 2022, https://sleepresearchsociety.org/about/conversations-our-founders/.

4. *As he inspected:* Dement, *The Sleepwatchers*, 136.

5. *Although no one had:* Ibid., 136–137.

6. *"What about sex?":* Ibid., 137.

7. *"I don't believe":* Ibid.

8. *About two weeks later:* Ibid.

9. *Dement declined:* Ibid., 137–138; "Sleepless Nights for Science," *San Francisco Chronicle*, September 7, 1970, 2.

10. *Thankfully, the Stanford:* Interview with Cathy Dement Roos, November 11, 2020.

11. *What truly set:* William C. Dement, "History of Sleep Medicine," *Neurologic Clinics* 23 (2005): 945–965; William C. Dement, "History of Sleep Medicine," *Sleep Medicine Clinics* 3 (2008): 147–156; Dement, "The Study of Human Sleep."

12. *More successful was:* "Ujamaa Celebrates 20 Years," *Stanford Daily*, October 15, 1990.

13. *About 40 percent:* "Minority Housing Program Begun," *Stanford Daily*, October 7, 1970; John Hennessy, "Should Stanford Expand the Freshman Class?," *Stanford Magazine*, September/October 2007, https://stanfordmag.org/contents/should-stanford

-expand-the-freshman-class; "Black Students at Stanford, 1970s," Stanford Publications, accessed April 12, 2022, https://exhibits.stanford.edu/stanford-pubs/catalog/xz351ht5726.

14. *In September 1970:* Interview with Cathy Dement Roos, November 2, 2020.

15. *Over the year:* Ibid.

16. *The dorm evolved:* Interview with Woody Myers, February 2, 2022.

17. *That winter, Dement:* "Courses and Degrees, 1973–1974," Stanford Publications, accessed May 3, 2022, https://exhibits.stanford.edu/stanford-pubs/catalog/qp502st4627.

18. *To his surprise:* Dement, *The Sleepwatchers,* 153–154; Richard Lyman, "At the Hands of the Radicals," *Stanford Magazine,* January/February 2009, accessed March 23, 2022, https://stanfordmag.org/contents/at-the-hands-of-the-radicals; SearchWorks catalog, Stanford Libraries, accessed April 5, 2022, https://searchworks-lb.stanford.edu/view/9719480.

19. *Although he wore:* Interview with Mary Carskadon, July 25, 2020.

20. *He had exchanged:* Rosanne Spector, "William Dement, Giant in Sleep Medicine, Dies at 91," Stanford Medicine, June 18, 2020, accessed October 5, 2022, https://med.stanford.edu/news/all-news/2020/06/william-dement-giant-in-field-of-sleep-medicine-dies-at-91.html; Richard Sandomir, "Dr. William Dement, Leader in Sleep Disorder Research, Dies at 91," *New York Times,* June 27, 2020, https://www.nytimes.com/2020/06/27/science/dr-william-dement-dead.html; Natalie Schwartz, "Prof Offers Final Year of 'Sleep & Dreams,'" *Stanford Daily,* January 13, 2002, 1; interview with Marc Miller, February 4, 2021; interview with Mark Rosekind, February 13, 2021; email message from Marc Miller, December 10, 2022.

21. *His approach, he explained:* Spector, "William Dement, Giant in Sleep Medicine"; "A Tribute for William C. Dement," YouTube video, 12:24, posted by Stanford, August 7, 2008, accessed April 5, 2022, https://www.youtube.com/watch?v=UWkeI45kbNU&t=706s; Dement, *The Sleepwatchers,* xxvi.

22. *Sleep and Dreams eventually:* Sarah DiGuilio, "The Surprising Way Colleges Are Helping Their Students Sleep More," HuffPost, April 20, 2016, accessed October 5, 2022, https://www.huffpost.com/entry/sleep-class-college-courses-teach-students-how-to-sleep_n_571578bae4b0060ccda425a2; interview with Mary Carskadon, September 5, 2020.

23. *While Dement's course:* Chokroverty and Billiard, *Sleep Medicine,* 105.

24. *The Bruges conference was:* APSS Bruges Conference program, Nathaniel Kleitman Papers, Box 12, Folders 9 and 12, Hanna Holborn Gray Special Collections Research Center, University of Chicago Library.

25. *What made the conference:* Chokroverty and Billiard, *Sleep Medicine,* 105.

26. *Guilleminault, like his:* Christian Guilleminault interview with Aly Sooyeon Suh, World Sleep Day, accessed March 31, 2022, https://worldsleepday.org/dr-christian-guilleminault.

27. *One of his current:* Dement, *The Sleepwatchers,* 62–63.

28. *Back at Stanford:* Chokroverty and Billiard, *Sleep Medicine,* 105; Christian Guilleminault CV, William C. Dement Papers, Box 53, Folder 10.7.2.

29. *He would serve:* Interview with Donald Bliwise, February 6, 2021; interview with Cathy Dement Roos, November 11, 2020; interview with Woody Myers, September 29, 2021.

30. *The new hire:* Chokroverty and Billiard, *Sleep Medicine*, 105; Guilleminault CV.

31. *Soon afterward, his team:* Dement, *The Sleepwatchers*, 63.

32. *After fitting the patient:* Christian Guilleminault, Frederic L. Eldridge, and William C. Dement, "Insomnia with Sleep Apnea: A New Syndrome," *Science* 181 (August 1973): 856–858.

33. *Guilleminault recognized:* Dement, *The Sleepwatchers*, 63.

34. *One of the clinic's psychiatrists:* Chokroverty and Billiard, *Sleep Medicine*, 106.

35. *The team soon found:* Christian Guilleminault, Ara Tilkian, and William C. Dement, "The Sleep Apnea Syndromes," *Annual Review of Medicine* 27 (1976): 465–484.

36. *Earlier researchers had found:* Christian Guilleminault et al., "Sleep Apnea Syndrome: Can It Induce Hemodynamic Changes?," *Western Journal of Medicine* 123 (July 1975): 7–16; Guilleminault et al., "The Sleep Apnea Syndromes"; Christian Guilleminault et al., "Apneas During Sleep in Infants: Possible Relationship with Sudden Infant Death Syndrome," *Science* 190 (November 14, 1975): 677–679.

37. *The team also found:* Dement, *The Sleepwatchers*, 63–64; William C. Dement and Christopher Vaughan, *The Promise of Sleep: A Pioneer in Sleep Medicine Explores the Vital Connection Between Health, Happiness, and a Good Night's Sleep* (New York: Delacorte, 1999), 183–184; Dement, "History of Sleep Medicine," *Sleep Medicine Clinics*.

38. *After the surgery:* Dement, *The Sleepwatchers*, 64; Dement and Vaughan, *The Promise of Sleep*, 183–184.

39. *Starting in 1973:* Christian Guilleminault, Frederic L. Eldridge, and William C. Dement, "Insomnia with Sleep Apnea: A New Syndrome," *Science* 181 (1973): 856–858; Guilleminault et al., "Sleep Apnea Syndrome"; Guilleminault, Tilkian, and Dement, "The Sleep Apnea Syndromes"; Sanjeet Bagcchi, "Prolific Sleep Researcher Christian Guilleminault Dies," *Scientist,* August 6, 2019, accessed April 5, 2022, https://www.the-scientist.com/news-opinion/prolific-sleep-researcher-christian-guilleminault-dies-66227.

40. *Another transformation took:* Dement, "History of Sleep Medicine," *Neurologic Clinics*.

41. *Unlike Dement:* Guilleminault CV.

42. *He was a gifted:* William C. Dement and Christian Guilleminault, "Sleep Disorders: The State of the Art," *Hospital Practice* 8, no. 11 (1973): 57–71.

43. *Under Guilleminault's guidance:* Association of Sleep Disorders Centers, *Diagnostic Classification of Sleep and Arousal Disorders,* 1st edition, prepared by the Sleep Disorders Classification Committee, H. P. Roffwarg, chairman, *Sleep* 2, no. 1 (1979): 1–137.

44. *And they learned:* Dement, *The Sleepwatchers*, 70; Association of Sleep Disorders Centers, *Diagnostic Classification of Sleep and Arousal Disorders.*

45. *In a startling:* Dement and Guilleminault, "Sleep Disorders"; Richard Lawrence Miller, *The Encyclopedia of Addictive Drugs* (Westport, CT: Greenwood Press, 2002), 175.

46. *The team also:* Christian Guilleminault, Mary Carskadon, and William C. Dement, "On the Treatment of Rapid Eye Movement Narcolepsy," *Archives of Neurology* 30 (January 1974): 90–93.

47. *For patients with parasomnias:* Roger J. Broughton, "Sleep Disorders: Disorders of Arousal?," *Science* 159 (March 1968): 1070–1077; Dement and Guilleminault, "Sleep Disorders."

48. *By late 1972:* William C. Dement, "The Study of Human Sleep."

Chapter 13

1. *Soon after Dement:* Jean Gayon, "From Mendel to Epigenetics: History of Genetics," *Comptes Rendus Biologies* 339 (2016): 225–230; Joseph S. Takahashi, "The 50th Anniversary of the Konopka and Benzer 1971 Paper in PNAS: 'Clock Mutants of Drosophila melanogaster,'" *PNAS* 118, no. 39 (2021), accessed March 30, 2022, https://doi.org/10.1073/pnas.2110171118; Ronald J. Konopka and Seymour Benzer, "Clock Mutants of *Drosophila melanogaster*," *Proceedings of the National Academy of Sciences, USA* 68, no. 9 (September 1971): 2112–2116.

2. *The discoverers were:* Takahashi, "The 50th Anniversary"; Konopka and Benzer, "Clock Mutants."

3. *The second breakthrough:* Lynne Lamberg, "Researchers Dissect the Tick and Tock of the Human Body's Master Clock," *JAMA* 278, no. 13 (October 1, 1997): 1049–1051; Lynne Lamberg, *Bodyrhythms: Chronobiology and Peak Performance* (New York: William Morrow, 1994), 24–26.

4. *Other researchers soon:* David R. Weaver, "The Suprachiasmatic Nucleus: A 25-Year Retrospective," *Journal of Biological Rhythms* 13, no. 2 (April 1998): 100–112; J. W. Shepard et al., "History of the Development of Sleep Medicine in the United States," *Journal of Clinical Sleep Medicine* 2, no. 1 (2005): 61–82.

5. *These studies proved:* Nathaniel Kleitman, *Sleep and Wakefulness as Alternating Phases in the Cycle of Existence* (Chicago: University of Chicago Press, 1939), 521.

6. *They also provided:* Elliot D. Weitzman et al., "Acute Reversal of the Sleep-Waking Cycle in Man: Effect on Sleep Stage Patterns," *Archives of Neurology* 22 (June 1970): 483–489.

7. *For Dement, one of:* Mary A. Carskadon and William C. Dement, "Sleep Studies on a 90-Minute Day," *Electroencephalography and Clinical Neurophysiology* 39 (1975): 145–155.

8. *Dement dubbed this method:* Ibid.

9. *He'd first tried it in 1971:* William C. Dement and Christopher Vaughan, *The Promise of Sleep: A Pioneer in Sleep Medicine Explores the Vital Connection Between Health, Happiness, and a Good Night's Sleep* (New York: Delacorte, 1999), 57–58; interview with Mary Carskadon, July 25, 2020.

10. *With a round face:* Mary Carskadon interview.

11. *As a child:* Ibid.

12. *At Elizabethtown Area High School:* Ibid.

13. *There, instead of:* Ibid.

14. *In the summer of 1970:* Mary Carskadon interview with Sonia Ancoli-Israel, "Conversations with Our Founders," Sleep Research Society, 2020, accessed April 6, 2022, https://sleepresearchsociety.org/about/conversations-our-founders/.

15. *Her first assignment:* Mary Carskadon interview.

16. *The lab had recently:* Interview with Grant Hoyt, February 14, 2022; interview with Bill Gonda, February 15, 2022; "The Stanford Family Carriage," Stanford Founders' Celebration, https://founders.stanford.edu/stanford-family-carriage; William C. Dement with Stuart Rawlings, *Dreamer: The Memoir of a Sleep Researcher* (unpublished manuscript, 2018), 119. In a bricked-up chamber, the crew found three century-old carriages that had belonged to university founder Leland Stanford; Dement displayed one of them in his driveway for several weeks before turning it over to the university.

17. *Carskadon was given:* Mary Carskadon interview.

18. *During her first months:* Ibid.; interview with William Dement, October 19, 2016.

19. *He also sometimes:* Mary Carskadon interview.

20. *Dement did, after all:* Bruges conference program, Nathaniel Kleitman Papers, Box 12, Folder 9, Hanna Holborn Gray Special Collections Research Center, University of Chicago Library; Eckard Weber et al., "a-N-Acetyl-B Endorphins in the Pituitary: Immunohistochemical Localization Using Antibodies Raised Against Dynorphin (1–13), *Journal of Neurochemistry* 36, no. 6 (June 1981), 1977–1985.

21. *Perhaps he was:* Carskadon interview.

22. *If so, the snub:* Ibid.

23. *The study's basic design:* Carskadon and Dement, "Sleep Studies on a 90-minute Day."

24. *The results held several:* Ibid.

25. *Although the details:* Ibid.

26. *For Carskadon, however:* Mary Carskadon interview.

27. *On December 10:* "Today's Television," *Democrat and Chronicle* (Rochester, New York), December 10, 1974, 4C.

28. *The host, Johnny Carson:* Johnny Carson interview with William Dement, *The Tonight Show Starring Johnny Carson,* December 10, 1974.

29. *"Nice to be here":* Johnny Carson interview.

30. *In reality, the book:* William C. Dement, *Some Must Watch and Some Must Sleep* (San Francisco: San Francisco Book Company, 1976), viii.

31. *Despite Carson's low-brow intro:* Johnny Carson interview.

32. *Dement's response was:* Ibid.

Chapter 14

1. *By the summer:* Interview with Mary Carskadon, July 25, 2020.

2. *her mother had died:* "Mrs. Carskadon, Secretary, Dies," *Lancaster New Era*, January 2, 1975, 10.

3. *Dement accompanied her:* Mary Carskadon, "Personal Memories and Reflections of Bill Dement," eulogy, June 18, 2020.

4. *To pursue an independent:* Mary Carskadon interview; email message from Mary Carskadon to author, April 10, 2022.

5. *Her mentor, too:* Interview with Cathy Dement Roos, November 2, 2020.

6. *When Dement wasn't:* William C. Dement, *The Sleepwatchers* (Menlo Park, CA: Nychthemeron Press, 1995), 146–149.

7. *Instead, he urged:* Mary Carskadon interview.

8. *Dement's motives:* Ibid.

9. *While waiting for:* Ibid.; Mary A. Carskadon, "Determinants of Daytime Sleepiness: Adolescent Development, Extended and Restricted Nocturnal Sleep," unpublished PhD dissertation, Stanford University, 1979, 13–23.

10. *In the spring of 1976:* Ibid.

11. *Carskadon named this:* Ibid.; Mary A. Carskadon, ed., *Encyclopedia of Sleep and Dreaming* (New York: Macmillan, 1993), 386–387; Mary A. Carskadon et al., "Pubertal Changes in Daytime Sleepiness," *Sleep* 2, no. 4 (1980): 453–460.

12. *usually ninety seconds:* Mary A. Carskadon and William C. Dement, "The Multiple Sleep Latency Test: What Does it Measure?," *Sleep* 5 (1982): 567–572.

13. *Each run was scored:* William C. Dement and Christopher Vaughan, *The Promise of Sleep: A Pioneer in Sleep Medicine Explores the Vital Connection Between Health, Happiness, and a Good Night's Sleep* (New York: Delacorte, 1999), 59–60.

14. *When the researchers:* Ibid.

15. *Over the next few years:* Gary S. Richardson et al., "Excessive Daytime Sleepiness in Man: Multiple Sleep Latency Measurement in Narcoleptic and Control Subjects," *Electroencephalography and Clinical Neurophysiology* 45 (1978): 621–627; Merrill Mitler et al., "REM Sleep Episodes During the Multiple Sleep Latency Test in Narcoleptic Patients," *Electroencephalography and Clinical Neurophysiology* 46 (1979): 479–481; Mary A. Carskadon and William C. Dement, "Effects of Total Sleep Loss on Sleep Tendency," *Perceptual and Motor Skills* 48 (1979): 495–506; Mary A. Carskadon et al., "Pubertal Changes in Daytime Sleepiness," *Sleep* 2, no. 4 (1980): 453–460; Mary A. Carskadon, Wesley Seidel, and William C. Dement, "Daytime Alertness, Insomnia, and Benzodiazepines," *Sleep* 5 (1982): S28–S45.

16. *By the mid-1980s:* Interview with Mary Carskadon, August 1, 2020.

17. *The idea for Sleep Camp:* Dement and Vaughan, *The Promise of Sleep*, 112–113.

18. *At a staff meeting:* Mary Carskadon interview; interview with Thomas Anders, October 21, 2022.

19. *Carskadon saw:* Mary Carskadon interview.

20. *She realized:* Ibid.; Thomas Anders interview; interview with Merrill Mitler, February 8, 2021.

21. *That April:* Carskadon, "Determinants of Daytime Sleepiness," 52–77.

22. *She would explore:* Ibid.; Patrick May, "Stanford Summer Camp Researchers Reunite and Recall a Long, Strange Trip," *Mercury News*, January 8, 2012, https://www.mercury news.com/2012/01/28/stanford-summer-camp-researchers-reunite-and-recall-a-long -strange-trip/; Stanford News Service Staff, "When Stanford's Lake Lagunita Had Water," Stanford News, February 25, 2020, accessed April 22, 2022, https://news.stanford .edu/2020/02/26/stanfords-lake-lagunita-water/; Michael Rosenbloom, "Recollections of the Stanford Summer Sleep Camp: A Conversation with Dr. Kim Harvey," End Your

Sleep Deprivation, Winter 2012, accessed April 22, 2022, http://www.end-your-sleep
-deprivation.com/dr-kim-harvey-stanford-summer-sleep-camp.html.

23. *The kids were recruited:* Mary Carskadon interview; Carskadon, "Determination of Daytime Sleepiness," 83–85; Christian Guilleminault and Elio Lugaresi, eds., *Sleep/Wake Disorders: Natural History, Epidemiology, and Long-Term Evolution* (New York: Raven Press, 1983), 202–203.

24. *When the camp's:* Mary Carskadon interview.

25. *Meanwhile, Carskadon continued:* Ibid.

26. *"Some of it was quite magical":* Ibid.

27. *For three years:* Ibid.

28. *And each summer:* Ibid.

29. *As Caskadon recorded:* Thorpy, "Elliot D. Weitzman MD and early sleep research and sleep medicine in New York."

30. *He spent part:* Thorpy; "Colin Pittendrigh, 'father of biological clock,' dies at 77," *Stanford News Service,* March 25, 1996, https://news.stanford.edu/pr/96/960325pittendrig.html (accessed April 17, 2022).

31. *Czeisler, then twenty-two:* Michael Thorpy, "Elliot D. Weitzman MD and Early Sleep Research Medicine in New York," *Sleep Medicine* 16 (2015): 1295–1300; Merrill Mitler interview.

32. *Weitzman became a mentor:* Charles Andrew Czeisler, "Human Circadian Physiology: Internal Organization of Temperature Sleep-Wake and Neuroendocrine Rhythms Monitored in an Environment Free of Time Cues," unpublished PhD dissertation, Stanford University, August 1978, v–vii; Thorpy, "Elliot Weitzman MD"; *helping to shift the latter's interest:* Interview with Charles Czeisler by Roger Bingham, Science Network, June 2009, accessed April 17, 2022, http://thesciencenetwork.org/programs/sleep-2009/charles-czeisler-2.

33. *Later that year:* Dement, *The Sleepwatchers,* 65.

34. *In early 1976:* Ibid., 88–90.

35. *Czeisler hypothesized:* Dement, *The Sleepwatchers,* 88–90; Elliot Weitzman et al., "Delayed Sleep Phase Syndrome," *Archives of General Psychiatry* 38 (July 1981): 737–746.

36. *In consultation with:* Dement, *The Sleepwatchers,* 88–90.

37. *Czeisler and his:* Weitzman et al., "Delayed Sleep Phase Syndrome"; Charles A. Czeisler, et al., "Chronotherapy: Resetting the Circadian Clocks of Patients with Delayed Sleep Phase Insomnia," *Sleep* 4, no. 1 (1981): 1–21; Elliott D. Weitzman et al., "Biological Rhythms in Man Under Non-Entrained Conditions and Chronotherapy for Delayed Sleep Insomnia," *Advances in Biological Psychiatry* 11 (1983): 136–149.

38. *In 1977, Weitzman built:* Steven Strogatz, *Sync: The Emerging Science of Spontaneous Order* (New York: Hyperion, 2003), 77.

39. *Weitzman and Czeisler hoped:* Ibid., 77–78.

40. *No clocks or watches:* Ibid.

41. *An initial report:* Czeisler, "Human Circadian Physiology"; Strogatz, *Sync,* 79.

42. *At the Montefiore isolation unit:* Strogatz, *Sync*, 79; Czeisler, "Human Circadian Physiology," 108–109, 119, 122, 128.

43. *Czeisler began searching:* Strogatz, *Sync*, 81–82; Czeisler, "Human Circadian Physiology," 280–283.

44. *Besides explaining:* Strogatz, *Sync*, 81–82; Czeisler, "Human Circadian Physiology," 307.

45. *Next, he turned:* Strogatz, *Sync*, 85–87; Czeisler, "Human Circadian Physiology," 294–298.

46. *Czeisler's preliminary research:* Strogatz, *Sync*, 85–87; Czeisler, "Human Circadian Physiology," 309–310.

47. *Among the few:* This stir was evidenced by the package of articles titled "REM Sleep: A Workshop on Its Temporal Distribution," which ran in the spring and summer 1980 issues of the journal *Sleep* (vol. 2, nos. 3 and 4). Edited by Czeisler and Christian Guilleminault, the series grew out of a conference held in Mexico City that March, where the papers—six coauthored by Czeisler, based on data from his dissertation research, and many of the rest citing his dissertation—were first presented.

48. *One of them:* Mary A. Carskadon, "Distribution of Sleep on a 90-Minute Sleep-Wake Schedule," *Sleep* 2, no. 3 (1980): 309–317.

49. *Carskadon, however:* Carskadon, "Determinants of Daytime Sleepiness."

50. *One came from:* Ibid., 155–167.

51. *The second bombshell:* Ibid., 78–144.

52. *But the study:* Ibid., 79.

53. *Like most experts:* Mary Carskadon interview.

54. *The campers' MSLTs:* Carskadon, "Determinants of Daytime Sleepiness," 78–80, 103–105.

55. *This finding had potentially:* Ibid., 80.

56. *That June, she received:* Carskadon, "Personal Memories and Reflections of Bill Dement."

57. *In April 1979:* Kenton Kroker, *The Sleep of Others and the Transformation of Sleep Research* (Toronto: University of Toronto Press, 2015), 379–380.

58. *The authors noted:* Ibid.

59. *The report asserted:* "Most of 25.6 Million Sleeping Pills Labeled Unnecessary," *Miami News*, April 5, 1979, 104.

60. *And it urged:* Kroker, *The Sleep of Others*, 381.

61. *In fact, this kind:* "Diagnostic Classification of Sleep and Arousal Disorders," *Sleep* 2, no. 1 (1979): 1–154; Kroker, *The Sleep of Others*, 387–390.

62. *The list was divided:* The dozen or so insomnias included those associated with psychoses, alcoholism, personality disorders, and restless legs syndrome. The somnolence disorders comprised daytime sleepiness caused by familiar ills such as narcolepsy and sleep apnea, as well as rarer conditions such as sleep drunkenness (confusion or combativeness upon awakening) and menstrual-associated syndrome (sleepiness during one's period). Under disorders of the sleep-wake schedule came delayed sleep phase syndrome (which drove people to go to sleep too late), advanced sleep phase syndrome (which compelled them to go to sleep too early), and troubles associated with rhythm-disrupting factors such as shift work and jet lag. The parasomnias included not only the commonplace

sleepwalking, sleep terrors, and bed-wetting but also such exotic miseries as sleep-related painful erections and sleep-related headbanging.

63. *Unlike manuals for:* Kroker, *The Sleep of Others*, 387–390.
64. *In response to:* "U.S. Embarks on 3-Year Effort to Help Insomniacs," *Los Angeles Times*, December 17, 1979, 8.
65. *The initiative would:* Wallace B. Mendelson, J. Christian Gillin, and William C. Dement, "Hypnotic Efficacy and Safety," *Sleep* 4, no. 2 (1981): 125–128.
66. *Project Sleep generated:* Kroker, *The Sleep of Others*, 419.
67. *One harbinger:* Charles A. Czeisler and Christian Guilleminault, "REM Sleep: A Workshop on Its Temporal Distribution," *Sleep* 2, no. 3 (1980): 285–286; Charles A. Czeisler, "Glossary of Standardized Terminology for Sleep Biological Rhythm Research," *Sleep* 2, no. 3 (1980): 287–288; Alexander A. Borbély, "Effects of Light and Circadian Rhythm on the Occurrence of REM Sleep in the Rat," *Sleep* 2, no. 3 (1980): 289–298; L. C. Johnson, "The REM Cycle is a Sleep-Dependent Rhythm," *Sleep* 2, no. 3 (1980): 299–307; Mary A. Carskadon and William C. Dement, "Distribution of REM Sleep on a 90-Minute Sleep-Wake Schedule," *Sleep* 2, no. 3 (1980): 309–317; Helmut Schultz et al., "The REM-NREM Sleep Cycle: Renewal Process or Periodically Driven Process?," *Sleep* 2, no. 3 (1980): 319–328; Charles A. Czeisler et al., "Timing of REM Sleep Is Coupled to the Circadian Rhythm of Body Temperature in Man," *Sleep* 2, no. 3 (1980): 329–346; Anita L. Weber, "Human Non-24-Hour Sleep-Wake Cycles in an Everyday Environment," *Sleep* 2, no. 3 (1980): 347–354; Ralph Lydic et al., "Suprachiasmatic Region of the Human Hypothalamus: Homolog to the Primate Circadian Pacemaker?," *Sleep* 2, no. 3 (1980): 355–361; Jurgen Zulley, "Distribution of REM Sleep in Entrained 24 Hour and Free-Running Sleep-Wake Cycles," *Sleep* 2, no. 4 (1980): 377–389; Elliot D. Weitzman et al., "Timing of REM and Stages 3+4 Sleep During Temporal Isolation in Man," *Sleep* 2, no. 4 (1980): 391–407; Janet C. Zimmerman et al., "REM Density Is Dissociated from REM Sleep Timing During Free-Running Sleep Episodes," *Sleep* 2, no. 4 (1980): 409–415; Kenneth I. Hume, "Sleep Adaptation After Phase Shifts of the Sleep-Wakefulness Rhythm in Man," *Sleep* 2, no. 4 (1980): 417–435.
68. *Czeisler and his:* Charles A. Czeisler et al., "Human Sleep: Its Duration and Organization Depend on Its Circadian Phase," *Science* 210, no. 4475 (December 12, 1980): 1264–1267.
69. *Another sign of:* Mary A. Carskadon et al., "Pubertal Changes in Daytime Sleepiness," *Sleep* 2, no. 4 (1980): 453–460.

Chapter 15

1. *When Colin Sullivan:* T. Scott Johnson, William A. Broughton, and Jerry Halberstadt, *Sleep Apnea: The Phantom of the Night* (Onset, MA: New Technology Publishing, 2003), xvii.
2. *To Sullivan, however:* Charles S. Barnes, "ResMed Origins: A Brief History of a Company Manufacturing Devices for the Diagnosis and Treatment of Sleep Disordered Breathing," ResMed, 2007, accessed April 27, 2022, https://document.resmed.com /en-us/documents/articles/resmed-origins.pdf.

3. *Phillipson's favored:* David K. Randall, *Dreamland: Adventures in the Strange Science of Sleep* (New York: W. W. Norton, 2013), 213–214.

4. *Like humans with:* "My Dog Snores: Does He Have Sleep Apnea?," ResMed, accessed April 27, 2022, https://www.resmed.com/en-us/sleep-apnea/sleep-blog/my-dog-snores-does-he-have-sleep-apnea/; Rio Dumitrascu et al., "Obstructive Sleep Apnea, Oxidative Stress and Cardiovascular Disease: Lessons from Animal Studies," Oxidative Medicine and Cellular Longevity, 2013, accessed April 27, 2013, https://www.ncbi.nlm.nih.gov/pmc/articles/PMC3603718/pdf/OXIMED2013-234631.pdf; Swati Chopra, Vsevolod Y. Polotsky, and Jonathan C. Jun, "Sleep Apnea Research in Animals: Past, Present, and Future," *American Journal of Respiratory Cell and Molecular Biology* 54, no. 3 (March 2016): 299–305.

5. *In Phillipson's lab:* Barnes, "ResMed Origins."

6. *Sullivan's next brainstorm:* William C. Dement, "My Nomination for the Nobel Prize in Physiology and Medicine," Free Library, July 1, 2009, accessed April 27, 2022, https://www.thefreelibrary.com/My+nomination+for+the+Nobel+Prize+in+Physiology+and+Medicine.-a0207360836.

7. *Sullivan tried this:* Barnes, "ResMed Origins."

8. *Sullivan's first subject:* Dement, "My Nomination"; Colin Sullivan, "Reversal of Obstructive Sleep Apnoea by Continuous Positive Airway Pressure Applied Through the Nares," *Lancet* 317, no. 8225 (April 18, 1981): 862–865.

9. *Sullivan—a tinkerer since childhood:* Tony Kirby, "Colin Sullivan: Inventive Pioneer of Sleep Medicine," *Lancet* 377 (April 30, 2011): 1485; ResMed Origins; Colin E. Sullivan, "Nasal Positive Airway Pressure and Sleep Apnea: Reflections on an Experimental Method That Became a Therapy," *American Journal of Respiratory Critical Care Medicine* 198, no. 5 (September 1, 2018): 581–587.

10. *One night soon:* Randall, *Dreamland*, 213–214.

11. *Over the following:* Sullivan, "Reversal of Obstructive Sleep Apnoea."

12. *Sullivan reported his:* Sullivan, "Reversal of Obstructive Sleep Apnoea."

13. *His technique, abbreviated:* Barnes, "ResMed Origins."

14. *At first, Sullivan:* Dement, "My Nomination."

15. *Sullivan secured a patent:* Barnes, "ResMed Origins."

16. *By 1985, about one hundred:* Randall, *Dreamland*, 215.

17. *Physicians who were:* Interview with Meir Kryger, February 3, 2021.

18. *That year, the:* Barnes, "ResMed Origins"; Jeffery Fraser, "Science's Role Grows in Area Economy," *Pittsburgh Press*, September 22, 1985, 82; "New Pulse-Gauging Gadgets Get Close to a Runner's Heart," *Minneapolis Star*, February 7, 1980, 18; "History of BiPAP—Respironics and Philips Respironics," CPAP.com, accessed October 10, 2022, https://www.cpap.com/blog/history-bipap-respironics-philips/.

19. *In 1988, Sullivan:* Barnes, "ResMed Origins."

20. *moved to San Diego:* "ResMed Inc.," Dunn & Bradstreet, accessed April 28, 2022, https://www.dnb.com/business-directory/company-profiles.resmed_inc.5dad8665978dd430e7cd15067ee08b39.html.

21. *overtook Respironics:* According to CompaniesMarketCap.com, ResMed's market capitalization in April 2022 was about $31 billion; Philips, now the parent company of Respironics, was valued at about $23 billion.

22. *The CPAP explosion:* Terry Young et al., "The Occurrence of Sleep-Disordered Breathing Among Middle-Aged Adults," *New England Journal of Medicine* 328, no. 17 (April 29, 1993): 1230–1235; William C. Dement, *The Sleepwatchers* (Menlo Park, CA: Nychthemeron Press, 1995), 170.

23. *And equipment makers:* Kenton Kroker, *The Sleep of Others and the Transformation of Sleep Research* (Toronto: University of Toronto Press, 2015), 415.

24. *CPAP altered the:* Ibid., 415–416.

25. *Early in the:* Mary Carskadon, ed., *Encyclopedia of Sleep and Dreaming* (New York: Macmillan, 1993), 634.

26. *They could also:* William A. Conway et al., "Adverse Effects of Tracheostomy for Sleep Apnea," *JAMA* 246, no. 4 (July 24/31, 1981): 347–350.

27. *In 1981, a team:* Shiro Fujita et al., "Surgical Correction of Anatomic Abnormalities in Obstructive Sleep Apnea: Uvulopalatopharyngoplasty," *Otolaryngology—Head and Neck Surgery* 89, no. 6 (November 1, 1981): 923–934; "Henry Ford Hospital— Historical Highlights: 1980s–1990s," Henry Ford, Health, accessed April 29, 2022, https://henryford.libguides.com/hfhsarchives_historicalhighlights/1980_90.

28. *First introduced in:* Meir H. Kryger, Thomas Roth, and William C. Dement, *Principles and Practice of Sleep Medicine*, 3rd edition (Philadelphia: Saunders, 2000), 917–918.

29. *Because people with:* Fujita et al., "Surgical Correction."

30. *Initial results were:* Chris Anne Raymond, "Popular, Yes, but Jury Still Out on Apnea Surgery," *JAMA* 256, no. 4 (July 25, 1986): 439–441.

31. *Eventually, other surgical:* Robert W. Riley, Nelson S. Powell, and Christian Guilleminault, "Maxillofacial Surgery and Obstructive Sleep Apnea: A Review of 80 Patients," *Otolaryngology—Head and Neck Surgery* 101, no. 3 (September 1989): 353–361; Kryger, Roth, and Dement, *Principles and Practice*, 3rd edition; Himanshu Wickramasinghe, "Obstructive Sleep Apnea (OSA) Treatment & Management," Medscape, accessed December 5, 2022, https://emedicine.medscape.com/article /295807-treatment.

32. *Although CPAP would:* Jay Summer, "Oral Appliances for Sleep Apnea," Sleep Foundation, April 27, 2022, accessed May 1, 2022, https://www.sleepfoundation.org /sleep-apnea/oral-appliance-for-sleep-apnea.

33. *That turned out:* Brian W. Rotenberg, Dorian Murariu, and Kenny P. Pang, "Trends in CPAP Adherence over 20 Years of Data Collection: A Flattened Curve," *Journal of Otolaryngology—Head and Neck Surgery* 45, no. 43 (August 19, 2016), accessed May 1, 2022, https://www.ncbi.nlm.nih.gov/pmc/articles/PMC4992257/; A. M. Sawyer et al., "A Systematic Review of CPAP Adherence Across Age Groups: Clinical and Empiric Insights for Developing CPAP Adherence Interventions," *Sleep Medicine Review* 15, no. 6 (2011): 343–356; Terri E. Weaver and Ronald R. Grundstein, "Adherence to Continuous Positive Airway Pressure Therapy: The Challenge to Effective Treatment," *Proceedings of the*

American Thoracic Society 5, no. 2 (2008): 173–178; Juan F. Masa and Jaime Corral-Peñafiel, "Should Use of 4 Hours Continuous Positive Airway Pressure Be Considered Acceptable Compliance?," *European Respiratory Journal* 44 (2014): 1119–1120.

34. *Another set of:* B. Gail Demo, "The Evolution of Oral Appliance Therapy for Snoring and Sleep Apnea: Where Did We Come From, Where Are We, and Where are We Going?," *Sleep Medicine Clinics* 13 (2018): 467–487; "Anti-Snoring Device May Have Sound Uses," *Herald and Review* (Decatur, IL), August 24, 1982, 19.

35. *In 1982, the pair:* Rosalind D. Cartwright and Charles F. Samelson, "The Effects of a Nonsurgical Treatment for Obstructive Sleep Apnea," *JAMA* 248, no. 6 (August 13, 1982): 705–709.

36. *Samelson's device:* Demo, "The Evolution of Oral Appliance Therapy."

37. *An even less invasive OSA treatment:* Nico de Vries, Madeline Ravesloot, and J. Peter can Maanen, eds., *Positional Therapy in Obstructive Sleep Apnea* (Cham, Switzerland: Springer, 2015), 225, 227; Rosalind D. Cartwright, "Effect of Sleep Position on Sleep Apnea Severity," *Sleep* 7, no. 2 (1984): 110–114.

38. *Cartwright was a:* Cartwright, "Effect of Sleep Position on Sleep Apnea Severity."

39. *The following year:* De Vries, Ravesloot, and Can Maanen, *Positional Training in Obstructive Sleep Apnea*, 230.

40. *Cartwright and her:* Rosalind D. Cartwright et al., "Sleep Position Training as Treatment for Sleep Apnea Syndrome: A Preliminary Study," *Sleep* 8, no. 2 (1985): 87–94.

41. *From these beginnings:* Grietje E. de Vries et al., "Usage of Positional Therapy in Adults with Obstructive Sleep Apnea," *Journal of Clinical Sleep Medicine* 11, no. 2 (2015): 131–137.

42. *As with oral apnea appliances:* Ibid.; Kenneth R. Casey, "Positional Therapy Is Worth a Try in Patients with Mild Obstructive Sleep Apnea," *Journal of Clinical Sleep Medicine* 11, no. 2 (2015): 89–90;

43. *It's worth noting:* Penelope Green, "Rosalind Cartwright, Psychologist and 'Queen of Dreams,' Dies at 98," *New York Times*, March 15, 2021, 22.

44. *For centuries, humanity's:* Sudhansu Chokroverty and Michel Billiard, eds. *Sleep Medicine: A Comprehensive Guide to Its Development, Clinical Milestones, and Advances in Treatment* (New York: Springer, 2015), 519–521.

45. *In the 1970s, some:* Ibid., 521; Ralph M. Turner and L. Michael Ascher, "Controlled Comparison of Progressive Relaxation, Stimulus Control, and Paradoxical Intention Therapies for Insomnia," *Journal of Consulting and Clinical Psychology* 47, no. 3 (1979): 500–508.

46. *Bootzin's simple method:* Turner and Ascher, "Controlled Comparison."

47. *In 1977, psychologist:* Chokroverty and Billiard, *Sleep Medicine*, 521.

48. *The more cumbersome:* An early example can be found in Marie de Manacéïne's 1897 book *Sleep: Its Physiology, Pathology, Hygiene, and Psychology.*

49. *Hauri, however, was:* Peter Hauri, *The Sleep Disorders* (Kalamazoo, MI: Upjohn Company, 1982), 22.

50. *Although studies failed:* Chokroverty and Billiard, *Sleep Medicine*, 521.

51. *The tipping point:* Ibid., 521–522.

52. *Over the next:* Ibid., 521–525.

53. *Soon afterward, psychologist:* Ibid., 523; Arthur J. Spielman, "Assessment of Insomnia," *Clinical Psychology Review* 6, no. 1 (1986): 11–25.

54. *Spielman, a cofounder:* Chokroverty and Billiard, *Sleep Medicine*, 523.

55. *In their initial:* Spielman's goal was to train patients to increase their sleep efficiency—the ratio between time spent sleeping and time spent trying to sleep. Participants in the study were asked to record their estimated sleep time for two weeks. The researchers then used the average time to calculate each patient's permitted time in bed (TIB). If a patient needed to get up at 7:00 a.m. to go to work and their TIB was five hours, their bedtime would be 2:00 a.m. Each morning, they reported their estimated sleep duration. If that time was at least 90 percent of their TIB for five nights in a row, they were rewarded with an extra fifteen minutes of TIB for the next five nights. If sleep efficiency was less than 85 percent, TIB was decreased to the average estimated sleep time. If sleep efficiency fell between 85 and 90 percent, the patient's TIB was left unchanged.

56. *the results were:* Spielman, "Assessment of Insomnia."

57. *Eventually, SRT, SCT:* Charles Morin et al., "Cognitive-Behavior Therapy for Late-Life Insomnia," *Journal of Consulting and Clinical Psychology* 61, no. 1 (February 1993): 137–146; Charles Morin, *Insomnia: Psychological Assessment and Management* (New York: Guilford, 1993); Chokroverty and Billiard, *Sleep Medicine*, 524–525, 536–537; Rob Newsom, "Cognitive Behavioral Therapy for Insomnia," Sleep Foundation, accessed October 10, 2022, https://www.sleepfoundation.org/insomnia/treatment/cognitive-behavioral-therapy-insomnia.

58. *Charles Czeisler was:* Lynne Lamberg, *Bodyrhythms: Chronobiology and Peak Performance* (New York: William Morrow, 1994), 202.

59. *The workers had:* Testimony of Preston Richey, *Biological Clocks and Shift Work Scheduling: Hearings Before the Investigations and Oversight Committee of the House Committee on Science and Technology, House of Representatives, Ninety-Eight Congress, First Session, March 23, 24, 1983* (Washington: US Government Printing Office, 1983), 161–162, accessed December 16, 2022, https://www.google.com/books/edition/Biological_Clocks_and_Shift_Work_Schedul/bF8VAAAAIAAJ?hl=en&gbpv=1&bsq=richey.

60. *Richey searched in vain:* Ibid.

61. *Czeisler flew to Utah:* Testimony of Charles A. Czeisler, *Biological Clocks and Shift Work Scheduling*, 171–175, accessed December 16, 2022, https://www.google.com/books/edition/Biological_Clocks_and_Shift_Work_Schedul/bF8VAAAAIAAJ?hl=en&gbpv=1&bsq=czeisler.

62. *The researchers soon recognized:* Martin Moore-Ede, "Evolution of Fatigue Risk Management Systems," Circadian white paper, 2010, https://www.circadian.com/images/pdf/evolution_of_frms.pdf; Keenan Mayo, "The Sleep Doctor to Elite Athletes, CEOs—Even Rock Stars," *Men's Fitness*, June 2014, accessed May 5, 2022, https://www.mensjournal.com/sports/sleep-doctor-elite-athletes-ceos-even-rock-stars/.

63. *The trouble with:* On the night shift, employees went to bed around 11:00 a.m. and set their alarm clocks for 7:00 p.m. When they rotated to the evening shift, they went to bed around 3:00 a.m., at which point, their internal clocks weren't ready for sleep;

they had to get up around 11:00 a.m., even if they'd spent hours lying awake. When they got to day shift, they went to bed around 10:00 p.m., which felt impossibly early, and then had to drag themselves upright around 6:00 a.m.

64. *They were in a state:* Mayo, "The Sleep Doctor"; Moore-Ede, "Evolution of Fatigue Risk Management Systems."

65. *Not only were many workers:* Charles A. Czeisler, Martin C. Moore-Ede, and Richard M. Coleman, "Rotating Shift Work Schedules That Disrupt Sleep Are Improved by Applying Circadian Principles," *Science* 217 (July 30, 1982): 460–463; Czeisler testimony, *Biological Clocks and Shift Work Scheduling*, 171–175.

66. *"It really disrupts your metabolism":* Testimony of Scott N. Nielsen, *Biological Clocks and Shift Work Scheduling*, 164–165, accessed December 16, 2022, https://www.google.com/books/edition/Biological_Clocks_and_Shift_Work_Schedul/bF8VAAAAIAAJ?hl=en&gbpv=1&bsq=nielsen.

67. *The company's shift:* Czeisler et al., "Rotating Shift Work Schedules"; Richard M. Coleman, *Wide Awake at 3:00 A.M.: By Choice or by Chance?* (New York: W. H. Freeman, 1986), 46–47.

68. *The direction of:* Czeisler et al., "Rotating Shift Work Schedules."

69. *After three months:* Mayo, "The Sleep Doctor"; Moore-Ede, "Evolution of Risk Management Systems"; Czeisler et al., "Rotating Shift Work Schedules."

70. *Unlike Nathaniel Kleitman's:* Moore-Ede, "Evolution of Risk Management Systems"; Czeisler et al., "Rotating Shift Work Schedules."

71. *Czeisler joined the:* Mayo, "The Sleep Doctor."

72. *Sadly, the scientist:* Michael Thorpy, "Elliot D. Weitzman MD and Early Sleep Research Medicine in New York," *Sleep Medicine* 16 (2015): 1295–1300; Walter H. Waggoner, "Dr. Elliot Weitzman, Expert on Sleep, Dies," *New York Times*, June 16, 1983, B13.

Chapter 16

1. *One of the paradoxes:* Chris Simms, "Occam's Razor: A Guiding Principle of Logic Exhorting Us to Keep Things as Simple as Possible," *New Scientist*, accessed October 11, 2022, https://www.newscientist.com/definition/occams-razor/.

2. *In March 1982:* "Prof. Dr. Med. Alexander Borbély," University of Zürich, accessed October 11, 2022, https://www.pharma.uzh.ch/static/sleep/borbely.htm; A. A. Borbély, "A Two Process Model of Sleep Regulation," *Human Neurobiology* 1, no. 195 (1982): 195–204; Stephen M. James et al., "Shift Work: Disrupted Circadian Rhythms and Sleep—Implications for Health and Well-Being," *Current Sleep Medicine Reports* 3, no. 2 (June 2017): 104–112; Jeff Mann, "Big Ideas: The Two-Process Model of Sleep," Sleep Junkies, accessed May 7, 2022, https://sleepjunkies.com/two-process-model/.

3. *Borbély offered an answer:* Borbély, "A Two Process Model of Sleep Regulation."

4. *Borbély described the first factor:* Ibid.

5. *The second factor in Borbély's model:* Ibid.

6. *Under ideal circumstances:* Ibid.

7. *The two-process model explained:* James et al., "Shift Work: Disrupted Circadian Rhythms and Sleep."

8. *Borbély's brainchild:* Sudhansu Chokroverty and Michel Billiard, eds. *Sleep Medicine: A Comprehensive Guide to Its Development, Clinical Milestones, and Advances in Treatment* (New York: Springer, 2015), 511–512.

9. *For Mary Carskadon:* Interview with Mary Carskadon, August 15, 2020.

10. *After receiving her PhD:* Mary Carskadon CV.

11. *Since then, she'd published:* Mary A. Carskadon and William C. Dement, "Cumulative Effects of Sleep Restriction on Daytime Sleepiness," *Psychobiology* 18, no. 2 (1981): 107–113; Mary A. Carskadon and William C. Dement, "Nocturnal Determinants of Daytime Sleepiness," *Sleep* 5, Supplement 2 (1982): S73–S81.

12. *Although the concept:* The earliest use of the term *sleep debt* that I've found is in D. Svorad and J. Wellnerová, "Temperature and Air Humidity as Factors Influencing Sleep and Wakefulness," *Human Bioclimatology* 3 (1959): 201–204. After that, the phrase doesn't turn up again on Google Scholar until 1966, after which there are approximately 100 references by 1982—an average of 7 per year. In the fifteen years after Carskadon's paper, the phrase appears 485 times, a nearly fivefold increase.

13. *Her study's findings:* Carskadon and Dement, "Nocturnal Determinants of Daytime Sleepiness."

14. *Borbély's two-process manifesto:* Mary Carskadon interview, August 15, 2020.

15. *At that point:* Interview with Mary Carskadon, August 22, 2020.

16. *At Stanford, only:* Interview with Mark Rosekind, May 29, 2022; email message from Donald Bliwise to author, April 13, 2022.

17. *Besides, she had:* Mary Carskadon interview, August 22, 2020.

18. *Carskadon was also:* Interview with Donald Bliwise, January 3, 2021; Mary Carskadon CV.

19. *In her dissertation:* Mary A. Carskadon, "Determinants of Daytime Sleepiness: Adolescent Development, Extended and Restricted Nocturnal Sleep," unpublished PhD dissertation, Stanford University, 1979, 168–185.

20. *One of her first:* Mary A. Carskadon, Ed Brown, and William C. Dement, "Sleep Fragmentation in the Elderly: Relationship to Daytime Sleep Tendency," *Neurobiology of Aging* 3 (1982): 321–327.

21. *She was soon:* Interview with Donald Bliwise, March 2, 2022. The uncle was William Offenkrantz, whose research focused on the relationship of REM sleep and dreaming.

22. *With Carskadon's help:* Donald Bliwise interview; D. L. Bliwise, "Sleep in a Nonagenarian: N. Kleitman," *Archives Italiennes de Biologie* 139 (2001): 9–21.

23. *He would publish:* Nathaniel Kleitman, "Basic Rest-Activity Cycle—22 Years Later," *Sleep* 5, no. 4 (1982): 311–317.

24. *He flew north:* Interview with Donald Bliwise, February 2, 2021.

25. *Kleitman stayed with:* Carskadon, "Personal Memories and Reflections of Bill Dement"; W. C. Dement, "Remembering Nathaniel Kleitman," *Archives Italiennes de Biologie* 139 (2001): 11–17.

26. *In part, this:* Willam C. Dement, "Rational Basis for the Use of Sleeping Pills," *Pharmacology* 27, Supplement 2 (1983): 3–38.

27. *One promising candidate:* Ibid.

28. *Those findings helped:* Institute of Medicine, *Halcion: An Independent Assessment of Safety and Efficacy Data* (Washington, DC: National Academies Press, 1997), accessed May 17, 2022, https://www.ncbi.nlm.nih.gov/books/NBK233849/.

29. *Marketed as Halcion:* Gina Kolata, "Maker of Sleeping Pill Hid Data on Side Effects, Researchers Say," *New York Times*, January 20, 1992, A1; Dawn House, "Judge Asked to Revive Halcion Controversy," *Salt Lake Tribune*, September 1, 1991, 15; Geoffrey Cowley, "Sweet Dreams or Nightmare," *Newsweek*, August 18, 1991, accessed February 19, 2023, https://www.newsweek.com/sweet-dreams-or-nightmare-203042; Kathy Slaughter, "Jury Seated in Petty Trial," *Battle Creek Enquirer*, March 11, 1987, 5. Note: Investigations by the FDA and the US Institute of Medicine ultimately found that Halcion was no more dangerous than other benzodiazepines when taken according to updated recommendations (i.e., in low doses and for limited periods of time).

30. *Perhaps a more compelling:* Interview with Donald Bliwise, May 16, 2022.

31. *That income was:* Matt Hourihan, "The Trump Administration's Science Budget: Toughest Since Apollo?," *AAAS News*, March 29, 2017, accessed May 17, 2022, https://www.aaas.org/news/trump-administrations-science-budget-toughest-apollo.

32. *Funds from other:* Interview with Donald Bliwise, May 16, 2022; William C. Dement to Tracey Loveridge et al. outlining financial challenges, September 1, 1985, William C. Dement Papers, (SC0633), Department of Special Collections and University Archives, Stanford University Libraries, Box 2, Folder 1.8.2; fundraising letter to Charles Butt, President of the H.E. Butt Grocery Company, Oct. 30, 1986, William C. Dement Papers, Box 1, Folder 1.8.1.

33. *Another source of financial strain:* Interview with Stuart Rawlings, May 13, 2022; John Walsh, "Indirect Cost Surge Prompts New Worries," *Science* 240 (June 10, 1988): 1400–1401; Marcia Barinaga, "Stanford Erupts over Indirect Costs," *Science* 248 (April 20, 1990): 292–294.

34. *To get around these hurdles:* Stuart Rawlings interview; "Sleep Disorders Foundation, Inc.," Connecticut Business Directory, accessed May 17, 2022, https://www.ctcompanydir.com/companies/sleep-disorders-foundation-inc/; William C. Dement, "Thirty Years of Narcolepsy Research (1953–1983)," William C. Dement papers, Box 5, Folder 4.1.2.1; Theodore Baker et al., "Canine Model of Narcolepsy: Genetic and Developmental Determinants," *Experimental Neurology* 75, no. 3 (March 1982): 729–742; Arthur S. Foutz et al., "Monoaminergic Mechanisms and Experimental Cataplexy," *Annals of Neurology* 19, no. 4 (October 1981): 369–376. *teetered perpetually at the brink of insolvency:* William C. Dement with Stuart Rawlings, *Dreamer: The Memoir of a Sleep Researcher* (unpublished manuscript, 2018), 66; Dement to Loveridge et al., September 1, 1985.

35. *In September 1982:* Stuart Rawlings interview.

36. *When Dement was:* Ibid.; Dement and Rawlings, *Dreamer*, 6.

37. *he faced the need:* William C. Dement to Michael Cowan, November 18, 1984, William C. Dement Papers, Box 22, Folder 5.3.5; "Early 1984," note from Dement to himself on frustration over negotiations for new clinic space, William C. Dement Papers, Box 53, Folder 10.8.

38. *There were tensions:* Interview with Merrill Mitler, February 4, 2021; interview with Sonia Ancoli-Israel, April 11, 2022. These tensions are also evident in a series of memos in which Dement upbraids Guilleminault for taking initiatives in personnel and research without prior consultation, and expresses concern over the lack of clarity around Guilleminault's official role: Dement to Guilleminault (subject: American Board of Sleep Medicine), February 3, 1986; Dement to Guilleminault (subject: SIDS Tech Recruitment), August 18, 1986; Dement to Guilleminault, November 12, 1986, all from William C. Dement Papers, Box 1, Folder 1.8.1.

39. *There were outbreaks:* Dement's conflicts with the DLAM over housing covered several years. They can be traced in documents such as Dement to Karl Grey (subject: Dispersal of the Narcoleptic Dog Colony), September 12, 1983; Dement to Thomas A. Gonda (subject: Canine Narcolepsy Colony), September 23, 1983; David H. Mendelow to Thomas Gonda, September 26, 1983; Dement to Theodore L. Baker, November 21, 1983; Tom Hamm to William Dement (subject: Narcolepsy dogs housed off-campus), November 21, 1984, all from William C. Dement Papers, Box 5, Folder 4.1.1.2.

40. *Meanwhile, the rise:* Bob Johnson and Stephen Korn, "In the Name of Humanity," *Up Front,* Winter 1983, 5–9.

41. *In recent years:* Benjamin Adams, "Legislative History of the Animal Welfare Act: Introduction," USDA National Agriculture Library, https://www.nal.usda.gov/legacy /awic/legislative-history-animal-welfare-act-introduction; *Animal Welfare and Scientific Research: 1985 to 2020: Recognizing 25 Years of Improving Animal Welfare, Advancing Science,* accessed May 20, 2022, https://grants.nih.gov/grants/olaw/seminar/docs /Booklet_AWSR.pdf.

42. *Such measures, however:* Christopher Wlach, "Animal Rights Extremism as Justification for Restricting Access to Government Records," *Syracuse Law Review* 67, no. 191 (2017): 190–215.

43. *The wave began:* "Flashback: 14 February 1982—ALF's Operation Valentine Raid," Red Black Green, accessed May 20, 2022, https://network23.org/redblackgreen/2015/02 /14/flashback-14-february-1982-alfs-operation-valentine-raid/; Wlach, "Animal Rights Extremism."

44. *In 1981, an activist:* Peter Carlson, "The Great Silver Spring Monkey Debate," *Washington Post,* February 24, 1991, accessed May 10, 2022, https://www.washingtonpost .com/archive/lifestyle/magazine/1991/02/24/the-great-silver-spring-monkey-debate /25d3cc06-49ab-4a3c-afd9-d9eb35a862c3/.

45. *Dement—whose relatively:* Cleveland Amory, "What You Can Do to Help the Animals," *Cosmopolitan,* undated clipping, 1968; Maurice Visccher to William C. Dement, June 26, 1968, William C. Dement Papers, Box 18, Folder 4.27; Bob Johnson and Stephen Korn, "In the Name of Humanity," *Up Front,* Winter 1983, 5–8.

46. *One morning, he:* William C. Dement, "Early 1984," undated manuscript, William C. Dement Papers, Box 53, Folder 10.8.

47. *It's not clear:* Ibid.

48. *Some of Dement's:* Interview with Sharon Keenan, April 7, 2022; Sharon Keenan CV; "History," School of Sleep Medicine, accessed May 18, 2022, https://www.sleepedu.net /the-school.

49. *In the spring:* German Nino-Murcia to Sleep Disorders Clinic employees, February 11, 1985, William C. Dement Papers, Box 4, Folder 3.15.2; William C. Dement, "Resources and Environment: Space, Facilities, and Equipment," undated, c. 1986, William C. Dement Papers, Box 8, Folder 4.8; "Hoover Pavilion—1931," Palo Alto Stanford Heritage, accessed May 18, 2022, https://www.pastheritage.org/Articles/HooverMF.html.

50. *In August, after:* "Lab Director Combines Animal Care with Research," press release, Stanford University Medical Center News Bureau, August 2, 1985; "Stanford's New Animal Care and Research Facility Opens," press release, Stanford University Medical Center News Bureau, August 2, 1985, William C. Dement Papers, Box 5, Folder 4.1.1.2.

51. *Satisfied with both:* Thomas A. Gonda to Thomas Hamm, August 12, 1985, William C. Dement Papers, Box 5, Folder 4.1.1.2; Dement and Rawlings, *Dreamer,* 67.

52. *Still, his sense:* "Testimony of William C. Dement, M.D., President of the Association of Sleep Disorders Centers, Before the House Appropriations Committee Subcommittee on Labor, Health and Human Services and Education Concerning Funding for Sleep Disorders Research on May 22, 1985," William C. Dement Papers, Box 1, Folder 1.8.1.

53. *To remedy the situation:* Interview with Merrill Mitler, April 9, 2022; Merrill Mitler CV; interview with Dale Dirks, November 3, 2021; "About," Health and Medicine Counsel, https://hmcw.org/about/.

54. *And so it was:* "Testimony of Merrill M. Mitler, Ph.D., Executive Treasurer of the Association of Sleep Disorders Centers, Before the Senate Appropriations Committee Subcommittee on Labor, Health and Human Services, and Education, Concerning Funding for Sleep Disorders Research on May 8, 1985," William C. Dement Papers, Box 1, Folder 1.8.1.

55. *Later that month:* "Testimony of William C. Dement, M.D., President of the Association of Sleep Disorders Centers, Before the House Appropriations Committee Subcommittee on Labor, Health and Human Services and Education Concerning Funding for Sleep Disorders Research on May 22, 1985," William C. Dement Papers, Box 1, Folder 1.8.1.

56. *As Dement made:* Interview with Mary Carskadon, August 22, 2020.

57. *She loved working:* Ibid.

58. *Yet job openings:* Ibid.

59. *Deliverance came that:* Ibid.; Thomas Anders CV; John Goncalves, "Bradley Hospital: A Place of Healing," Joukowsky Institute for Archaeology, accessed May 20, 2022, https://www.brown.edu/Departments/Joukowsky_Institute/courses/placesofhealing2012/17387.html.

60. *In September 1985, after:* Mary Carskadon interview; email message from Mary Carskadon to author, April 25, 2022.

61. *Soon afterward:* Email message from Mary Carskadon to author, August 22, 2021.

Chapter 17

1. *Emma Pendleton Bradley was:* "Emma Pendleton Bradley," Find a Grave, https://www.findagrave.com/memorial/61102167/emma-pendleton-bradley; "Bradley Hospital,

Leaders in Mental Health Care for Children for 90 Years: Our History," Lifespan, https://www.lifespan.org/locations/bradley-hospital/about-bradley-hospital/our-history.

2. *Her father had passed away:* Henry H. Work, "George Lathrop Bradley and the War Over Ritalin," Cosmos, 2001, accessed June 3, 2022, http://www.cosmosclub.org /journals/2001/work.html; Madeleine P. Strohl, "Bradley's Benzedrine Studies on Children with Behavioral Disorders," *Yale Journal of Biology and Medicine* 84 (2011): 27–33. Quote is from "Bradley Hospital Turns 90," Bradley Hospital, accessed June 4, 2022, https://giving.lifespan.org/Bradley/bradley-hospital-turns-90.

3. *The Emma Pendleton Bradley Home:* Thomas Anders CV; John Goncalves, "Bradley Hospital: A Place of Healing," Joukowsky Institute for Archaeology, accessed May 20, 2022, https://www.brown.edu/Departments/Joukowsky_Institute/courses/placesofhealing 2012/17387.html.

4. *Psychologist Herbert H. Jasper:* Frederick Andermann, "In Memoriam: Herbert Henri Jasper, 1906–1999," *Epilepsia* 41, no. 1 (2000): 113–120; D. B. Lindsley and C. E. Henry, "The Effects of Drugs on Behavioral and the Electroencephalograms of Children with Behavior Disorders," *Psychosomatic Medicine* 4 (1942): 140–149.

5. *His successor, Donald B. Lindsley:* Norman M. Weinberger, "Donald Benjamin Lindsley: 1907–2003," *Biographical Memoir* (Washington, DC: National Academy of Sciences, 2009), accessed June 4, 2022, http://www.nasonline.org/publications /biographical-memoirs/memoir-pdfs/lindsley-donald.pdf.

6. *Medical director Charles Bradley:* Walter A. Brown, "Images in Psychiatry: Charles Bradley, M.D., 1902–1979," *American Journal of Psychiatry* 155, no. 7 (July 1998): 968; Work, "George Lathrop Bradley."

7. *Bradley Hospital:* "Today's News: Medical School, Providence, R.I.," *Naugatuck Daily News*, December 14, 1973, 7.

8. *By the time Mary Carskadon:* Interview with Mary Carskadon, September 5, 2020; Mary Carskadon CV; interview with Thomas Anders, October 21, 2022.

9. *Carskadon was far:* Mary Carskadon interview.

10. *The scene that:* Mary Carskadon interview; Mary Carskadon email message to author, June 19, 2022.

11. *Once she got physically:* Mary Carskadon interview.

12. *It occurred to Carskadon:* Ibid.; Mary Carskadon CV; "Richard Millman, MD," Lifespan, accessed June 12, 2022, https://www.lifespan.org/providers/richard-millman-md.

13. *Carskadon was flying:* Interview with Mary Carskadon, September 16, 2016.

14. *As a teenager:* Interview with Mary Carskadon, October 3, 2022.

15. *For women:* Betty M. Vetter, "Women's Progress," *Mosaic* 18, no. 1 (Spring 1987): 7.

16. *Carskadon aspired to:* Mary Carskadon interview.

17. *In the autumn of:* "H.J. Res. 372—Balanced Budget and Emergency Deficit Control Act of 1985," US Congress, accessed June 15, 2022, https://www.congress.gov /bill/99th-congress/house-joint-resolution/372; Darrell M. West, "Gramm-Rudman-Hollings and the Politics of Deficit Reduction," *Annals of the American Academy of Political and Social Science* 499 (September 1988): 90–100.

18. *Fearing that federal:* William C. Dement to Robert Purpura, January 17, 1986, William C. Dement Papers (SC033), Department of Special Collections and University Archives, Stanford University Libraries, Box 1, Folder 8.1.

19. *Partnerships between university:* H. Garrett deYoung, "University and Industry in Agreement," *Biotechnology* 6 (August 1988): 906–910. Note: The growth of such partnerships was fueled by the Bayh-Dole Act of 1980, which permitted researchers to patent inventions they developed with federal funding and to transfer the technology to the private sector.

20. *Dement wasn't content:* William C. Dement to Ted Cooper, November 21, 1985, William C. Dement Papers, Box 1, Folder 8.1.

21. *His anxiety was apparent:* William C. Dement to Robert Purpura, January 17, 1986, William C. Dement Papers, Box 1, Folder 8.1.

22. *On January 28:* Elizabeth Howell, "Space Shuttle Challenger and the Disaster That Changed NASA Forever," Space.com, February 1, 2022, accessed May 22, 2022, https://www.space.com/18084-space-shuttle-challenger.html; Matt Haskell, "Wings of Exploration: Reflecting on the 40th Anniversary of the Space Shuttle," Spaceflight Insider, https://www.spaceflightinsider.com/space-flight-history/wings-of-exploration -reflecting-on-the-40th-anniversary-of-the-space-shuttle/. Note: Haskell and others cite reports that 17 percent of Americans watched the liftoff; at the time, the US population was 240 million.

23. *In just over a minute:* "Space Shuttle," NASAfacts, accessed June 13, 2022, https://www .nasa.gov/centers/johnson/pdf/167751main_FS_SpaceShuttle508c.pdf; "Space Shuttle Challenger Explosion (1986)," YouTube, 4:22, posted by CNN, January 27, 2011, accessed May 21, 2022, https://www.youtube.com/watch?v=AfnvFnzs91s.

24. *The seven astronauts:* Jay Barbree, "An Eternity of Descent," NBC News, accessed June 13, 2022, https://www.nbcnews.com/id/wbna3078062.

25. *Debris was scattered:* John J. Glitch and Lynne Bumpus-Hooper, "Coverage from the Day Space Shuttle Challenger Exploded: Searchers Seek Clues, Not Survivors from the Outset, NASA Held Out Little Hope," *Orlando Sentinel*, January 29, 1986, accessed June 13, 2022, https://www.orlandosentinel.com/space/orl-challenger-25-day-of-the -search-story.html.

26. *That June:* Andrew J. Dunar and Stephen P. Waring, *The Power to Explore: A History of Marshall Space Flight Center, 1960–1990* (Washington, DC: US Government Printing Office, 1999), 339–387; *Report of the Presidential Commission on the Space Shuttle Challenger Accident, Volume 2, Appendix G—Human Factor Analysis*, NASA. https://history.nasa.gov/rogersrep/v2appg.htm; Leo Hickman, "The Hidden Dangers of Sleep Deprivation," *Guardian*, February 9, 2011, accessed June 13, 2022, https://www .theguardian.com/lifeandstyle/2011/feb/09/dangers-sleep-deprivation; Merrill Mitler et al., "Catastrophes, Sleep, and Public Policy: Consensus Report," *Sleep* 11, no. 1 (February 1988): 100–109.

27. *The* Challenger *explosion:* Lenore C. Terr et al., "Children's Symptoms in the Wake of *Challenger:* A Field Study of Distant-Traumatic Effects and an Outline of Related

Conditions," *American Journal of Psychiatry* 156, no. 10 (October 1999): 1536–1544; Howell, "Space Shuttle Challenger."

28. *At the 1986 APSS meeting:* Mitler et al., "Catastrophes, Sleep, and Public Policy."

29. *University of Pennsylvania psychologist:* David F. Dinges CV; "David F. Dinges, MS, MA, PhD," Penn Psychiatry, accessed June 14, 2022, https://www.med.upenn.edu/uep /faculty_dinges.html; David F. Dinges and Roger J. Broughton, *Sleep and Alertness: Chronobiological, Behavioral, and Medical Aspects of Napping* (New York: Raven Press, 1989).

30. *neuropsychologist R. Curtis Graeber:* "Raymond Curtis Graeber," ResearchGate, https:// www.researchgate.net/profile/Raymond-Graeber; "Board of Directors: R. Curtis Graeber," Washington State Academy of Sciences, accessed June 14, 2022, http://depts .washington.edu/washacad/about/board/graeber_c.html.

31. *In May 1986:* Peretz Lavie, "Ultrashort Sleep-Waking Schedule.III. 'Gates' and 'Forbidden Zones' for Sleep," *Electroencephalography and Clinical Neurophysiology* 63 (1986): 414–425.

32. *Lavie's findings echoed:* Interview with Mary Carskadon, August 8, 2020; Steven Strogatz, *Sync: The Emerging Science of Spontaneous Order* (New York: Hyperion, 2003), 94–95.

33. *Carskadon's study proved:* Strogatz, 94–95.

34. *In Czeisler's lab:* Ibid., 92–97; Steven H. Strogatz, *The Mathematical Structure of the Human Sleep-Wake Cycle* (Berlin: Springer-Verlag, 1986), 90–97; Steven H. Strogatz, Richard E. Kronauer, and Charles A. Czeisler, "Circadian Pacemaker Interferes with Sleep Onset at Specific Times Each Day: Role in Insomnia," *American Journal of Physiology* 253, no. 1 (July 1, 1987): R172–R178.

35. *Soon after reaching:* Strogatz, *Sync*, 94, 302n94; Merrill M. Mitler and James C. Miller, "Methods of Testing for Sleeplessness," *Behavioral Medicine* 21, no. 4 (1996): 171–183.

36. *In a 1987 paper:* Strogatz, Kronauer, and Czeisler, "Circadian Pacemaker."

37. *One such therapy:* Charles A. Czeisler et al., "Bright Light Resets the Human Circadian Pacemaker Independent of the Timing of the Sleep-Wake Cycle," *Science* 233, no. 4764 (August 8, 1986): 667–671; Alfred J. Lewy et al., "Light Suppresses Melatonin Secretion in Humans," *Science* 210 (December 1980): 1267–1269. *about five times the level:* "Illuminance—Recommended Light Level," Engineering Toolbox, accessed June 18, 2022, https://www.engineeringtoolbox.com/light-level-rooms-d_708.html.

38. *In August 1986:* Czeisler et al., "Bright Light Resets."

39. *Further tests showed:* Charles A. Czeisler et al., "Bright Light Induction of Strong (Type 0) Resetting of the Human Circadian Pacemaker," *Science* 244, no. 4910 (June 16, 1989): 1328–1333; Charles A. Czeisler et al., "Exposure to Bright Light and Darkness to Treat Physiologic Maladaptation to Night Work," *New England Journal of Medicine* 322, no. 18 (May 3, 1990): 1253–1259.

40. *Around the time:* H. R. Lieberman et al., "Effects of Melatonin on Human Mood and Performance," *Brain Research* 323, no. 2 (December 1984): 201–207; David Evered and Sarah Clark, eds., *Ciba Foundation Symposium 117—Photoperiodism,*

Melatonin and the Pineal (London: Pitman, 1985), 266–283; H. R. Lieberman, "Behavior, Sleep and Melatonin," *Journal of Neural Transmission,* Supplementum, no. 21 (January 1, 1986): 233–241; S. P. James et al., "The Effect of Melatonin on Normal Sleep," *Neuropsychopharmacology* 1, no. 1 (December 1, 1987): 41–44.

41. *Dement and his team:* "CFIR Studies 1977–1988," William C. Dement Papers, Box 5, Folder 4.4.1.

42. *a condition, surveys:* Mary A. Carskadon et al., "Guidelines for the Multiple Sleep Latency Test (MSLT): A Standard Measure of Sleepiness," *Sleep* 9, no. 4 (1986): 519–524.

43. *For the field's:* William C. Dement, "The Neuropsychopharmacology of Daytime Alertness," *Pharmacology Bulletin* 23, no. 3 (1987): 435; Merrill M. Mitler, "Alerting Drugs: Do They Really Work?," *Pharmacology Bulletin* 23, no. 3 (1987): 435–439; Richard Saltus, "Waking Up to a Modern Plague: Drowsiness," *Boston Globe,* December 15, 1986, 41–42.

44. *In a* Boston Globe *feature:* Saltus, "Waking Up."

45. *The* Globe *story also:* Ibid.

46. *He continued to:* Dement to Kathryn Dement (letter reporting that the deal has been finalized), February 5, 1988, William C. Dement Papers, Box 53, Folder 10.8.

47. *He was pressing:* William C. Dement to David Korn, "Sleep Disorders Program," May 13, 1986, William C. Dement Papers, Box 1, Folder 1.8.1; William C. Dement to David Korn, "Administrative Support," October 24, 1986, William C. Dement Papers, Box 1, Folder 1.8.1.

48. *He was launching:* William C. Dement to Ken and Barbara Oshman, April 2, 1986, William C. Dement Papers, Box 1, Folder 8.1; William C. Dement to Donald Kennedy, May 5, 1986, William C. Dement Papers, Box 1, Folder 8.1.

49. *He was advising the Black Pre-Medical Association:* Interview with Woody Myers, September 29, 2021; interview with Debra Myers, November 15, 2021.

50. *He was traveling:* Dement to program chiefs, "Upcoming Trip—FYI," November 11, 1986, William C. Dement Papers, Box 1, Folder 8.1.

51. *He was writing:* William C. Dement CV; William C. Dement Papers, passim; William C. Dement to Tracey Loveridge, "Time Pressures," April 23, 1986, Wiliam C. Dement Papers, Box 1, Folder 1.8.

52. *In February 1987:* Interview with Cathy Dement Roos, November 6, 2020; William C. Dement to Robert E. Neger, January 14, 1988, William C. Dement Papers, Box 1, Folder 1.8.3.

53. *Dement, Pat, and their children:* "History" (intake form for Dr. Neger), December 1987, William C. Dement Papers, Box 1, Folder 1.8.

54. *The family set up:* Ibid.

55. *For Dement, however:* Interviews with Woody Myers and Debra Myers.

56. *"As you know":* William C. Dement to Woodrow Myers and Debbie Hammond, October 13, 1987, William C. Dement Papers, Box 1, Folder 1.8.3.

57. *In a draft memo:* William C. Dement to Hugh McDevitt, Carl Grumet, Larry Steinman, Kym Faull, Roland Ciaranello, undated. William C. Dement Papers, Box 1, Folder 1.8.3.

58. *At the ASDC meeting:* J. W. Shepard et al., "History of the Development of Sleep Medicine in the United States," *Journal of Clinical Sleep Medicine* 2, no. 1 (2005): 61–82. Today, the organization is known as the American Academy of Sleep Medicine (AASM).

59. *The event also:* William C. Dement to Nathaniel Kleitman, October 9, 1987, William C. Dement Papers, Box 1, Folder 1.8.3.

60. *Although he returned:* William C. Dement to staff, November 2, 1987, William C. Dement papers, Box 1, Folder 1.8.3.

61. *"I don't know how it seems to you":* Ibid.

Chapter 18

1. *In March 1988:* Merrill Mitler et al., "Catastrophes, Sleep, and Public Policy: Consensus Report," *Sleep* 11, no. 1 (February 1988): 100–109.

2. *The committee explained:* Ibid.

3. *Accidents were not:* Ibid.

4. *But the paper's:* Ibid.

5. *"The committee recognizes":* Ibid.

6. *With the space shuttle:* Anthony Perry, "Alarming Link Ties Disaster to Drowsiness," *Des Moines Register*, February 10, 1988, 7; Anthony Perry, "Make the World Safer: Take a Nap," *Los Angeles Times*, February 12, 1988, 61; Peg Byron (United Press International), "Accidents Linked to Sleep Patterns," *Olathe Daily News*, March 13, 1988, 23; Associated Press, "Drowsiness Linked to Disasters: Study," *Merced Sun-Star*, February 22, 1988, 2.

7. *"Society has got to be aware":* Perry, "Make the World Safer."

8. *That summer:* J. W. Shepard et al., "History of the Development of Sleep Medicine in the United States," *Journal of Clinical Sleep Medicine* 2, no. 1 (2005): 61–82; interview with Dale Dirks, November 3, 2021.

9. *Led by Dement:* Tom Roth, "An Overview of the Report of the National Commission on Sleep Disorders Research," *European Psychiatry* 10, Supplement 3 (1995): 109s–113s.

10. *Witnesses with narcolepsy:* Paul Recer, "Millions Suffer from Sleep Disorders That Cause Accidents, Experts Say," AP News, September 25, 1990, accessed July 21, 2022, https://apnews.com/article/2ecd06561e22eed14e3a6c541e7722b2; Robert Baird, "Sleep Disorders a Nightmare for up to 40% of Us," *Pittsburgh Press*, May 10, 1991, 37; transcript, National Commission on Sleep Disorders Research Public Hearing, Charleston, West Virginia, May 11, 1991, William C. Dement Papers, Series 2, Box 4.

11. *A father spoke:* Mark Pressman and William C. Orr, *Understanding Sleep: The Evaluation and Treatment of Sleep Disorders* (Washington, DC: American Psychological Association, 1997), 118.

12. *In September 1992:* Dement, *The Sleepwatchers*, 188; Shepherd et al., "History of the Development."

13. *And the following May:* "S-1—National Institutes of Health Revitalization Act of 1993," US Congress, accessed July 26, 2022, https://www.congress.gov/bill/103rd-congress/senate-bill/1/actions.

14. *In June:* Shepherd et al., "History of the Development"; interview with James Walsh (sleep researcher who lobbied Congress with Dement), September 15, 2022.

15. *It took two or three:* Shepherd et al., "History of the Development"; J. P. Kiley et al., "The National Center on Sleep Disorders Research—Progress and Promise," Sleep 42, no. 6 (June 2019): zsz105, https://doi.org/10.1093/sleep/zsz105; email message from Dale Dirks to author, November 4, 2021; H. R. Colten and B. M Altvogt, eds., *Sleep Disorders and Sleep Deprivation: An Unmet Public Health Problem,* Institute of Medicine (US) Committee on Sleep Medicine and Research (Washington, DC: National Academies Press, 2006), Table G-2, accessed October 20, 2022, https://www.ncbi.nlm.nih.gov/books/NBK19960/.

16. *Meanwhile, Mary Carskadon:* Mary Carskadon CV; interview with Mary Carskadon, August 22, 2020; Steven J. Zottoli, "The Origins of the Grass Foundation," *Biological Bulletin* 201 (October 2001): 218–226.

17. *By then, Carskadon:* Mary Carskadon interview; Mary A. Carskadon and Joan Mancuso, "Reported Sleep Habits in Boarding School Students: Preliminary Data," *Sleep Research* 16 (1987): 173; Mary A. Carskadon and Joan Mancuso, "Sleep Habits in High School Adolescents: Boarding Versus Day Students," *Sleep Research* 17 (1988): 74; Mary A. Carskadon and Joan Mancuso, "Daytime Sleepiness in High School Adolescents: Influence of Curfew," *Sleep Research* 17 (1988): 75; Mary A. Carskadon, Joan Mancuso, and Mark R. Rosekind, "Impact of Part-Time Employment on Adolescent Sleep Patterns," *Sleep Research* 18 (1989). *questionnaires distributed to more than 4,500 students:* Mary A. Carskadon, "Patterns of Sleep and Sleepiness in Adolescents," *Pediatrician* 17 (1990): 5–12; David Streitfeld, "And So to Bed…But Not Necessarily to Sleep," *Washington Post*, April 26, 1988, accessed July 5, 2022, https://www.washingtonpost.com/archive/lifestyle/1988/04/26/and-so-to-bed-but-not-necessarily-to-sleep/d251a95e-248c-4ccc-ba76-be94d97d195b/.

18. *Carskadon was aided:* Interview with Mark Rosekind, February 13, 2021.

19. *Rosekind arrived in Providence:* Ibid.

20. *The summer of 1988:* Mary Carskadon CV; "'Sleep Deficit Is an Eye-Opener,' Say Researchers," *Democrat and Chronicle*, April 28, 1988, 1.

21. *In early 1990:* Carskadon, "Patterns of Sleep."

22. *The big news had:* Ibid.

23. *Teenagers with after-school jobs:* Ibid.

24. *"By far the most striking":* Ibid.

25. *Carskadon also revealed:* Ibid.

26. *She expanded on:* Mary A. Carskadon, "Adolescent Sleepiness: Increased Risk in a High-Risk Population," *Alcohol, Drugs and Driving* 5/6, no. 1 (1990): 317–328.

27. *This paper was:* Ibid.

28. *Carskadon found that:* Ibid.

29. *The last section:* Ibid.

30. *Carskadon concluded with:* Ibid.

31. *Soon afterward, Carskadon:* Mary A. Carskadon, Cecilia Vieira, and Christine Acebo, "Association Between Puberty and Delayed Phase Preference," *Sleep* 16, no. 3 (1993): 258–262.

32. *To test the latter:* Ibid.

33. *The children's questionnaires:* Ibid.

34. *When Carskadon and her team:* Ibid.

35. *The study was published:* Ibid.

36. *Carskadon's paper made:* Gannett News Service, "Puberty May Push Teens to Sleep Late," *Chillicothe Gazette*, April 22, 1993, 15; Gannett News Service, "Study Finds Teens May Have Later Body Clocks," *Daily Item*, April 22, 1993, 6; Marie Ellis, "The Young and the Restless: What Makes Teenagers Tick?," *Honolulu Advertiser*, April 22, 1993, 1.

37. BODY CLOCK BLAMED: Roger Highfield, "Body Clock Blamed for Lie-In Teenagers," *Daily Telegraph*, April 15, 1993, 7.

38. *Over the next couple:* Mary Carskadon CV; Orna Tzichinsky et al., "Sleep Habits and Salivary Melatonin Onset in Adolescents," *Sleep Research* 24 (1995): 543; Mary A. Carskadon et al., "An Approach to Studying Circadian Rhythms of Adolescent Humans," *Journal of Biological Rhythms* 12, no. 3 (June 1997): 278–289.

39. *She also launched:* Mary A. Carskadon et al., "Early school Schedules Modify Adolescent Sleepiness," *Sleep Research* 24 (1995): 92; Mary A. Carskadon et al., "Adolescent Sleep Patterns, Circadian Timing, and Sleepiness at a Transition to Early School Days," *Sleep* 21, no. 8 (1998): 871–881.

40. *"The students may be":* Mary A. Carskadon, ed., *Adolescent Sleep Patterns: Biological, Social, and Psychological Influences* (Cambridge, England: Cambridge University Press, 2002), 19.

41. *Carskadon reported on her:* Mary Carskadon CV.

42. *But her most consequential:* Lisa L. Lewis, *The Sleep-Deprived Teen: Why Our Teenagers Are So Tired, and How Parents and Schools Can Help Them Thrive* (Coral Gables, FL: Mango Publishing, 2022), 54–55; Hawley Montgomery-Downs, *Sleep Science* (New York: Oxford University Press, 2020), 175.

43. *In 1994, when the association:* Lewis, *The Sleep-Deprived Teen*, 55.

44. *Getting other stakeholders:* Ibid., 46–52, 55–56; interview with Terra Ziporyn Snider (cofounder of the nonprofit Start School Later), September 8, 2022.

45. *Dragseth led an outreach:* Lewis, *The Sleep-Deprived Teen*, 55–56.

46. *The immediate results:* Ibid., 56; Duschene Paul Drew, "High School Start Times Under Review," *Star Tribune*, December 10, 1996, 1.

47. *Hoping to spread:* Lewis, *The Sleep-Deprived Teen*, 57–58; Drew, "High School Start Times"; interview with Mary Carskadon, September 19, 2020.

48. *By May 1997:* Lewis, *The Sleep-Deprived Teen*, 57; "Minneapolis Schools to Start 2 1/2 Hours Later," *St. Cloud Times*, May 7, 1997, 2.

49. *A movement had been:* Lewis, *The Sleep-Deprived Teen*, 54–61, 200, 218–235; Terra Ziporyn Snider interview.

50. *In June 1995:* Chip Brown, "The Stubborn Scientist Who Unraveled a Mystery of the Night," *Smithsonian*, October 2003; interview with Jill Aserinsky Buckley, February 24, 2022; interview with Mary Carskadon, September 19, 2020; interview with Sonia Ancoli-Israel, January 19, 2021; interview with Hortense Kleitman Snower, September 10, 2016.

51. *Too frail to stand:* Video of symposium, collection of Jill Aserinsky Buckley.

52. *What was stunning:* Ibid.

53. *Next, Dement invited:* Ibid.; William C. Dement, "Dr. Aserinsky Remembered," *SRS Bulletin* 4, no. 2 (September 1998): 14.

54. *Still, that gesture:* Interview with Armond Aserinsky, November 6, 2021; Jill Aserinsky Buckley interview; Lynne Lamberg, "Scientists Never Dreamed Finding Would Shape a Half-Century of Sleep Research," *JAMA* 290, no. 20 (November 26, 2003): 2652–2654; Wallace Mendelson, "In Memory of Eugene Aserinsky (1921–1998), *Journal of the History of the Neurosciences* 7, no. 3 (1996): 250–251.

55. *"He spent the rest of his life":* Mallika Rao, "Third State," *HiberNation* (podcast), July 15, 2021, accessed November 7, 2021, https://hibernation.simplecast.com/episodes/third-state.

56. *"Just as Dr. Kleitman":* Video of symposium.

57. *The room erupted:* Ibid.

58. *Jouvet had been:* Ibid.

59. *After Jouvet finished:* Interview with sleep researcher Jerome Siegel (who organized the symposium), April 21, 2021.

60. *"It was one of the highlights":* Brown, "The Stubborn Scientist."

61. *Aserinsky's life ended:* "Eugene Aserinsky, Sleep Researcher, 77," *New York Times*, August 7, 1998, D17; Brown, "The Stubborn Scientist."

62. *Kleitman followed him:* Vicki Cheng, "Nathaniel Kleitman, Sleep Expert, Dies at 104," *New York Times*, August 19, 1999, B8; County of Los Angeles, Registrar-Recorder/County Clerk, Certificate of Death, Nathaniel Kleitman, August 13, 1999.

63. *He had never quite accepted:* Brown, "The Stubborn Scientist."

64. *The annual APSS meeting:* Sleep 2019 meeting program, accessed December 18, 2022, https://sleepmeeting.org/wp-content/uploads/2018/11/SLEEP2019_EP.pdf.

65. *There are approximately:* "Sleep Medicine in America: Infographic," AASM Sleep Education, https://sleepeducation.org/sleep-medicine-america-infographic/. *three times the figure:* My estimate is based on the membership of the ASDA in 1995 (2,500) and a letter from AASM president Conrad Iber to Steve Phurrough at the US Centers for Medicare and Medicaid Services, dated May 5, 2004, in which he gives the number of board-certified specialists at that time as 2,656. See: "Initial Public Comments for NCA for Continuous Positive Airway Pressure (CPAP) Therapy for Obstructive Sleep Apnea," Centers for Medicare & Medicaid Services, https://www.cms.gov/Medicare/Coverage/DeterminationProcess/Downloads/id110a-1.pdf. *Membership in the American Academy:* "About American Academy of Sleep Medicine," AASM, accessed December 18, 2022, https://aasm.org/about/; Shepard et al., "History of the Development."

66. *As Jouvet put it:* Michel Jouvet, *The Paradox of Sleep: The Story of Dreaming* (Cambridge, MA: MIT Press, 1999), 177.

67. *Therapeutic puzzles persist:* Sudhansu Chokroverty and Michel Billiard, eds., *Sleep Medicine: A Comprehensive Guide to Its Development, Clinical Milestones, and Advances in Treatment* (New York: Springer, 2015), 205–335, 357–405, 519–571.

68. *And there still:* Nathaniel F. Watson, Ilene M. Rosen, and Ronald D. Chervin, "The Past Is Prologue: The Future of Sleep Medicine," *Journal of Clinical Sleep Medicine* 13, no. 1 (2017): 127–135.

69. *Our growing attachment:* Rob Newsom, "How Blue Light Affects Sleep," Sleep Foundation, October 18, 2022, accessed October 27, 2022, https://www.sleepfoundation.org /bedroom-environment/blue-light; *Among low-income Americans:* National Center on Sleep Disorders Research, "Goal 4: Sleep and Circadian Disruptions and Health Disparities," *National Institutes of Health Sleep Research Plan* (2021), accessed June 9, 2023, https://www.nhlbi.nih.gov/sleep-research-plan/health-disparities.

70. *Around 20 percent:* Ryan Fiorenzi, "Sleep Statistics: Understanding Sleep and Sleep Disorders," Start Sleeping, accessed October 27, 2022, https://startsleeping.org/statistics/; Megan McHugh, Diane Farley, and Adovich S. Rivera, "A Qualitative Exploration of Shift Work and Employee Well-Being in the US Manufacturing Environment," *Journal of Occupational and Environmental Medicine* 62, no. 4 (April 2020): 303–306.

71. *And despite continued:* Terra Ziporyn Snider interview.

Epilogue

1. *In August 2020:* Author's notes. I was among the virtual attendees.

2. *Between sets:* Author's notes; interview with Rafael Pelayo, February 24, 2021; "Clete A. Kushida, MD, PhD," Stanford Profiles, accessed October 29, 2022, https://profiles .stanford.edu/clete-kushida.

3. *Dement's daughter Cathy:* Author's notes.

4. *Now, the family:* By "fathers," I mean the researchers (all of them male) responsible for the creation of modern sleep science. Among the earliest investigators of REM sleep, Howard Roffwarg and Allan Rechtschaffen outlived Dement (they died, respectively, in October and November 2021), but his work helped inspire both to enter the field.

5. *she'd just been awarded:* "Carney Researcher to Lead New Sleep, Circadian Rhythms Center," Robert J. and Nancy D. Carney Institute for Brain Science, April 28, 2021, accessed October 29, 2022, https://www.brown.edu/carney/news/2021/04/28/carney -researcher-lead-new-sleep-circadian-rhythms-center.

6. *for three decades:* Email message from Mary Carskadon to author, July 10, 2022. Her lab moved into this facility in the spring of 1990.

7. *"I mean, sleep is pretty hot":* Interview with Mary Carskadon, August 2, 2021.

8. *The rationale behind the new Center:* COBRE grant proposal.

9. *The NIH award:* Mary Carskadon interview.

10. *So far, the center:* COBRE Grant proposal; Mary Carskadon interview.

11. *The second study:* Ibid.

12. *But the objects he was carrying:* Interview with Jared Saletin, August 2, 2021.

13. *"We went to this tile-making studio":* Mary Carskadon interview.

14. *"My favorite place in the lab":* Ibid.

15. *The sleeping rooms themselves:* Ibid.

16. *"Jared's project will use":* Mary Carskadon; "How to Wear ZMax," Hypnodyne, accessed October 30, 2022, https://hypnodynecorp.com/howtowear.php.

17. *Our next stop:* Mary Carskadon interview.

18. *The latter picture:* Ibid.

19. *In the classes she taught:* Ibid.

20. *As a parting gift:* Ibid.
21. *"Bill said, 'Here, Mary,'":* Ibid.
22. *An answer came:* Interview with Armond Aserinsky, November 6, 2021.

Appendix A

1. *~500 BC—Alcmaeon:* Gaston G. Celesia, "Alcmaeon of Croton's Observations on Health, Brain, Mind, and Soul," *Journal of the History of the Neurosciences* 21, no. 4 (2012): 409–426; Douglas Benjamin Kirsch, "There and Back Again," *Chest* 139, no. 4 (April 2011): 939–946.

2. *~350 BC—Aristotle:* Kirsch, "There and Back Again"; Sudhansu Chokroverty and Michel Billiard, eds., *Sleep Medicine: A Comprehensive Guide to Its Development, Clinical Milestones, and Advances in Treatment* (New York: Springer, 2015), 50–51.

3. *1729—Jean-Jacques d'Ortous de Mairan:* Lynne Lamberg, *Bodyrhythms: Chronobiology and Peak Performance* (New York: William Morrow, 1994), 20.

4. *1807—London introduces:* "Jan. 28, 1807: Flickering Gaslight Illuminates Pall Mall," *Wired,* January 28, 2008, https://www.wired.com/2008/01/dayintech-0128/.

5. *1832—Augustin Pyramus de Candolle:* Vinod Kumar, ed., *Biological Timekeeping: Clocks, Rhythms and Behavior* (New Delhi: Springer, 2017), 5–6.

6. *1849—Observing cerebral circulation:* Michael J. Thorpy, "History of Sleep Medicine," *Handbook of Clinical Neurology* 98 (2011): 3–25.

7. *1862—Ernst Kohlschütter publishes:* Mathias Basner, "Arousal Threshold Determination in 1862: Kohlschütter's Measurements on the Firmness of Sleep," *Sleep Medicine* 11 (2010): 417–422.

8. *1864—Bromide salts:* R. H. Balme, "Early Medicinal Use of Bromides," *Journal of the Royal College of Physicians* 10, no. 2 (January 1976): 205–208.

9. *1869—Chloral hydrate becomes:* Wallace B. Mendelson, *Nepenthe's Children: The History of the Discoveries of Medicines for Sleep and Anesthesia* (New York: Pythagoras Press, 2020), 60–64.

10. *1880—Thomas Edison patents:* Callum McKelvie and Elizabeth Peterson, "Who Invented the Lightbulb?," LiveScience, November 2, 2022, accessed December 6, 2022, https://www.livescience.com/43424-who-invented-the-light-bulb.html.

11. *1880—Jean-Baptiste-Édouard Gélineau:* Kenton Kroker, *The Sleep of Others and the Transformation of Sleep Research* (Toronto: University of Toronto Press, 2015), 86–87.

12. *1880s—The term* insomnia: Ibid., 76–79.

13. *1890—Ludwig Mauthner theorizes:* Ibid., 194, 196; James E. Lebensohn, "The Eye and Sleep," *Archives of Ophthalmology* 25, no. 3 (1941): 401–411.

14. *1891—Eduard Michelson shows:* Matthias M. Weber and Wolfgang Burgmair, " 'The Assistant's Bedroom Served as a Laboratory': Documentation in 1888 of Within Sleep Periodicity by the Psychiatrist Eduard Robert Michelson," *Sleep Medicine* 10 (2009): 378–384.

15. *1894—Maria Mikhailovna Manaseina:* M. Bentivoglio and G. Grassi-Zucconi, "The Pioneering Experimental Studies on Sleep Deprivation," *Sleep* 20, no. 7 (July 1997): 570–576, doi: 10.1093/sleep/20.7.570.

16. *1895—Lord Rosebery resigns:* Jonathan Davidson, *Downing Street Blues: A History of Depression and Other Mental Afflictions in British Prime Ministers* (Jefferson, NC: McFarland, 2011), 96–97.

17. *1896—George Patrick and:* Bentivoglio and Grassi-Zucconi, "The Pioneering Experimental Studies."

18. *1899—Sigmund Freud publishes:* Kroker, *The Sleep of Others,* 156–159.

19. *1903—Barbiturates are introduced:* Francisco Lopez-Muñoz, Ronaldo Ucha-Udabe, and Cecilio Alamo, "The History of Barbiturates a Century After Their Clinical Introduction," *Neuropsychiatric Disease and Treatment* 1, no. 4 (2005): 329–343.

20. *1909—Kuniomi Ishimori publishes:* Kisou Kubota, "Kuniomi Ishimori and the First Discovery of Sleep-Inducing Substances in the Brain," *Neuroscience Research* 6 (1989): 497–518.

21. *1917—Constantin von Economo identifies:* Kroker, *The Sleep of Others,* 193–196; Lazaros Triarhou, "The Percipient Observations of Constantin von Economo on Encephalitis Lethargica and Sleep Disruption and Their Lasting Impact on Contemporary Sleep Research," *Brain Research Bulletin* 69 (2006): 244–258.

22. *1923—Ivan Pavlov theorizes:* Y. Popov and L. Rokhin, eds., *I. P. Pavlov: Psychopathology and Psychiatry, Selected Works* (Moscow: Foreign Languages Publishing House, n.d.), 109–125, 513–514.

23. *1923—Kleitman publishes his first:* Nathaniel Kleitman, "Studies on the Physiology of Sleep: I. The Effects of Prolonged Sleeplessness on Man," *American Journal of Physiology* 66 (1923): 67–92.

24. *1926—Von Economo identifies:* Triarhou, "The Percipient Observations."

25. *1929—Kleitman disputes:* Kroker, *The Sleep of Others,* 217.

26. *1929—Hans Berger publishes:* T. J. La Vaque, "The History of EEG: Hans Berger," *Journal of Neurotherapy: Investigations in Neuromodulation, Neurofeedback and Applied Neuroscience* 3, no. 2 (1999): 1–9.

27. *1929—Von Economo theorizes:* Triarhou, "The Percipient Observations."

28. *1932—Kleitman publishes:* Claude Gottesmann, *Henri Piéron and Nathaniel Kleitman: Two Major Figures of 20th Century Sleep Research* (New York: Nova Science Publishers, 2013), 60–64; Nathaniel Kleitman, *Sleep and Wakefulness as Alternating Phases in the Cycle of Existence* (Chicago: University of Chicago Press, 1939): 138–140.

29. *1933—Kleitman publishes the first study:* Gottesmann, *Henri Piéron,* 60–61.

30. *1933—Kleitman publishes the first controlled:* Ibid., 65–66.

31. *1935—Alfred Lee Loomis:* Kroker, *The Sleep of Others,* 287–290; Alfred L. Loomis, E. Newton Harvey, Garret Hobart, "Potential Rhythms of the Cerebral Cortex During Sleep," *Science* 81, no. 2111 (June 14, 1935): 597–598.

32. *1935—Frédéric Bremer:* Bremer, "Cerveau 'Isolé' et Physiologie du Sommeil," *Comptes Rendus des Séances de la Société de Biologie* 118 (1935): 1235–1241.

33. *1948—Kleitman spends two weeks:* Matthew J. Wolf-Meyer, *The Slumbering Masses: Sleep, Medicine, and Modern American Life* (Minneapolis: University of Minnesota Press, 2016), 133–134.

34. *1949—Kleitman's experimental:* Ibid.

35. *1949—Horace Magoun and Giuseppe Moruzzi:* Giuseppe Moruzzi and Horace Magoun, "Brain Stem Reticular Formation and Activation of the EEG," *EEG and Clinical Neurophysiology* 1 (1949): 455–473.

36. *1953—Aserinsky and Kleitman:* Eugene Aserinsky and Nathaniel Kleitman, "Regularly Occurring Periods of Eye Motility, and Concomitant Phenomena, During Sleep," *Science* 118, no. 3062 (September 4, 1953): 273–274.

37. *1957—Dement and Kleitman:* William Dement and Nathaniel Kleitman, "Cyclic Variations in EEG During Sleep and Their Relation to Eye Movements, Body Motility, and Dreaming," *Electroencephalography and Clinical Neurophysiology* 9 (1957): 673–690.

38. *1957—Thalidomide, prescribed:* "Remind Me Again, What Is Thalidomide and How Did It Cause So Much Harm?," Conversation, December 6, 2015, https://theconversation .com/remind-me-again-what-is-thalidomide-and-how-did-it-cause-so-much-harm -46847.

39. *1958—Dement describes REM:* William Dement, "The Occurrence of Low Voltage, Fast, Electroencephalogram Patterns During Behavioral Sleep in the Cat," *Encephalography and Clinical Neurophysiology* 10, no. 2 (May 1958): 291–296.

40. *1959—Michel Jouvet reports:* M. Jouvet, F. Michel, and J. Courjon, "Sur un Stade d'Activité Électrique Cérébrale Rapide au Cours du Sommeil Physiologique," *Comptes Rendus des Séances de l'Académie de la Société Biologique et de Ses Filiales* 153 (1959): 1024–1028; M. Jouvet and F. Michel, "Corrélations Électromyographiques du Sommeil Chez le Chat Décortiqué et Mésencephalique Chronique," *Comptes Rendus des Séances de la Société de Biologie et de Ses Filiales* 153 (1959): 422–425.

41. *1959—Dement identifies:* William C. Dement and Christopher Vaughan, *The Promise of Sleep: A Pioneer in Sleep Medicine Explores the Vital Connection Between Health, Happiness, and a Good Night's Sleep* (New York: Delacorte, 1999), 43.

42. *1959—Kleitman retires:* Nathaniel Kleitman autobiography, dictated to Paulena Kleitman circa 1970s, Nathaniel Kleitman Papers, Box 1, Folder 7, Hanna Holborn Gray Special Collections Research Center, University of Chicago Library; Richard Sandomir, "Allan Rechtschaffen, Eminent Sleep Researcher, Dies at 93," *New York Times*, December 19, 2021, A3.

43. *1960—Gerald Vogel describes:* Gerald Vogel, "Studies in the Psychophysiology of Dreams, III: The Dream of Narcolepsy," *Archives of General Psychiatry* 3, no. 4 (October 1960): 421–482.

44. *1960—Dement publishes:* William Dement, "The Effect of Dream Deprivation," *Science* 131, no. 3415 (June 10, 1960): 1705–1707.

45. *1960—The Ciba Conference:* Adrian R. Morrison, "Coming to Grips with a 'New' State of Consciousness: The Study of Rapid-Eye-Movement Sleep in the 1960s," *Journal of the History of the Neurosciences* 22, no. 4 (2013): 392–407; G. E. W. Wolstenhome and Maeve O'Connor, eds., *Ciba Foundation Symposium on the Nature of Sleep* (Boston: Little, Brown, 1961).

46. *1960—Cold Spring Harbor:* Kumar, *Biological Timekeeping*, 12.

47. *1960—Benzodiazepine sleep aids:* Jeanette Y. Wick, "The History of Benzodiazepines," *Consultant Pharmacist* 9 (September 2013): 538–548.

48. *1961—Jürgen Aschoff conducts:* Rütger A. Wever, *The Circadian System of Man: Results of Experiments Under Temporal Isolation* (New York: Springer-Verlag), 9–10.

49. *1961—Dement and Rechtschaffen establish:* "Historical Note," Association for the Psychophysiological Study of Sleep Records, Hanna Holborn Gray Special Collections Research Center, University of Chicago Library, accessed February 11, 2022, https://www.lib.uchicago.edu/e/scrc/findingaids/view.php?eadid=ICU.SPCL.SLEEP; "Some Selected Details from the Earliest Sleep Conferences (1961–1964)," Association for the Psychophysiological Study of Sleep Records, Box 1, Folder 2.

50. *1963—Dement and Rechtschaffen report:* Allan Rechtschaffen et al., "Nocturnal Sleep of Narcoleptics," *Electroencephalography and Clinical Neurophysiology* 15, no. 4 (August 1963): 599–609.

51. *1966—Dement and Roffwarg report:* Howard P. Roffwarg, Joseph N. Muzio, and William C. Dement, "Ontogenetic Development of the Human Sleep-Dream Cycle," *Science* 152, no. 3722 (April 29, 1966): 604–619.

52. *1968—Rechtschaffen and Anthony Kales:* Allan Rechtschaffen and Anthony Kales, *A Manual of Standardized Technology, Techniques and Scoring System for Sleep Stages of Human Subjects* (Washington, DC: Public Health Service, US Government Printing Office, 1968).

53. *1969—Franz Halberg:* Franz Halberg, "Chronobiology," *Annual Review of Physiology* 31 (March 1969): 675–726.

54. *1971—Ronald Konopka and Seymour Benzer:* Ronald J. Konopka and Seymour Benzer, "Clock Mutants of *Drosophila melanogaster*," *Proceedings of the National Academy of Sciences, USA* 68, no. 9 (September 1971): 2112–2116.

55. *1972—Researchers identify:* Lynne Lamberg, "Researchers Dissect the Tick and Tock of the Human Body's Master Clock," *JAMA* 278, no. 13 (October 1, 1997): 1049–1051.

56. *1972—Richard Bootzin introduces:* Chokroverty and Billiard, *Sleep Medicine*, 521.

57. *1973—Carskadon and Dement publish:* Mary A. Carskadon and William C. Dement, "Sleep Studies on a 90-Minute Day," *Electroencephalography and Clinical Neurophysiology* 39 (1975): 145–155.

58. *1974—Dement starts:* Dement and Vaughan, *The Promise of Sleep*, 203; Henry Nicholls, "Dreaming of a Cure: The Battle to Beat Narcolepsy," *Guardian*, October 22, 2017, accessed April 3, 2022, https://www.theguardian.com/science/2017/oct/22/dreaming-cure-battle-to-beat-narcolepsy; Merrill Mitler et al., "Narcolepsy-Cataplexy in a Female Dog," *Experimental Neurology* 45, no. 2 (November 1974): 332–340.

59. *1975—Guilleminault describes:* Christian Guilleminault et al., "Sleep Apnea Syndrome: Can It Induce Hemodynamic Changes?," *Western Journal of Medicine* 123 (July 1975): 7–16.

60. *1975—The Association of Sleep Disorders Centers:* Dement and Vaughan, *The Promise of Sleep*, 48–49.

61. *1976—Steven Glotzbach:* Steven F. Glotzbach and H. Craig Heller, "Central Nervous Regulation of Body Temperature During Sleep," *Science* 194, no. 4264 (October 29, 1976): 537–539.

62. *1977—Peter Hauri introduces:* Chokroverty and Billiard, *Sleep Medicine*, 521.

63. *1978—The APSS begins:* J. W. Shepard et al., "History of the Development of Sleep Medicine in the United States," *Journal of Clinical Sleep Medicine* 2, no. 1 (2005): 61–82.

64. *1979—The first* Diagnostic Classification: "Diagnostic Classification of Sleep and Arousal Disorders," *Sleep* 2, no. 1 (1979): 1–154.

65. *1980—Carskadon reports:* Mary A. Carskadon et al., "Pubertal Changes in Daytime Sleepiness," *Sleep* 2, no. 4 (1980): 453–460.

66. *1981—Carskadon and Dement publish:* Mary A. Carskadon and William C. Dement, "Cumulative Effects of Sleep Restriction on Daytime Sleepiness," *Psychobiology* 18, no. 2 (1981): 107–113.

67. *1981—Colin Sullivan introduces:* Colin Sullivan, "Reversal of Obstructive Sleep Apnoea by Continuous Positive Airway Pressure Applied through the Nares," *Lancet* 317, no. 8225 (April 18, 1981): 862–865.

68. *1981—Shiro Fujita introduces:* Shiro Fujita et al., "Surgical Correction of Anatomic Abnormalities in Obstructive Sleep Apnea: Uvulopalatopharyngoplasty," *Otolaryngology—Head and Neck Surgery* 89, no. 6 (November 1, 1981): 923–934.

69. *1981—Czeisler describes delayed:* Elliot Weitzman et al., "Delayed Sleep Phase Syndrome," *Archives of General Psychiatry* 38 (July 1981): 737–746; Elliott D. Weitzman et al., "Biological Rhythms in Man Under Non-Entrained Conditions and Chronotherapy for Delayed Sleep Insomnia," *Advances in Biological Psychiatry* 11 (1983): 136–149.

70. *1982—Alexander Borbély publishes:* A. A. Borbély, "A Two Process Model of Sleep Regulation," *Human Neurobiology* 1, no. 195 (1982): 195–204.

71. *1982—Czeisler et al.:* Charles A. Czeisler, Martin C. Moore-Ede, and Richard M. Coleman, "Rotating Shift Work Schedules That Disrupt Sleep Are Improved by Applying Circadian Principles," *Science* 217 (July 30, 1982): 460–463.

72. *1982—Rosalind Cartwright:* Rosalind D. Cartwright and Charles F. Samelson, "The Effects of a Nonsurgical Treatment for Obstructive Sleep Apnea," *JAMA* 248, no. 6 (August 13, 1982): 705–709.

73. *1982—Kleitman publishes:* Nathaniel Kleitman, "Basic Rest-Activity Cycle—22 Years Later," *Sleep* 5, no. 4 (September 1982): 311–317.

74. *1983—Rechtschaffen theorizes:* Allan Rechtschaffen et al., "Physiological Correlates of Prolonged Sleep Deprivation in Rats," *Science* 221, no. 4606 (July 8, 1983): 182–184.

75. *1986—Zopiclone:* Karen L. Goa and Rennie C. Heel, "Zopiclone: A Review of Its Pharmacodynamic and Pharmacokinetic Properties and Therapeutic Efficacy as an Hypnotic," *Drugs* 32 (1986): 48–65; "Eszoplicone for Insomnia," *Cochrane Database Systemic Review* 10 (2018), accessed December 6, 2022, https://www.ncbi.nlm.nih.gov/pmc/articles/PMC6492503/.

76. *1986—Investigators report:* Merrill Mitler et al., "Catastrophes, Sleep, and Public Policy: Consensus Report," *Sleep* 11, no. 1 (February 1988): 100–109.

77. *1986—Peretz Lavie reports:* Peretz Lavie, "Ultrashort Sleep-Waking Schedule. III. 'Gates' and 'Forbidden Zones' for Sleep," *Electroencephalography and Clinical Neurophysiology* 63 (1986): 414–425.

78. *1987—Steven Strogatz:* Steven Strogatz, *Sync: The Emerging Science of Spontaneous Order* (New York: Hyperion, 2003), 92–97; Steven H. Strogatz, *The Mathematical*

Structure of the Human Sleep-Wake Cycle (Berlin: Springer-Verlag, 1986); Steven H. Strogatz. Richard E. Kronauer, and Charles A. Czeisler, "Circadian Pacemaker Interferes with Sleep Onset at Specific Times Each Day: Role in Insomnia," *American Journal of Physiology* 253, no. 1 (July 1, 1987): R172–R178.

79. *1987—Arthur Spielman introduces:* Chokroverty and Billiard, eds., *Sleep Medicine*, 523.

80. *1988—Merrill Mitler:* Mitler et al., "Catastrophes, Sleep, and Public Policy."

81. *1988—Congress passes:* Shepherd et al., "History of the Development."

82. *1989—Rechtschaffen reports:* Clete A. Kushida, Bernard A. Bergmann, and Allan Rechtschaffen, "Sleep Deprivation in the Rat: IV. Paradoxical Sleep Deprivation," *Sleep* 12, no. 1 (1989): 22–30; Jerome Siegel, *The Neural Control of Sleep and Waking* (New York: Springer-Verlag, 2002), 139–141.

83. *1989—Publication of* Principles: Interview with Meir Kryger, February 3, 2021; Meir H. Kryger, Thomas Roth, and William C. Dement, *Principles and Practice of Sleep Medicine*, 1st edition (Philadelphia: Saunders, 1989).

84. *1990—Publication of the first:* Diagnostic Classification Steering Committee and Michael J. Thorpy, *International Classification of Sleep Disorders: Diagnostic and Coding Manual* (Rochester, NY: American Sleep Disorders Association, 1990).

85. *1990—Czeisler and colleagues:* Charles A. Czeisler et al., "Exposure to Bright Light and Darkness to Treat Physiologic Maladaptation to Night Work," *New England Journal of Medicine* 322, no. 18 (May 3, 1990): 1253–1259.

86. *1990—NASA uses light therapy:* Charles A. Czeisler, August J. Chasera, and Jeanne F. Duffy, "Research on Sleep, Circadian Rhythms and Aging: Applications to Manned Spaceflight," *Experimental Gerontology* 26 (1991): 217–232.

87. *1990—The National Commission:* Mark R. Pressman and William C. Orr, *Understanding Sleep: The Evaluation and Treatment of Sleep Disorders* (Washington, DC: American Psychological Association, 1997), 114.

88. *1993—The first epidemiological study:* Terry Young et al., "The Occurrence of Sleep-Disordered Breathing Among Middle-Aged Adults," *New England Journal of Medicine* 328, no. 17 (April 29, 1993): 1230–1235.

89. *1993—Carskadon finds evidence:* Mary A. Carskadon, Cecilia Vieira, and Christine Acebo, "Association Between Puberty and Delayed Phase Preference," *Sleep* 16, no. 3 (1993): 258–262.

90. *1993—President Clinton signs:* "S-1—National Institutes of Health Revitalization Act of 1993," US Congress, accessed July 26, 2022, https://www.congress.gov /bill/103rd-congress/senate-bill/1/actions.

91. *1993—Charles Morin introduces:* Charles Morin et al., "Cognitive-Behavior Therapy for Late-Life Insomnia," *Journal of Consulting and Clinical Psychology* 61, no. 1 (February 1993): 137–146; Charles Morin, *Insomnia: Psychological Assessment and Management* (New York: Guilford Press, 1993); Chokroverty and Billiard, *Sleep Medicine*, 524–525, 536–537.

92. *1995—The American Medical Association:* Shepherd et al., "History of the Development"; Markus Helmut Schmidt, "In Memoriam: A Tribute to an Unsung Hero," *Sleep* 34, no. 3 (March 1, 2011): 403–404.

93. *1995—Joel H. Benington:* Ann K. Finkbeiner, "Getting Through the Sleep Gate," *Sciences*, September/October 1998, 14–18; Joel H. Benington, Susheel K. Kodali, and H. Craig Heller, "Stimulation of A1 Adenosine Receptors Mimics the Electroencephalographic Effects of Sleep Deprivation," *Brain Research* 692 (1995): 79–85.

94. *1995—Kleitman addresses:* Chip Brown, "The Stubborn Scientist Who Unraveled a Mystery of the Night," *Smithsonian*, October 2003.

95. *1996—Edina, Minnesota:* Lisa L. Lewis, *The Sleep-Deprived Teen: Why Our Teenagers Are So Tired, and How Parents and Schools Can Help Them Thrive* (Coral Gables, FL: Mango Publishing, 2022), 55–56.

96. *1997—Carskadon reports:* Orna Tzichinsky et al., "Sleep Habits and Salivary Melatonin Onset in Adolescents," *Sleep Research* 24 (1995): 543.

97. *1997—Minneapolis pushes back:* Lewis, *The Sleep-Deprived Teen*, 57; "Minneapolis Schools to Start 2 1/2 Hours Later," *St. Cloud Times*, May 7, 1997, 2.

98. *1998—Death of Aserinsky:* "Eugene Aserinsky, Sleep Researcher, 77," *New York Times*, August 7, 1998, D17.

99. *1998—Researchers at the Scripps:* Alberto K. de la Herrán-Arita, Magdalena Guerra-Crespo, and René Drucker-Colin, "Narcolepsy and Orexins: An Example of Progress in Sleep Research," *Frontiers in Neurology* 2, no. 26 (April 2011), https://www.frontiersin.org/articles/10.3389/fneur.2011.00026/full; Henry Nicholls, "Dreaming of a Cure: The Battle to Beat Narcolepsy," *Guardian*, October 22, 2017, accessed October 25, 2022, https://www.theguardian.com/science/2017/oct/22/dreaming-cure-battle-to-beat-narcolepsy; J. M. Siegel et al., "A Brief History of Hypocretin/Orexin and Narcolepsy," *Neuropsychopharmacology* 25, no. S5 (2001): S14–S20; L. de Lecea et al., "The Hypocretins: Hypothalamus-Specific Peptides with Neuroexcitatory Activity," *Proceedings of the National Academy* 95 (January 1998): 322–327; Takeshi Sakurai et al., "Orexins and Orexin Receptors: A Family of Hypothalamic Neuropeptides and G Protein-Coupled Receptors that Regulate Feeding Behavior," *Cell* 92 (February 20, 1998): 572–585.

100. *1999—Emmanuel Mignot:* De la Herrán-Arita et al., "Narcolepsy and Orexins"; Nicholls, "Dreaming of a Cure"; Ling Lin et al., "The Sleep Disorder Canine Narcolepsy Is Caused by a Mutation in the *Hypocretin (Orexin) Receptor 2* Gene," *Cell* 98 (August 6, 1999): 365–376; "Emmanuel Mignot," Stanford Medicine CAP Profiles, accessed October 25, 2022, https://med.stanford.edu/profiles/emmanuel-mignot.

101. *1999—Death of Kleitman:* Brown, "The Stubborn Scientist."

102. *1999—Carskadon becomes:* "Past Presidents," Sleep Research Society, accessed October 26, 2022, https://sleepresearchsociety.org/about/leadership/past-presidents/; interview with Mary Carskadon, October 3, 2020.

103. *1999—Eve Van Cauter:* Karine Spiegel, Rachel Leproult, and Eve Van Cauter, "Impact of Sleep Debt on Metabolic and Endocrine Function," *Lancet* 354 (October 23, 1999): 1435–1439.

104. *2000—Carol Everson:* Carol A. Everson and Linda A. Toth, "Systemic Bacterial Invasion Induced by Sleep Deprivation," *American Journal of Physiology: Regulatory, Integrative and Comparative Physiology* 278 (2000): R905–R916; Siegel, *The Neural Control*, 141–142.

105. *2007—The American Board:* Interview with Sharon Keenan, February 11, 2021; Kingman P. Strohl, "Sleep Medicine Training Across the Spectrum," *Chest* 138, no. 5 (May 2011): 1221–1231.

106. *2009—The NIH's Sleep Heart Health:* Gail G. Weinmann et al., "A Perspective: Division of Lung Diseases at Fifty," *American Journal of Respiratory Critical Care Medicine* 12 (December 15, 2019): 1466–1471.

107. *2011—A poll by:* Ruthann Richter, "Among Teens, Sleep Deprivation an Epidemic," Stanford Medicine, October 8, 2015, accessed October 26, 2022, https://med.stanford .edu/news/all-news/2015/10/among-teens-sleep-deprivation-an-epidemic.html.

108. *2011—Two Maryland mothers:* Interview with Terra Ziporyn Snider, September 8, 2022; Lewis, *The Sleep-Deprived Teen*, 59–60.

109. *2012—Neuroscientist Maiken Nedergaard:* Kenneth Miller, "Getting Enough Sleep Can Be a Matter of Life and Death," *Discover,* February 25, 2015, accessed December 6, 2022, https://www.discovermagazine.com/mind/getting-enough-sleep-can-be-a -matter-of-life-and-death; Jeffrey J. Iliff et al., "A Paravascular Pathway Facilitates Paravascular Flow Through the Brain Parenchyma and the Clearance of Interstitial Solutes, Including Amyloid-B," *Science Translational Medicine* 4, no. 147 (August 15, 2012); Lulu Xie et al., "Sleep Drives Metabolite Clearance from the Adult Brain," *Science* 342, no. 18 (October 18, 2013): 373–377.

110. *2014—The FDA approves:* Patrick Strollo Jr. et al., "Upper Airway Stimulation for Obstructive Sleep Apnea," *New England Journal of Medicine* 370, no. 2 (January 9, 2014): 139–147; Harneet K. Walia et al., "Upper Airway Stimulation vs. Positive Airway Pressure Impact on BP and Sleepiness Symptoms in OSA," *Chest* 157, no. 1 (January 1, 2020): 173–183; Saif Mashaqi et al., "The Hypoglossal Nerve Stimulation as a Novel Therapy for Treating Obstructive Sleep Apnea—a Literature Review," *International Journal of Environmental Research and Public Health* 18, no. 1642 (2021), https://pubmed .ncbi.nlm.nih.gov/33572156/; "Obstructive Sleep Apnea: How Do the Leading Therapies Stack Up in Effects on Sleepiness and Blood Pressure?," Consult QD, October 7, 2019, accessed October 26, 2022, https://consultqd.clevelandclinic.org/obstructive-sleep-apnea -how-do-the-leading-therapies-stack-up-in-effects-on-sleepiness-and-blood-pressure/.

111. *2014—The FDA also approves:* Ashok J. Dubey, Shailendra S. Hanru, and Pramod K. Mediratta, "Suvorexant: The First Orexin Antagonist to Treat Insomnia," *Journal of Pharmacology and Pharmacotherapeutics* 6, no. 2 (April–June 2015): 118–121; "Belsomra," Drugs.com, https://www.drugs.com/belsomra.html; Daniel F. Kripke, "Is Suvorexant a Better Choice Than Alternative Hypnotics?," *F1000Research* 4, no. 456 (August 3, 2015), accessed October 26, 2022, https://www.ncbi.nlm.nih.gov/pmc /articles/PMC4648222/.

112. *2014—Two large studies:* Kyla Wahlstrom et al., "Examining the Impact of Later School Start Times on the Health and Academic Performance of High School Students: A Multi-Site Study," Center for Applied Research and Educational Improvement, St. Paul, Minnesota, University of Minnesota, accessed October 26, 2022, https:// conservancy.umn.edu/handle/11299/162769; Judith Owens et al., "School Start Time

Change: An In-Depth Examination of School District in the United States," *Mind, Brain, and Education* 8, no. 4 (2014): 182–213.

113. *2014—The American Academy:* Rhoda Au et al., "School Start Times for Adolescents," *Pediatrics* 134, no. 3 (September 2014): 642–649; Lewis, *The Sleep-Deprived Teen*, 61.

114. *2014–2016—The American Psychological Association:* Lewis, *The Sleep-Deprived Teen*, 61; "Summary of Policy Statements on School Start Time," Minnesota Sleep Society, https://www.mnsleep.net/school-start-time-toolkit/appendix/summary-of-policy -statements-on-school-start-time/?doing_wp_cron=1671398036.7194778919219970 703125; "Later School Start Times Promote Adolescent Well-Being," American Psychological Association, https://www.apa.org/pi/families/resources/school-start-times.pdf; "Position Statements and Resolutions on Sleep and School Start Times," Start School Later, accessed December 18, 2022, https://www.startschoollater.net/position-statements .html.

115. *2019—California becomes:* Richard Cano, "California Just Pushed Back School Start Times—You Weren't Dreaming. Now What?," CalMatters, October 28, 2019, accessed October 27, 2022, https://calmatters.org/education/k-12-education/2019/10 /how-school-start-time-law-will-work-in-california/.

116. *2020—Death of Dement:* Richard Sandomir, "Dr. William Dement, Leader in Sleep Disorder Research, Dies at 91," *New York Times,* June 27, 2020, https://www.nytimes .com/2020/06/27/science/dr-william-dement-dead.html.

117. *2021—Mary Carskadon launches:* "Carney Researcher to Lead New Sleep, Circadian Rhythms Center," Robert J. and Nancy D. Carney Institute for Brain Science, April 28, 2021, accessed October 29, 2022, https://www.brown.edu/carney/news/2021/04/28 /carney-researcher-lead-new-sleep-circadian-rhythms-center; Brian Amaral, "New Center in R.I. Will Research Sleep, and How It Affects Kids' Mental Health," *Boston Globe,* April 30, 2021, accessed December 15, 2022, https://www.bostonglobe.com/2021/04/30 /metro/new-center-ri-will-research-sleep-how-it-affects-kids-mental-health/.

Appendix B

1. *The roots of chronobiology:* Vinod Kumar, ed., *Biological Timekeeping: Clocks, Rhythms and Behavior* (New Delhi: Springer, 2017), 5–6.

2. *The first scientist:* Ibid.

3. *Charles Darwin endorsed:* Urs Albrecht, ed., *The Circadian Clock*, 12th edition (New York: Springer, 2010), 3.

4. *In the 1920s:* Kumar, ed., *Biological Timekeeping*, 7–8; William K. Stevens, "Curt Richter, Credited with Idea of Biological Clock, Is Dead at 94," *New York Times*, December 22, 1988, section D, 23.

5. *In the '30s:* M. K. Chandrashekaran, "Erwin Bünning (1906–1990): A Centennial Homage," *Journal of the Biosciences* 31, no. 1 (March 2006): 5–12; Kumar, ed., *Biological Timekeeping*, 7–8.

6. *Many experts, meanwhile:* Albrecht, *The Circadian Clock*, 2–4.

7. *Not long after Bünning:* Serge Daan, *Die innere Uhr des Meschen: Jürgen Aschoff (1913–1998), Wissenschaftler in einem beweigten Jahrnhundert* (Wiesbaden, Germany: Reichert Verlag, 2017), 44–80.

8. *Aschoff's experiments involved:* Jürgen Aschoff, "Sources of Thoughts: From Temperature Regulation to Rhythm Research," *Chronobiology International* 7, no. 3 (1990): 179–186.

9. *After the war:* Ibid.

10. *Wondering whether diurnal:* Jürgen Aschoff, "Exogenous and Endogenous Components in Circadian Rhythms," *Cold Spring Harbor Symposia on Quantitative Biology* 25 (1960): 11–28; Jürgen Aschoff, "Die 24-Stunden-Periodik der Maus unter konstanten Umgebungsbedingungen," *Naturwissenschaften* 138 (1951): 506–507; J. Aschoff and J. Meyer-Lohmann, "Angeborene 24-Stunden-Periodik beim Kücken, *Pflügers Archiv* 260 (1954): 170–176.

11. *For Aschoff, this posed:* Aschoff, "Sources of Thoughts."

12. *In 1954, Aschoff published:* Albrecht, *The Circadian Clock*, 10; Aschoff, "Sources of Thoughts"; Jürgen Aschoff, "Zeitgeber der Tierischen Tagesperiodik," *Naturwissenschaften* 41 (1954): 49–56.

13. *Soon afterward, Princeton:* Albrecht, *The Circadian Clock*, 9; Adriano A. Buzzati-Traverso, ed., *Perspectives in Marine Biology* (Berkeley: University of California Press, 1958), 239–268.

14. *Aschoff and Pittendrigh met:* Serge Daan, "Colin Pittendrigh, Jürgen Aschoff, and the Natural Entrainment of Circadian Systems," *Journal of Biological Rhythms* 15, no. 3 (June 2000): 195–207.

15. *In the meantime:* Thomas C. Erren, "A Chronology of Chronobiology," *Physiology News Magazine* 113 (Winter 2018): 32–35.

16. *More than 150 scientists:* "List of Those Attending the Symposium," *Cold Spring Harbor Symposia on Quantitative Biology* 25 (1960): ix–xi, http://library.cshl.edu/symposia/1960/participants.html.

17. *Erwin Bünning:* "Opening Address: Biological Clocks," *Cold Spring Harbor Symposia on Quantitative Biology* 25 (1960): 1–9.

18. *Aschoff discussed the evidence:* Aschoff, "Exogenous and Endogenous Components."

19. *Pittendrigh hypothesized:* Colin S. Pittendrigh, "Circadian Rhythms and the Circadian Organization of Living Systems," *Cold Spring Harbor Symposia on Quantitative Biology* 25 (1960): 159–184.

20. *Nearly 50 other experts:* Contents, *Cold Spring Harbor Symposia on Quantitative Biology* 25 (1960): xii-xiii, accessed December 6, 2022, http://symposium.cshlp.org/content/25.toc.pdf.

21. *The Cold Spring Harbor symposium:* Kumar, *Biological Timekeeping*, 12.

Index

Page numbers in *italics* indicate illustrations; n indicates a note.